Silicon Technologies

Silicon Technologies

Ion Implantation and Thermal Treatment

Edited by
Annie Baudrant

iSTE

WILEY

First published 2011 in Great Britain and the United States by ISTE Ltd and John Wiley & Sons, Inc.

ISTE Ltd
27-37 St George's Road
London SW19 4EU
UK

www.iste.co.uk

John Wiley & Sons, Inc.
111 River Street
Hoboken, NJ 07030
USA

www.wiley.com

Library of Congress Cataloging-in-Publication Data

Silicon technologies : ion implantation and thermal treatment / edited by Annie Baudrant.
p. cm.
Includes bibliographical references and index.
ISBN 978-1-84821-231-2
1. Semiconductor doping. 2. Ion implantation. 3. Semiconductors--Heat treatment. I. Baudrant, Annie.
TK7871.85.S5485 2011
621.3815'2--dc22

2011008131

British Library Cataloguing-in-Publication Data
A CIP record for this book is available from the British Library
ISBN 978-1-84821-231-2

Printed and bound in Great Britain by CPI Antony Rowe, Chippenham and Eastbourne.

FSC
www.fsc.org
MIX
Paper from responsible sources
FSC® C013604

Table of Contents

Preface

Electronic components based on semiconductors are at the core of all electrical and electronic equipment. They are stakeholders of an increasing share of the objects surrounding and accompanying us. They give to these objects various functions: capture, storage, transmission or information restoration (cameras, phones, TV, etc.), control-command, aid to decision-making, safety, etc.

Nowadays, the components aggregate more than 20% of the value of electrical or electronic equipment, against 7% in 1985. Therefore, there is a global market of the "semiconductor" sector of 270 billion dollars.

These components are indeed at the heart of all applications, for the "digital society", of which they ensure both the engine and the memory, or for the improvements expected in the fields of transport, aeronautics, health, safety, and electrical energy management close to the user or within distribution systems.

Broadband available to all, the intelligent and efficient management of energy in buildings, telehealth, the intelligent road and zero emission cars or even the replacement of the electrical infrastructure, go inevitably through an increasing use of electronic components with more complex and powerful semiconductors.

The number of delivered components amounts to billions of parts per week (3.25 billion of units/week, at the end of December 2009).

This increasing and continuous penetration of electronic components in equipment was made possible by two other key features of the sector, which have been permanent over the last four decades:

– the monolithic and collective fulfillment or the increasing integration, of complex electronic functions (combination of elementary components) on hardware support, which constitutes an integrated silicon circuit, with a fast renewal of products and technologies;

– the continuous reduction of manufacturing costs and unit prices.

For the driving markets of the memories, of the specific processors and digital circuits, the production is primarily based on CMOS technologies. They are characterized by the most critical dimension of the transistors constituting them. Thus, the "40 nm node" corresponds to the technology, where this lower limit measures 40 nm.

Nowadays, the most aggressive generation of production is the 40 nm node, but the 28 nm node will be proposed *in production* from the next half-year.

The dimensions' reduction, for which the transistors (elementary components of the integrated circuits) are produced, is at the heart of two important product evolutions:

– the rise in performance (transistor speed, related to the minimal size of these elementary components);

– the integration of a growing number of transistors, today exceeding a billion per circuit.

The evolution of technologies leading to this increasing integration is fast. It follows an empirical prediction made by Gordon Moore (cofounder of Intel), of an integration doubled every 18 months. Unequalled fact in any industrial sector, this technical evolution of the products has lasted for four decades (see Figure 1).

The manufacturing technologies of electronic components have evolved a lot since the invention of the first transistor: we went from an element the size of a 1€ coin to submicron dimensions (45 nm previewed at IBM in spring 2006, while in December 2007, the TSMC group sampled SRAMs in 32 nm technology. Nowadays, we are heading for 22 nm and below).

Nevertheless, some principles have been preserved, and of course the technology is subjected to constraints, resulting from the fundamental properties of the semiconductors, such as they appear in the chapters on the components' physics. Let us note however that 32 nm noticeably represents the thickness of a hundred atomic lines, and that we imperceptibly come near a limit below which we will no longer be able to go, except by completely changing the principles.

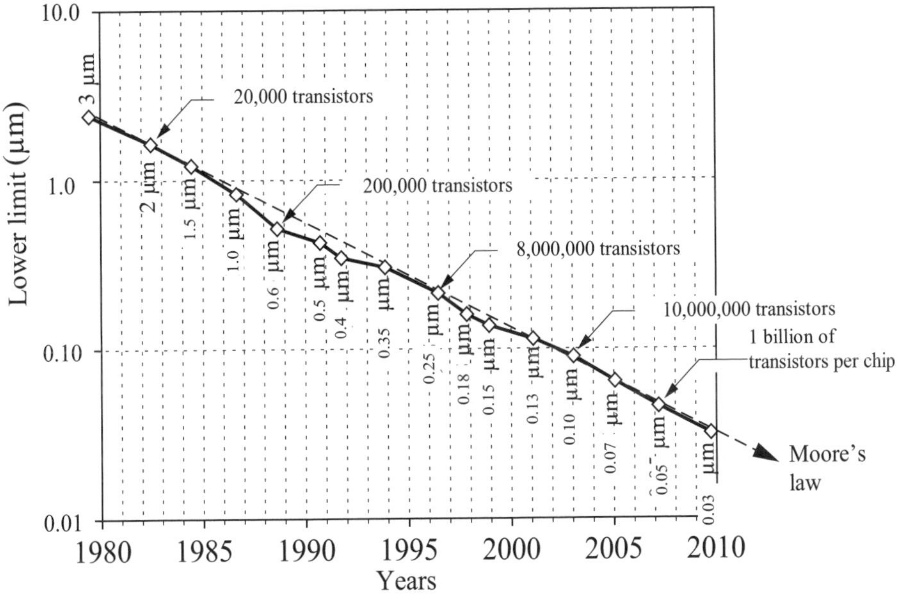

Figure 1. *Moore's law*

Surprisingly, there are no works written in the French language describing these 30 years of history and evolution of technological methods. We can remember:

– that the evolution of CMOS technologies is made up of incremental and radical evolutions;

– that the control of these radical evolutions gives place to the highest costs and are the object of strategic decisions.

We often find a good number of precise publications on each topic, but little or no scientific summaries outlining the "why and how" of the various technical choices.

Among these key points, we have often quoted, over the last ten years:

– around the year 2000, for 130 to 90 nm nodes:

- introduction of 193 nm optical lithography;

- introduction of the Copper technology: with this method, we can accelerate the interconnections between the transistors and reduce the losses;

– around the year 2005, for 65 to 45 nm nodes:

- dipping lithography: with this, we can go down below 65 nm. Its unit piece of equipment exceeds 50 million dollars;

- the process-induced strain, introduced in 2003 by Intel, and taken over since then by almost all of the other microprocessor manufacturers (i.e. the strain induced in a localized way into the channel of the transistors;

– around 2010, for 32 to 22 nm nodes:

- introduction of the technology of metal gate and of high permittivity dielectric, which reduces the leak of transistors during their miniaturization;

- transition to substrates of thin silicon films (SOI), to improve the performances of transistors of the smallest sizes;

– around 2014, for the 18-12 nm node:

- EUV (*extreme ultraviolet*) lithography.

All these stages use new techniques and thus new investments. We forget to name the advances made in the course of time *by elementary methods*, with the constant objective to decrease all dimensions (width, length, thickness) and to increase all the electrical performances (current, speed, low consumption, etc.).

This book thus proposes to the reader a timeline, of the development of active zones, modifying the surface structure of the massive silicon substrate, the basic principles, the implementations according to the technological fields, the limits and constraints, with a focus on the recent advances.

Its objective is not to go through the issue exhaustively for all the elementary processes of a technological assembly, but to point out the essential and fundamental pieces of data, only for the methods aiming to use and to improve the properties of the material or of the silicon semiconductor, for the electronic components, as well as for the microsystems (integrated sensors for example).

Below, a few paragraphs introduce each of the four matters evoked, but leave to the authors (J.J. Ganem, I. Trimaille, J.J. Grob, D. Mathiot and J.M. Hartmann, the best French-speaking specialists on the topic), the pleasure of elaborating on the manufacturing methods and scientific materials (physical and chemical laws),which consolidate or predict the outcomes.

Silicon oxidation

Oxidation is a very important stage in the achievement of integrated silicon circuits, because it is thanks to this specific property that the silicon, which is not *a priori* a very good semiconductor, has become the most frequently used material in microelectronics. This operation is necessary throughout the modern manufacturing methods of integrated circuits. It is thus essential to know how to carry out an oxide of good quality.

The oxide can be used as:

– an implantation and diffusion mask of dopants;

– a passivating layer on the silicon surface;

– insulation zones between various components of an integrated structure;

– an active zone in MOS (gate oxide) transistors;

– electrical insulation between the adjacent layers to improve the integration and the reduction of dimensions ("spacer" for example, see further);

– electrical insulation between the various levels of metallization or of conducting layers in strongly doped polycrystalline silicon;

– sacrificial layers that can improve the circuit performances and integration. These sacrificial layers can also be used to manufacture microstructures containing polycrystalline silicon and to intervene in integrated microsystems (MEMS: micro electro-mechanical systems).

Ion implantation

Ion implantation is a low-temperature process. The implantation process takes the ions of the chosen species, accelerates them with an electrical field, and then makes them scan the surface of the slice, to carry out a uniform pre-deposition.

This method of introducing doping atoms into silicon was developed in the 1960s: it is ion implantation (an ion is a loaded atom). Very briefly, the ion implantation consists of projecting the ions of the adequate doping species towards the slice, through the openings of an oxide mask or of hardened resin.

The ions used for the doping, such as boron, phosphorus or arsenic, are generally produced from a gas source, guaranteeing a great purity of the source. These gases have a tendency to be very dangerous. When they are implanted in a semiconductor, each doping atom creates a charge carrier (hole or electron according to whether it is a type p or n dopant), thus locally altering the conductivity of the semiconductor.

The ion implantation is also a method used for the preparation of SOI (silicon-on-insulator) substrates from conventional silicon substrates.

Diffusion

Diffusion is the term used to describe the movement of atoms, molecules or particles from a high concentration zone towards a lower concentration zone.

Diffusion is a phenomenon, depending on time and temperature. The diffusion velocity of an atom, molecule or compound from an area of high concentration towards a low concentration zone is a function of time and temperature. The parameter connecting the diffusion velocity to the temperature at a given time, is known as the diffusion or diffusivity coefficient.

The atoms of the dopant must move the silicon atoms of the crystalline structure and take their place to become electrically active. The diffusion process is used in the manufacture of integrated circuits, in order to introduce a controlled quantity of a specific dopant into a specific area of the semiconductor crystal. The diffusion process used to achieve this substitution is divided into distinct stages.

Epitaxy of strained Si/Si1-xGex hetero-structures

The term epitaxy is of Greek origin and means "to build above". Deposition by epitaxy is in general the construction of a single-crystal silicon layer on a slice (also a single-crystal). The layer deposited is a crystallographic extension of the substrate, from the point of view of the atomic arrangement (i.e. identical crystalline structure). The substrate can thus be regarded as the "seed" essential to generate a single-crystal growth.

Deposition by epitaxy is a process of CVD. The first use of CVD was the single-crystal silicon deposition at the end of the 1950s. This technique then played a crucial role in this industry, but this chapter will not develop the aspect "material - growth and crystallography of silicon substrates", instead focusing on the use of the epitaxy technique for the manufacture of active SiGe zones in nano-CMOS new technologies.

We hope that by introducing a comprehensive overview of these techniques to readers, this book will answer the expectations of those students, professors, technicians, engineers or researchers, closely interested in the manufacture of silicon micro nanostructures.

This preface would not be complete without extending to the authors, J.J. Ganem, I. Trimaille, J.J. Grob, D. Mathiot and J.M. Hartmann, my warmest thanks for their rich contributions to this book.

Annie BAUDRANT
May 2011

Chapter 1

Silicon and Silicon Carbide Oxidation

This chapter is devoted to the physics of silicon and silicon carbide oxidation. We will find in this chapter, an examination of the main techniques of deposition and growth of thin films. The reader will then discover how the laws governing oxidation are established, in particular those concerning silicon oxidation. This remains nowadays the most widespread method in the manufacture of integrated circuits and of MEMS in the broad sense. This chapter has a double purpose. First, it is written to expose in detail the theoretical principles that are particularly interesting for researchers, and secondly, to review a certain number of experimental results, useful in the practice of any process engineer.

1.1. Introduction

The substantial improvement of the electrical and physical characteristics of the SiO_2/Si interface leads to an impressive development of integrated circuits. That was made possible by a better understanding, over time, of the way in which silica is manufactured by deposition or growth on silicon.

Although thermal growth is nowadays one of the most frequently used methods, other techniques have also been developed. In section 1.2, we review the main passivation techniques employed today in industry as well as in research laboratories.

Chapter written by Jean-Jacques GANEM and Isabelle TRIMAILLE.

In the semiconductor industry, silica can be manufactured by thermal growth from silicon substrates placed at atmospheric pressure in a flux of water vapor, oxygen or a mix of oxidizing gases. Chemical vapor deposition is also usually used to obtain thick passivation films.

In laboratories, more often aiming at the study of physical phenomena governing the production of layers, the techniques used will be based on the physics of the methods (anodic oxidation in an electrolytic environment or by oxygen plasma) or on specific conditions of thermal oxidation (low or high pressure, etc.).

In section 1.3, we detail the silica properties useful for understanding the growth phenomena. We pay particular attention to showing the properties of atomic transport and the solubility of gases in silica. For a more complete review of the physical characteristics of silica, we refer the reader to the works of Bruckner [BRU 70]. For self-scattering and chemical diffusion in glasses, we recommend the works of Frischat [FRI 75].

In section 1.4, we develop the general equations of transport taking place during oxide growth, and the borderline cases arising from various approximate assumptions that can be made on the flux of mobile species in the case of oxidation. Thus, we also could deduct the expressions of the flux of the transported ion species, in the case of anodic oxidation.

In section 1.5, we show how it is possible to give a satisfactory answer to the two following questions: Which are the mobile species in silicon and in silicon carbide oxidation? How do they move? For this purpose, isotopic labeling techniques are also presented, as well as the experimental results.

In section 1.6, we state the transport equations after the simplifying hypotheses. Two cases are considered, where the transported species are either neutral or charged. When they are charged, two cases still emerge. For the films with a thickness lower than 3 nm, where the tunneling currents are established, the Cabrera and Mott theory predicts the growth laws. For the larger thicknesses, equations become very complex.

The classic Deal and Grove theory, describing silicon oxidation is detailed in section 1.7. It is based on the fact that oxidizing species (O_2, H_2O) are dissolved in silica in interstitial positions and migrate towards the SiO_2/Si interface, to form there the new oxide by a chemical reaction assumed to be of the first order. This theory leads to the *linear-parabolic* growth kinetics, which we largely resort to in industry. They give a good account of the growth kinetics in the case of oxidation under water vapor, and in the case of oxidation under O_2, for thicknesses higher than 20-30 nm.

Whereas the Deal and Grove theory assumes that species diffuse through oxide without interacting with it, Breed and Doremus have proposed a theory developed in section 1.8, assuming the interaction of the diffusing species with the silica network. The concentration profiles of the various species that we can expect from it are compared to the experimental results.

In section 1.9, we examine the growth kinetics of film slower than 30 nm and those of very thin films (< 3 nm) obtained by thermal growth, for which the Deal and Grove theory is no longer satisfactory. Several alternative models are stated to explain the initial regime of fast growth.

Finally, in section 1.10, we show how the important oxidation parameters, k_P and k_L, are affected by the experimental oxidation conditions such as the pressure, temperature and crystalline direction of the substrate and doping.

Several English journals, already published, explain the classical theories linked to the subjacent oxidation mechanisms. Those show the crucial interest for the industry to control and understand silicon oxidation, which must produce increasingly thin functional oxide layers. This chapter relies on the work of Rigo [RIG 86], while updating them by taking into account all the latest breakthroughs of the field.

1.2. Overview of the various oxidation techniques

1.2.1. *General information*

There are many techniques used to form silica films on silicon. They are included in two categories, according to whether they rely on an oxidation or deposition process.

1.2.1.1. *Oxidation process*

In oxidation processes, the film growth depends on the injection phenomena taking place at the two interfaces: oxide/gas and oxide/substrate. Those are *a priori* rather well defined. In all cases, the oxide film itself always plays a fundamental role. The atomic transport, from one interface to another, of at least one component through oxide is necessary to the growth. The oxide composition is determined by the external environment and by the laws of thermodynamics. Oxidation can be carried out in two ways:

– *Thermal oxidation*: heating a silicon sample at high temperatures under an oxidizing atmosphere, increases the hopping probability of the atoms. Indeed, the hopping probability being proportional to $\exp(-W/k_BT)$, raising the temperature thus increases the transport of most mobile species. Moreover, fluxes at the interfaces are then increased. At this stage, it is necessary to stress that mobile species can be either electrically charged or neutral.

– *Anodic oxidation*: the injection and transport of charged species (anions and cations) are caused by an important electrical field (10 MV.cm - 1), which decreases the potential barrier that the ions must overcome. This type of growth can be carried out at room temperature.

1.2.1.2. *Deposition process*

In deposition processes, the films' growth depends on phenomena taking place outside the silica film or its surface. The main part of the film already formed does not play any role in the following growth stages (if the film is maintained at a sufficiently low temperature). The composition of the formed film should vary considerably according to the techniques used: evaporation, sputtering of a material or chemical vapor deposition. We will find a comparison of the physical and chemical properties of these films in relation to their formation process in the summary article by Pliskin [PLI 77].

1.2.2. *Most frequently used methods in the semiconductor industry*

1.2.2.1. *Thermal oxidation*

Among all the methods employed in the silicon industry, thermal oxidation is the most widely used. It can be done either with dry oxygen (dry oxidation), or with water vapor (wet oxidation).

The intermediate case, i.e. oxidation in an O_2+H_2O mixture, will not be discussed in this chapter. The two types of thermal oxidation will be developed in distinct sections. However, they are sometimes used sequentially, during the manufacture of integrated circuits (for example, wet oxidation followed by a stage of dry oxidation), to grow thick field oxides.

1.2.2.1.1. Oxidation under dry oxygen

Typically, dry oxidation is carried out in a device schematized in Figure 1.1b. The furnace is a quartz tube heated by the Joule effect. After being cleaned, the silicon wafers are placed in a socket. The latter is slowly introduced in the center of the furnace, in an oxygen flux, so as to avoid thermal shocks. When we use a pure O_2 flux, the oxidation is carried out with a P_{O2} oxygen pressure equal to the atmospheric pressure.

By using a mixture of O_2 and of inert gases, we can oxidize with lower partial oxygen pressures.

To form very thin oxides (a few nm), we will use the rapid processing systems (RTF – rapid thermal furnace) principle. They rely on a rapid heating system (halogen lamps), in order to increase the temperature of a silicon wafer from the room temperature to 1,000°C in a fraction of a second. These furnaces allow very short processing times at high temperature, to obtain thin oxides with good electrical properties. A rapid thermal processing system under controlled atmosphere is schematized in Figure 1.1a.

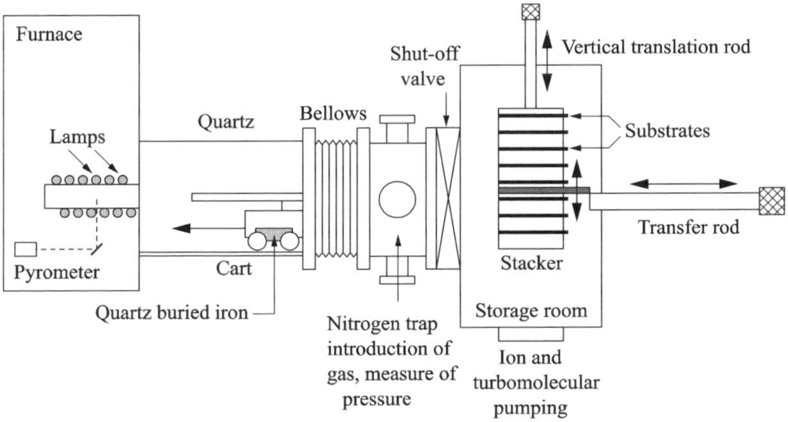

Figure 1.1a. *Schematic diagram of a rapid thermal oxidation furnace under a controlled atmosphere (according to Ganem [GAN 92])*

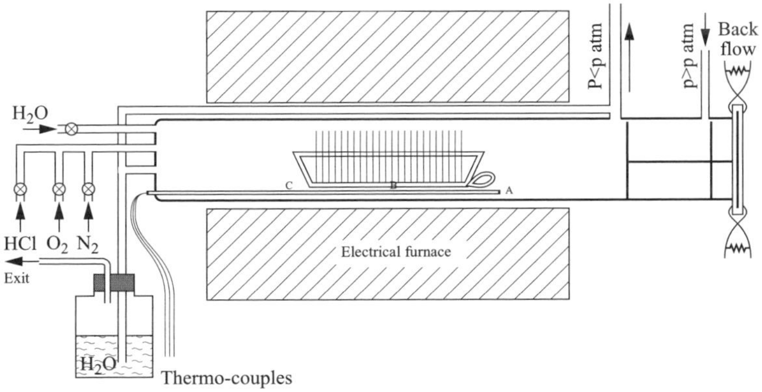

Figure 1.1b. *Schematic diagram of a of thermal oxidation furnace (according to Rigo [RIG 86])*

Dry oxidation, although very simple, has some disadvantages:

– the process can become complicated if we need to grow several films of different thicknesses;

– to obtain a Si-SiO_2 interface of good quality, the oxide must be annealed (for example, under a N_2 flux, after stopping the O_2 flux).

A pre-oxidation stage can be necessary (for example, under a flux of chlorinated compounds).

To oxidize under an atmosphere of pure oxygen, we must reduce the contamination sources to a minimum. They were listed by Revesz [REV 79] and are primarily CO_2 and H_2O. The gas flux always contains traces of contaminants, which are mainly water vapor and hydrocarbons. The latter will react with oxygen in the hot area of the furnace, to form CO_2 and H_2O. It is thus necessary to burn these hydrocarbons (for example by injecting the gas into a small quartz tube containing quartz wool heated to 900°C) [IRE 78a]. H_2O and CO_2 are then eliminated by sending gas through zeolites and/or by the means of a low temperature trap.

A third possible contaminant is H_2. Above 1,000°C, the hydrogen diffuses through the quartz tube from the outside. Various techniques have already been conceived (in the laboratory), in order to avoid this phenomenon. We then use a tube with double or triple inner surfaces, in between which an inert gas circulates [REV 69, VAN 72].

Lastly, the native oxide and the adsorbed impurities are also contamination sources. The silicon substrate is covered at room temperature with a native oxide layer (thickness ~1 nm) and with adsorbed impurities. This layer can be removed before the oxidation, through 10 minute processing under H_2 at 1,200°C [REV 69], followed by a neutral gas annealing.

Another means reported by several authors [KAM 77, HOR 78] would consist of carrying out oxidations with low partial oxygen pressures. This is made possible by mixing O_2 with inert gases (N_2 or Ar). However, knowing that neutral gases always contain traces of impurity, it is almost impossible to simultaneously obtain dry oxygen and low values of P_{O2}.

1.2.2.1.2. Oxidation under water vapor or wet oxidation

The vapor flux with atmospheric pressure is obtained by heating a balloon containing very high purity water. The entry channel of the vapor into the furnace is itself heated to avoid any condensation. By mixing the vapor with O_2, we can also carry out the oxidation under wet oxygen. In order to improve the purity of the vapor, we resort to H_2O produced by pyrogenic synthesis [DEA 78]. It is obtained by the reaction of O_2 with H_2, simultaneously introduced and heated, before they reach the oxidation furnace.

1.2.2.2. *Chemical vapor deposition*

At the same time as oxidation, there are other methods widely spread in the semiconductor industry, including CVD (chemical vapor deposition), evaporation and the sputtering of targets. We will limit ourselves to the case of the CVD technique.

Chemical vapor deposition can be carried out by pyrolysis of silane SiH_4 in the presence of oxygen. The basic equations are then as follows:

$$SiH_4 + O_2 \rightarrow SiO_2 + H_2$$

$$\rightarrow SiO_2 + H_2O$$

Derivatives of silane such as $Si(OR)_4$ and the oxidizing compounds such as N_2O and CO_2 can also be used.

Whereas the high partial pressures of silane P_{SiH4} and the high temperatures (600-1,000°C) support the formation of H_2O, low pressures and low temperatures lead to the formation of H_2. Silica films are formed by a heterogeneous reaction on the surface. However, in the gas phase, due to nucleation process, a homogeneous cloud of colloidal particles also occurs. This is a significant contamination source. Indeed, some of these particles condense in a white powdery coating on the surfaces at low temperatures, such as the walls of the reactor.

There are several types of CVD techniques at room temperature:

– low temperature with normal pressure (low temperature CVD –LTCVD);

– low pressure (low pressure CVD – LPCVD);

– enhanced plasma (plasma enhanced CVD – PECVD).

1.2.2.2.1. Low temperature with normal pressure (LTCVD)

LTCVD reactors function under normal pressure conditions and at temperatures lower than 500°C. In most commercial facilities, the substrate is heated by the Joule effect, whereas the walls and entry points of gas are cooled to remove heterogeneous nucleation in the gas phase. The complete elimination of the contaminating particles constitutes one of the most serious difficulties of this technique.

1.2.2.2.2. Low pressure (LPCVD)

The pressure can vary between 0.5 and 1 Torr and the temperature between 700 and 800°C. This way, we manage to increase the mass transfer of the gas phase towards the substrate of an order of magnitude. It is the surface reaction that determines the oxide growth rate. Thus, we decrease the effects related to mass transfer the variations. Since the walls of the low pressure reactor are hot, the films deposit there in the same way as on the substrate. The schematic diagram of a LPCVD reactor is presented in Figure 1.2.

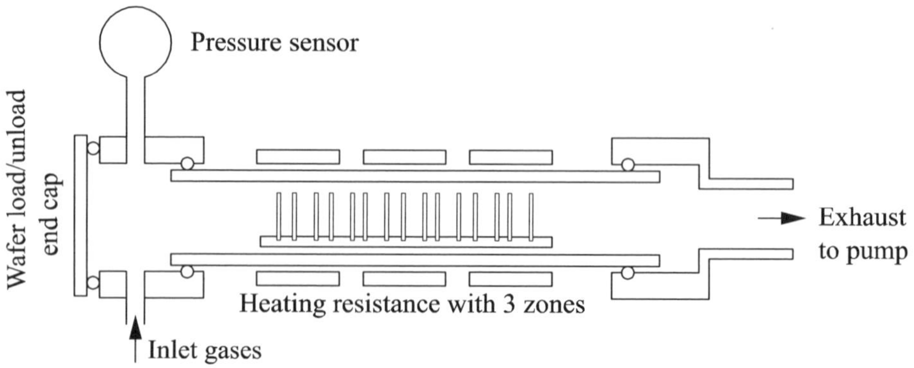

Figure 1.2. *Schematic diagram of a LPCVD reactor (according to Kern and Rosler [KER 77])*

1.2.2.2.3. Plasma enhanced (PECVD)

A glow discharge plasma is generally created with pressures ranging between 0.01 and 1 Torr. This plasma splits up the gas molecules to form very reactive species, in order to carry out a deposition in chemical phase of low temperature films. J.R. Hollahan mentions the silica film deposition for temperatures ranging between 250 and 350°C [HOL 79].

An article by W. Kern and R.S. Rosler [KER 77] reviews the basic principles, the deposition parameters, the advantages and the limitationss of various CVD techniques depositing dielectric passivation films.

1.2.3. *Other methods*

We mention here other methods used in the laboratory to grow silica films on silicon.

1.2.3.1. *Anodic oxidation*

An oxide is grown in an electrolytic cell by implementing a potential difference between the silicon wafer, acting as an anode, and a cathode plate immersed in an electrolyte or a gas plasma [DEL 71].

The aqueous solutions containing salt traces are not desirable for silicon oxidation, because they cause its corrosion. The oxide thus formed is porous.

The following non-aqueous solutions have been used to form compact SiO_2 films [CRO 71a, DRE 66, SCH 57]:

Ethylene glycol [DRE 66] + NO_3K (0.1 - 1%) +H_2O (0.1 - 2%)

Di-ethylene glycol + NO_2K (0.1 - 1%) +H_2O (0.1 - 2%)

N-methylacetamide [SCH 57] + NO_2P (0.1 - 1%) +H_2O (0.1 - 2%)

Tetrahydrofurfuryl alcohol + NH_4NO_3 (0.1 - 1%) +H_2O (0.1 - 2%)

A schematic representation of an anodic oxidation cell is given in Figure 1.3.

Oxidations are performed at a constant current (typically 1 to 10 $mA.cm^{-2}$). As the oxide thickness grows, the cell voltage increases up to a pre-established value. This value is then maintained constant until the density of the current decreases, to reach a value of about 10 $\mu A.cm^{-2}$.

The oxidation process can last from 1 to 10 hours, depending on the desired oxide thickness. With these experimental parameters, we obtain SiO_2 films of 0.6 nm per volt. In practice, it is rather easy to carry out oxidations up to 300 V (180 nm of silica). For values of higher voltages, it is necessary to be freed from the corrosion phenomena of impurities, such as fluorine ions [CRO 71b].

Voltages going up to 500 V (i.e. X = 300 nm) can be reached using a bath of dihydroxydiethyl ether containing 6.7×10^{-3} mole/liter of $Al(NO_3)_3$, $9H_2O$.

With this type of oxidation method, the film growth is due to the ionization current, representing unfortunately only 1% of the total current. Indeed, the main part of the current (99%) is an electronic current circulating through the oxide during growth.

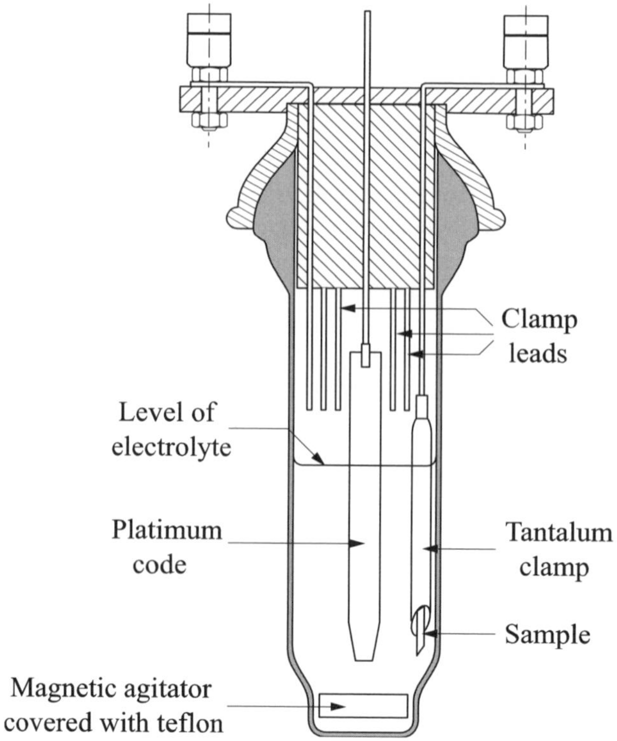

Figure 1.3. *Anodic oxidation cell (according to Croset et al. [CRO 71a])*

We can also dope the oxides during the anodic growth thanks to suitable salts, such as diethylphosphate [SCH 64]. The doped oxides can be used as diffusion sources of controlled composition [SCH 65].

This technique has some advantages:

– oxidations are carried out at room temperature;

– we can produce homogeneous films and repeatable thicknesses even in the case of low thicknesses, of a few nm.

It however involves disadvantages:

– the method is performed in a wet environment;

– the insulating properties of the obtained materials are not as good as those reached by thermal oxidation. They are deteriorated by the presence of water, coming from the external environment [DRE 66, NAN 70], as well as by the possible contamination of the electrolyte by alkaline ions.

1.2.3.2. *Oxidation assisted by plasma*

In 1963, Miles and Smith [MIL 63] suggested using plasmas to make oxides grow. Since then, this technique was applied to various metals (such as Ta, Nb, Al, Mg, etc.) and to semiconductors (Ge, Si, GaAs, etc.). A review discusses the physical mechanisms implied for this type of oxidation [GOU 81].

In this method, the electrolyte is substituted by a gaseous plasma of oxygen. The oxide surface is bombarded by a large number of charged and neutral species (molecules, atoms, ions, free radical and electrons), whose states of charge and concentrations depend on the various phenomena taking place in the plasma. These will vary with the type of discharge and with the experimental parameters, such as geometry, pressure, etc.

This method is called "plasma oxidation", when the substrate is electrically floating, with respect to the plasma, or "plasma anodization", when the substrate is positively biased. During plasma oxidation, no current circulates through oxide, following the example of what occurs during thermal oxidation. On the other hand, in the case of plasma anodization, an electrical current goes through the oxide, as in the case of the electrolytic anodic oxidation. The latter is the only way to obtain important thicknesses. Several types of discharge and geometry can be used.

1.2.3.2.1. Discharge with constant current

This is obtained by applying a continuous voltage (around 1 kV) between the anode and the cathode immersed in a gas at low pressure (10^{-2} to 1 Torr). The electronic density is of the order of $10^{10}cm^{-3}$. A common device is presented in Figure 1.4 [OHA 70].

The highly energetic ions due to the high voltage can produce sputtering of the electrodes. This can then involve a strong contamination of the oxide film by the sputtered cathode material, even if the sample is not facing the electrode [LES 78]. This contamination affects the oxide characteristics, as well as the growth kinetics. The ion bombardment on the substrate can also involve an ablation of the oxide film itself. In order to reduce this effect, the sample must be located as far as possible from the electrodes. Moreover, the electrode must be made of a material with low sputtering rates (Al, Si, etc.).

Copeland and Pappu [COP 71] reported a low growth rate of oxide (100 nm in 4-5 hours) at low temperatures (32-55°C).

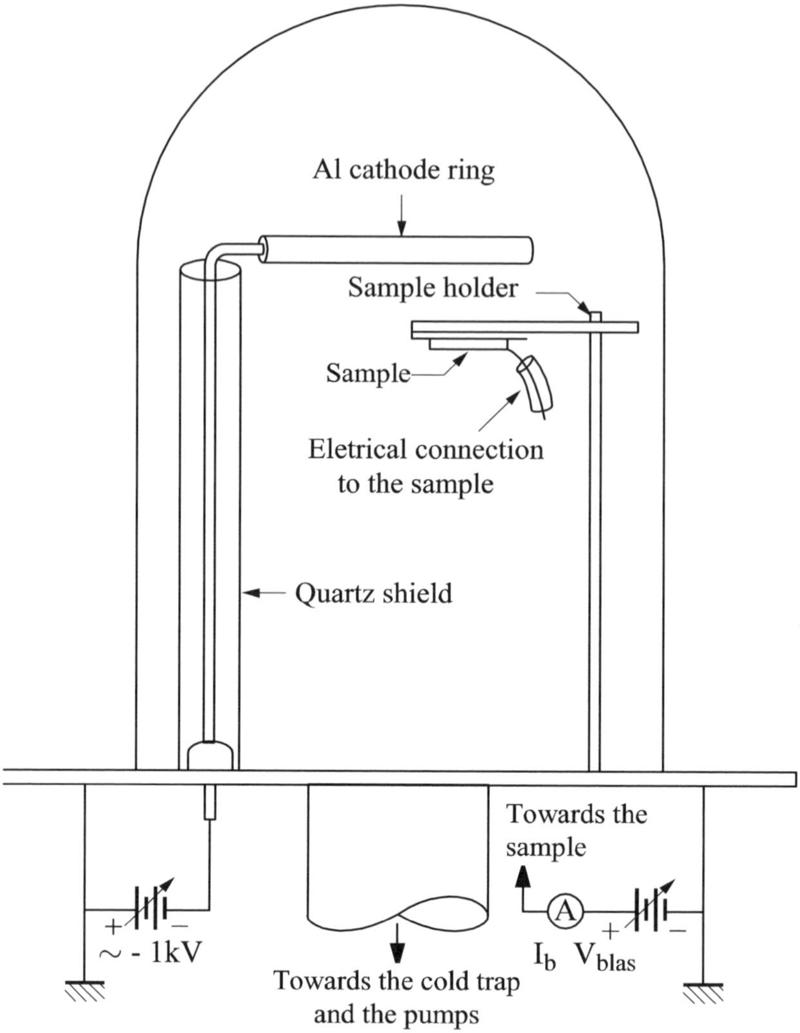

Figure 1.4. *Typical device of a plasma reactor (according to O'Hanlon [OHA 70])*

1.2.3.2.2. Hot cathode discharge

The hot cathode emits an electron flux producing a plasma with a higher electron density (10^{10}-10^{11} cm^{-3}), when the plasma is magnetically confined [GOU 80]. This discharge can be maintained with a low DC voltage (lower than 50 V) and the plasma is thus free from very energetic particles. The main problem is the sputtering of the hot cathode in the oxygen atmosphere. High quality SiO_2 films have been produced by this method, with fast growth rates (90 nm in 10 minutes), by setting the substrate at a temperature of 225°C [LIG 70].

1.2.3.2.3. High frequency discharge

This discharge is obtained by applying a RF voltage (a few MHz) [PUL 73, PUL 74] or a microwave (a few GHz) [KRA 67, LIG 65]. The oxygen pressure can vary between 0.01 and 1 Torr. The advantage of this technique is that the electron density can reach $10^{13}cm^{-3}$ [LIG 65]. The major disadvantage lies in the fact that the energetic particles can damage the oxide.

By using the microwave discharge, we can obtain strong oxidation rates: 200 nm in 5 minutes [LIG 65] and 600 nm in 2 hours [KRA 67], at a substrate temperature of 300-400°C.

In the case of the RF discharge, we reach 3 nm/min [PUL 73] or 5.2 nm/min [PUL 74], with a substrate temperature around 200°C. Chang *et al.* [CHA 80] have shown that, by mixing a small quantity of CF_4 with O_2, we considerably increased growth rates with a contamination of fluorine not exceeding 0.5 atom for 100.

Anodic plasma has the following characteristics (for silicon, as well as for other materials):

– very low efficiency of the ionic current (less than 1%);

– if the total current is constant, the oxidation rate generally decreases as growth proceeds;

– the electron bombardment on the substrate surface plays a crucial role in oxidation. The growth rate is enhanced with the electron temperature, but strongly decreases, if we prevent the electrons from reaching the sample [OHA 73].

This technique offers the following advantages:

– this is a process in a dry environment;

– growth rates are significant at low temperatures.

It also has disadvantages, the main ones being:

– the oxides thus produced have in general high densities of electron traps and thus require annealing, which can be at high temperature (approximately 1,000°C) [LIG 65]. This disadvantage can be mitigated by reducing the bombardment of the energetic ions on the surface of the substrate. That was done [CHA 77, TSU 78] by removing the plasma source of the process chamber, where the high frequency discharge is taking place.

1.2.3.3. *Low pressure oxidation under an ultra-high vacuum*

Ultra-thin layers (1-10 nm) have a very important advantage, because they illustrate the first stages of silicon oxidation. They are also very important practically in the manufacture of very high integration silicon components. These films are often grown under a low oxygen pressure by a mixture of O_2 with an inert gas. However, as mentioned previously, this oxidation mode is very sensitive to water vapor contamination. With a set up under ultra-high vacuum, we can use very low partial oxygen pressures, while minimizing contaminations. Moreover, a very clean silicon surface, free from natural oxide, is easily obtained by heating under an ultra-high vacuum.

This type of technique was described and used by Hopper *et al.* [HOP 75] and Smith *et al.* [SMI 82]. The wafers were electrically heated, thanks to a current crossing them from a sample-holder made of platinum or molybdenum. The basic vacuum of the device was tiny until approximately 10^{-10} Torr. The reaction chamber was then drained three times with pure oxygen containing at maximum 3 ppm of water vapor. The pressure there was then reduced to a value lower than 10^{-7} Torr. Later, the sample was heated up to 1,200°C – 1,350°C for 5-10 minutes, to remove the superficial oxide and the carbonaceous contaminants. Immediately after this cleaning process, the substrate temperature is reduced to the desired value and the oxygen is introduced from bakeable UHV leakage valves, until we reach the desired pressure.

The adjustable experimental parameters have been examined between the following values: 700 - 950°C, 50 - 1,200 Torr in the case of Hopper *et al.* [HOP 75] and 890 - 1,130°C, 5.10^{-5} - 5.10^{-2} Torr for Smith *et al.* [SMI 82]. The latter have determined the critical values of oxygen pressure and substrate temperature, for which the oxide growth is made possible. These critical conditions are illustrated in Figure 1.5 [SMI 82].

Derrien *et al.* [DER 82] measured the oxidation kinetics for pressures ranging between 10^{-4} - 10^{-2} Torr and 700 – 1,050°C. The silicon surface is thermally cleaned under an ultra-high vacuum at 1,100°C, sometimes leaving very little carbon traces. Figure 1.6 [ROC 84] shows a furnace containing a quartz tube heated by the Joule effect, connected to an ultra-high vacuum room. It enables oxidations in a very wide range of pressures (10^{-6} - 60 Torr), under an ultra-dry oxygen atmosphere (less than 3 ppm H_2O).This oxygen can be strongly enriched in ^{18}O (beyond 99%), to study the oxidation mechanisms. The purity and the labeling of the gas are controlled by a mass spectrometer. However, in this device, the water vapor contamination is always possible, if the gas oxygen reacts with hydrogenated species diffusing through the inner surface of the quartz tube.

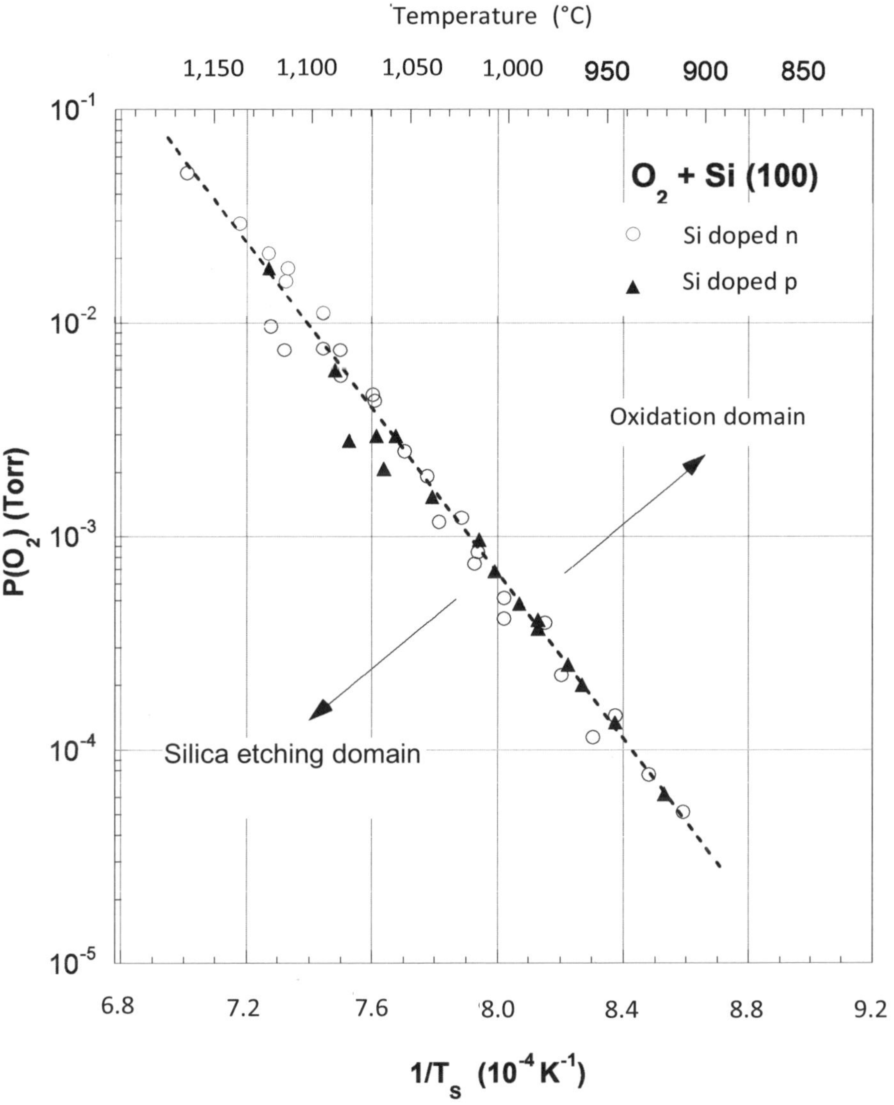

Figure 1.5. *Diagram representing the critical values of P_{O2} oxygen pressure and of T_S substrate temperature beyond which the oxide growth is feasible. If the experimental data of P_{O2} and T_S are placed in the upper right corner of the graphic, oxidation can take place. If they are located in the lower left corner, the silicon surface is cleaned and oxidation is inhibited (according to Smith and Ghidini [SMI 82])*

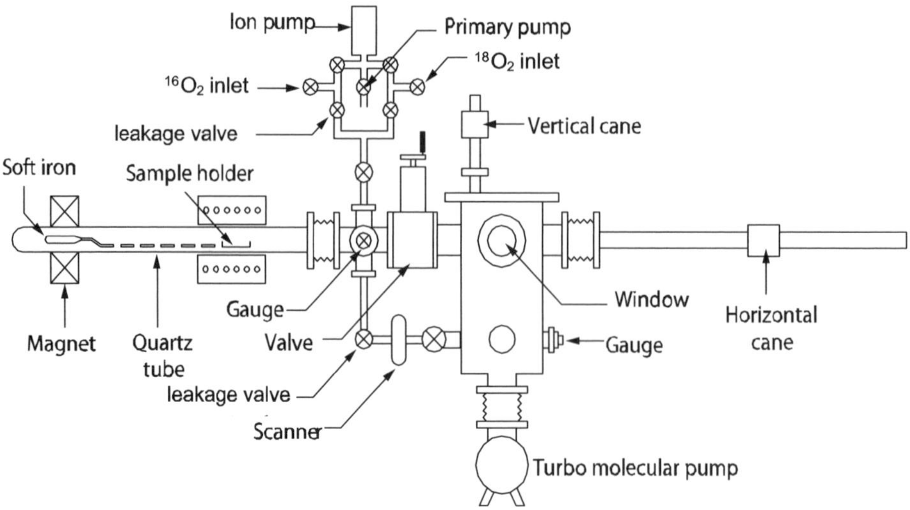

Figure 1.6. *Schematic diagram of the ultra-high vacuum furnace for processing under dry gas of $^{18}O_2$ and $^{16}O_2$ (according to Rochet et al. [ROC 84])*

1.2.3.4. *High pressure oxidation*

Katz and Howells [KAT 79] described a thermal scanning processing system under flowing water vapor at high pressure (see Figure 1.7).The metal walls of the furnace are covered with quartz. Due to their coefficient of thermal expansion, they limit the processing temperature to a maximal value of 750°C. High purity deionized steam is injected into the furnace with a pressure of 20 atm. The pressure regulation is ensured by the opening of a load shedding valve, thanks to a solenoid. The steam flow rate at 20 atm is maintained at 4 liters/min. The oxidation kinetics carried out under such conditions at 725°C were published in [KAT 79]. The latter shows results identical to those obtained at 1,050°C under water vapor at atmospheric pressure.

Thus, 1 μm of oxide is obtained after a 5-hour processing. The oxides carried out at high pressure present higher refractive indexes (1.474 against 1.469), higher densities (2.31 g.cm^{-3} against 2.21 g.cm^{-3}) and lower etching speed rates in buffered HF solutions (855 Å/min against 922 Å/min), than oxides obtained at 900°C under atmospheric pressure of water vapor. Nevertheless, this type of oxide presents similar electrical characteristics, after being carefully submitted to suitable annealing processing (the measured interface defects are about 3×10^{10} cm^{-2}).

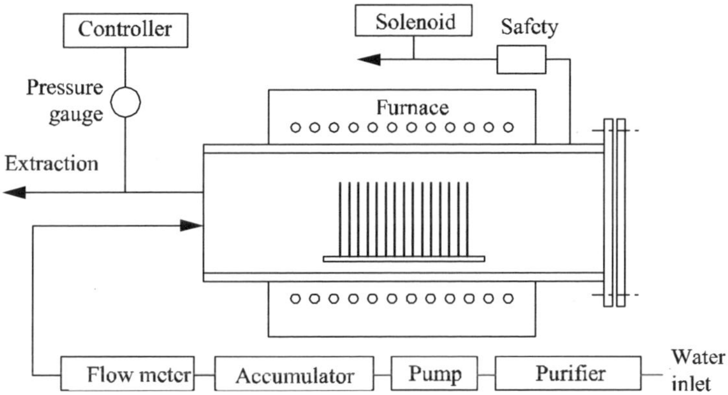

Figure 1.7. *Schematic diagram of a high pressure oxidation furnace (according to Katz and Howells [KAT 79])*

1.3. Some physical properties of silica

In this section, we present some properties of silica, useful for a better understanding of silicon oxidation and the growth models developed in this chapter.

During silicon oxidation, the oxide growth proceeds because a phenomenon of atomic transport takes place through the film (as shown in the following sections).

Thus, we must consider the transport properties of silica, which should depend in theory on its atomic structure, and thus on its physical characteristics.

1.3.1. *The silica structure*

The basic silica structure is represented in Figure 1.8. It is a tetrahedron, with in its center, a silicon atom linked to 4 oxygen atoms located at the corners. Each oxygen atom belongs to 2 adjacent tetrahedrons, while being linked to 2 silicon atoms. The mean value of inter-atomic distances was measured by Mozzi and Warren [MOZ 69] in the fused quartz with a 2.20 g/cm^3 density.

They report the following values:

– 1.62 Å for the Si-O bond;

– 2.62 Å for the O-O bond;

– 3.12 Å for the Si-Si bond.

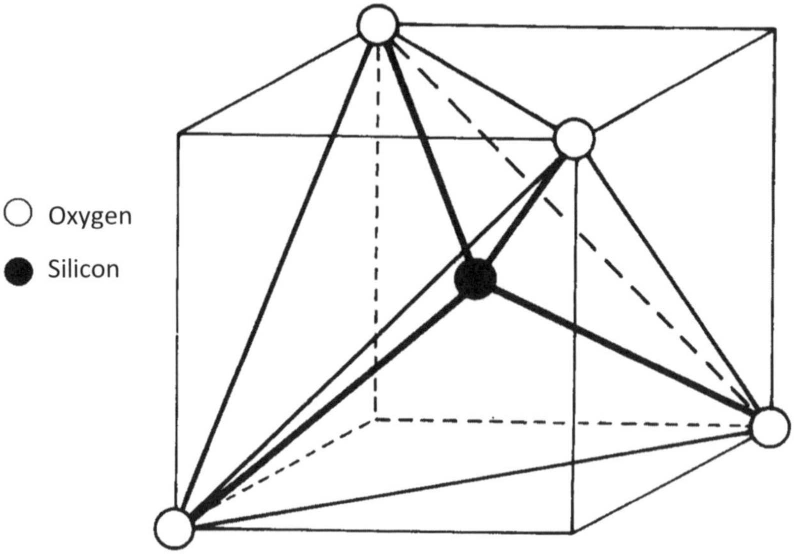

Figure 1.8. *Basic structure of SiO_2*

Crystalline form	System	Angle of Si-O-Si bond (deg.)	Molar volume (cm^3)
Quartz - α	trigonal	143.9	22.7
Quartz - β	hexagonal	146	24.0
Cristobalite - α	tetragonal	148	25.8
Cristobalite - β	cubic	152	27.3
Tridymite - α	orthorhombic		
Tridymite - β	hexagonal		
Silica glass	amorphous	120-180	27.2

Table 1.1. *Structural characteristics of the most widespread crystalline forms of SiO_2 (according to Revesz [REV 70, REV 80])*

The stability zones of silica and polymorphic transformations are represented in Figure 1.9; and the various structures in Figures 1.10, 1.11 and 1.12.

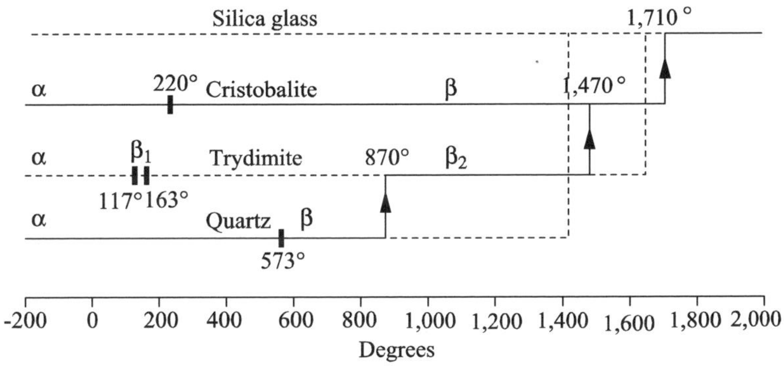

Figure 1.9. *Stability zones of silica at atmospheric pressure and its polymorphic transformations (according to Pascal [PAS 62])*

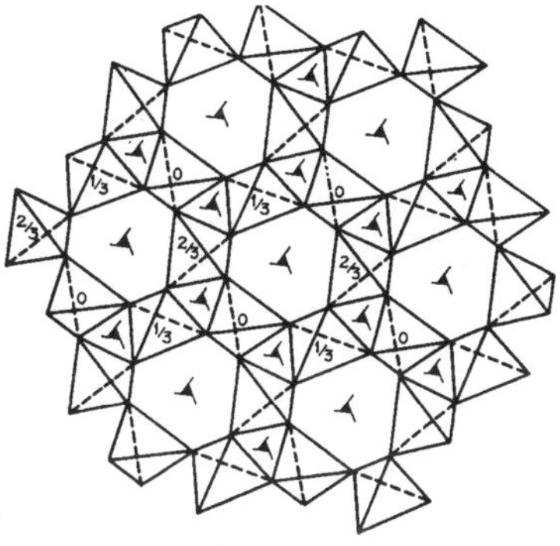

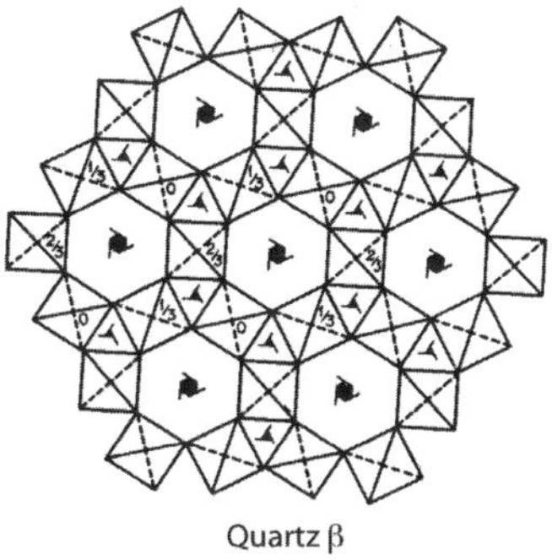

Figure 1.10. *Projection in the (001) plane of structures α and β of the quartz (according to Pascal [PAS 62])*

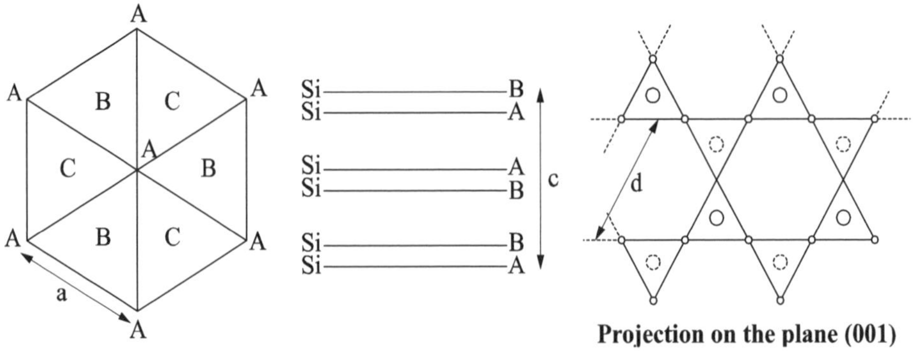

Figure 1.11. *Structure of tridymite-β (according to Pascal [PAS 62])*

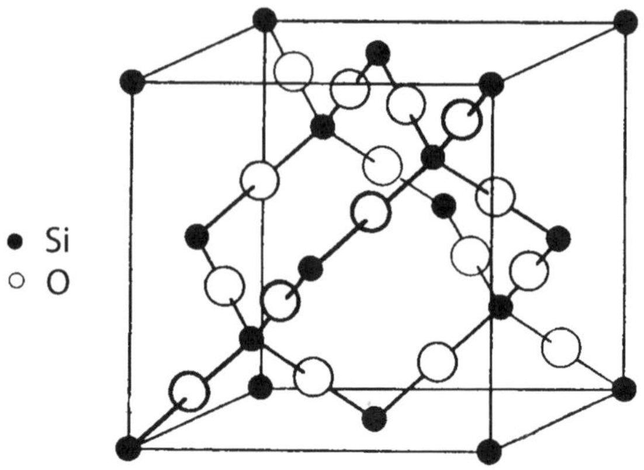

Figure 1.12. *Structure of cristobalite-β (according to Pascal [NOT 62])*

Revesz [REV 80] allocated the wide distribution of measured angles, Φ, for the Si-O-Si bond to the flexibility of the Si-O bond that has three components (covalent-σ, covalent-π and ion (40%)).The contribution of each of its components would vary with Φ without significantly changing the total bonding energy.

Indeed, several configurations of the same energies could coexist (see Table 1.1).

Moreover, it was shown that in the crystalline structures, the π bonding was enhanced along the directions parallel to axis c.

1.3.2. *Three useful parameters of silica*

The properties of vitreous silica depend on its thermal history and on the impurities it can contain [HET 62] (OH, H, Al, etc.). That could explain the wide range of parameters given in Table 1.2.

Refraction index n (at 546 nm)	**Density** ρ ($g.cm^{-3}$)	**Coefficient of thermal expansion** ($10^{-6}.K^{-1}$)
1.460	2.200-2.206	0.5-1.2

Table 1.2. *Some important parameters of silica glass (according to Bruckner [BRU 70])*

The porosity and the transparency of the structure are two of the most important physical characteristics of silica films. We can quantify them by using mass density values. However, for thin films this is not easily directly measurable, contrary to the refraction index. Two equations connect the density ρ, with the refraction index n, of an environment.

The Gladston-Dale equation:

$$\rho = k_1(n-1) \qquad [1.1]$$

The Lorentz-Lorentz equation:

$$\rho = k_2 \frac{n^2 - 1}{n^2 + 2} \qquad [1.2]$$

with k_1 and k_2, constants where k_2 can also be expressed according to the polarizability α and to the molar mass M of silica:

$$k_2 = \frac{3}{4\pi} \frac{M}{6.02.10^{23}} \alpha$$

These equations (see Figure 1.13) are based on the hypothesis that the basic material remains identical in its atomic structure, but with different porosities and optical transparency.

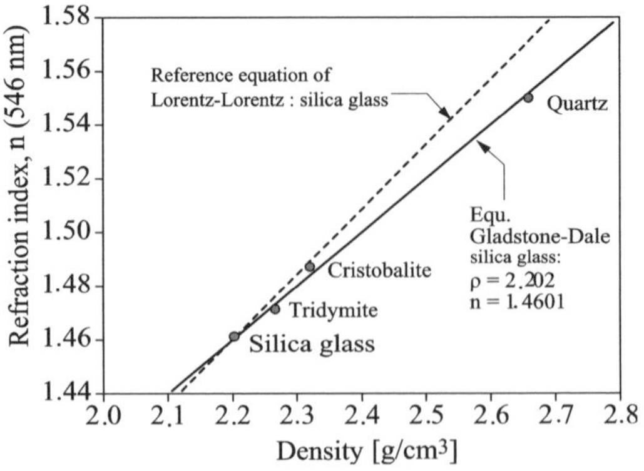

Figure 1.13. *Refraction index of silica according to the density for various SiO_2 forms (according to Pliskin [PLI 77])*

1.3.3. *Transport properties in silica*

To understand the transport mechanisms taking place during oxidation, we should not only know the SiO_2 structure, but also examine the way in which the species diffuse and/or possibly react in amorphous silica. For example, the rare gases diffuse without reacting, whereas hydrogen reacts with its diffusive environment for temperatures higher than 500°C. The species that will be examined are those that we generally encounter during oxidation and annealing processes.

The diffusion of rare gases helps us to define two important concepts: solubility and diffusivity.

1.3.3.1. *Diffusion of rare gases in silica*

The rare gases (He, Ne, Ar, Xe) diffuse interstitially without reacting with the silica network.

1.3.3.1.1. Solubility

Solubility is the aptitude of a network to dissolve a gas which is immersed there at a given temperature and pressure. It is best described by the C* parameter. The latter is the maximum concentration of gas dissolved in the material, after having reached the thermodynamic equilibrium.

The solubility process can be described by two distinct models: the free volume model, suggested by Doremus [DOR 94] and the complete statistical thermodynamics model, developed by Shackelford and other scientists [SHA 72, SHE 77].

The free volume model relies on the idea that the solid network contains a certain accessible volume Vs, remaining constant with the temperature and the pressure, as long as the structure is not modified. In that case, C^* can be expressed as follows:

$$C^* = C_g \frac{V_S}{V_{ox}} \quad [1.3]$$

with C_g, the atomic concentration per unit of gas volume, and V_{ox}, the volume of the solid.

In the model derived from statistical mechanics, the equilibrium between the gaseous and dissolved states of the same species requires us to know that the Gibbs free energies in the gaseous state Gg equals that in the dissolved state Gs. Thus, C^* can be expressed as follows:

$$C^* = K(T)pC_0 \quad [1.4]$$

where p is the gas pressure and C_0 is the concentration of the solubility sites (2.3×10^{21} and 1.3×10^{21} cm^{-3} respectively for He and Ne). K(T) is expressed:

$$K(T) = f(T)\exp(-W_{SG}/RT) \quad [1.5]$$

where W_{SG} is the difference between the potential energy that an atom has at rest, on the one hand, in the dissolved state, and on the other hand, in its gaseous environment
(-1.5 kcal/mole and -2.9 kcal/mole respectively for He and Ne). The expression of f(T) depends on the type and on the degree of freedom assumed [STU 70].

1.3.3.1.2. Diffusivity

Diffusivity is the aptitude of a species to diffuse in an environment. This physical characteristic is the best represented by the diffusion coefficient D, given by the Arrhenius equation:

$$D = D_0 \exp(-W_0/k_B T)$$

However, the experimental results show that D_0 and W_0 would also depend on T, in the range of temperatures considered [PER 71, SHE 77, SWE 61]:

– Sweets [SWE 61] suggests that it could be due to the fact that the considered crystalline forms of silica present strong phase reversal in the range of temperatures [160°C - 280°C].

– Perkins and Begeal [PER 71] expose models derived from the statistical model. They predict that the pre-exponential factor D_0 should be independent from the temperature, if the partition function *fq* corresponding to the displacement in the direction of the diffusion is the same as that of a linear oscillator. If *fq* is the partition function of a linear translation, then D_0 must be proportional to T1/2.

– By taking an interest in the amorphous structure of glass, Shelby and Keeton [SHE 74] suggest that the energy barrier to be crossed to go from one site to another can have several different values with a Gaussian distribution. The adjustment of the experimental results for the helium diffusion (24°C - 100°C) confirms the model, and validates the Doremus hypothesis [DOR 62], predicting a linear dependence in T of D_0.

– The activation energy W_0, of the molecular diffusion increases with the radius r, of the diffusing molecule (see Table 1.3). This behavior is described by the model suggested by Anderson and Stuart [AND 54]: W_0 is the potential energy of elasticity necessary to stretch an orifice of radius r_D to the radius r of the diffusing molecule, so that:

$$W_0 = 8\pi G(r - r_D)^2 \qquad [1.6]$$

G being the shear modulus. A reasonable agreement is obtained with experience by taking r_D= 0.6 Å and $G = 1.33\ 10^{10}$Pa [PER 71] as values.

1.3.3.2. *Molecular diffusion in silica*

Gas molecules such as H_2, O_2, H_2O, diffuse in a way similar to rare gases.

As we can see in Table 1.3, the activation energy of the molecules diffusion is similar to that of rare gases with the same radius. However, the interactions and chemical reactions with the environment can complicate the observation and extraction of parameters. Moreover, the presence of impurities traces can strongly influence chemical reactions.

Diffusing part	Temp.[°C]	C* x 10^{17} at.cm^3.atm^{-1}	V_s/V_{ox}	D_0x 10^{-4} cm^2.s^{-1}	W_0 Kcal/mole	Diam. [Å]	Ref.
He	0-1,000	7-2	~0.026			2	[STU 70]
	380-1,034	3.6-2	~0.0311	7.4	6.6		[SWE 61]
	84-300	6.2-3.6	~0.028	3.04	5.6		[SWE 61]
	52-540	4.6-1.9					[SHA 72]
Ne	450-1,000	2.2-1.2	~0.018			2.4	[STU 70]
	80-180	4.6-2.8	~0.019	0.51	9.5		[PER 71]
	400-780	2.1-1.3					[SHA 72]

H_2	310-490 5-560	4.3-2.8 1.2-2.2	~0.021 0.044-0.025	 1.2	 9.3-10.3	2.5	[SHA 72] [SHE 77]
O_2	950-1,078 800-1,200	0.61-0.51 1.7-3.4	~0.01	2.7 0.45-0.67	27 21-23	3.2	[NOR 61] [KAJ 05]
N_2	700-1,000			1	26	3.23	
Ar	670-900	0.8-0.5	~0.009	1.21	28.7	3.2	[PER 71]
D_2	134-144 20-180	 13-10	 ~0.005	7.9 1.5	11.6 9.2		[SHE 74] [PER 71]
Kr					>44b)	3.8	[PER 71]
H_2O Hydroxyl	1,000 ~600-1,200	0,6* 2,000-600	0.01 0.01	2.5 0.0027-0.01	~18 18.3-18.3	3.3 3.3	[MOU 61] [BUR 70] [DOR 69] [MOU 61] [BUR 70]
O-O exchange	925-1,225 900-1,200 1,010-1,220			151 4.3 10^{-3} 3.7 10^{-5}	71.2 56 55		[SUC 63] [HAU 69] [HAU 69]
O-O exchange in quartz	850-1,250			2 10^{-5}	29		[WIL 65]
Na Silica type 1	170-250 250-570 570-1,000			2.1-300 3,980 344	18 25.8 21.1	1.9	[FRI 68] [FRI 68] [FRI 68]
Na Silica type 3	740-1,200			59.4	35.2	1.9	[FRI 69]
CO	1,100	$DxC^* = 0.9 \times 10^8$ at.cm-1.s-1.atm-1					[CAV 09]

a) Deposited films
b) 54 is the value of the activation energy for permeability
*Value estimated for Vs/Vox ~ 0.01

Table 1.3. *Values of the solubility parameters (C^*, Vs/Vox) and of the diffusivity parameters (D_0, W_0) for various elements. Vs/Vox is calculated according to the hypothesis of free volume*

1.3.3.2.1. Hydrogen diffusion

The hydrogen primarily diffuses without reacting up to 500°C [SHE 77, VAN 75]. At higher temperatures, Si-OH groupings are formed and the equilibrium concentration of these hydroxyls varies as the square root of the molecular gas pressure [BEL 62]. Hetherington *et al.* [BEL 62, HET 64] propose the following reactive mechanism:

$$\bullet\underset{\bullet}{\overset{\bullet}{\mathrm{Si}}}{}^{4+}-\mathrm{O}-\underset{\bullet}{\overset{\bullet}{\mathrm{Si}}}{}^{4+}\bullet+\frac{1}{2}\,\mathrm{H_2}\rightarrow\ \bullet\underset{\bullet}{\overset{\bullet}{\mathrm{Si}}}{}^{3+}+\ \mathrm{HO}\text{-}\underset{\bullet}{\overset{\bullet}{\mathrm{Si}}}{}^{4+}\bullet \qquad [1.7]$$

Whereas Van der Steen and Papanikolau [VAN 75] propose the following:

$$\bullet\underset{\bullet}{\overset{\bullet}{\mathrm{Si}}}-\mathrm{O}-\underset{\bullet}{\overset{\bullet}{\mathrm{Si}}}\bullet+\mathrm{H_2}\rightarrow\ \bullet\underset{\bullet}{\overset{\bullet}{\mathrm{Si}}}-\mathrm{H}+\ \mathrm{HO}\text{-}\underset{\bullet}{\overset{\bullet}{\mathrm{Si}}}\bullet \qquad [1.8]$$

This hypothesis cannot be checked by measurement of infrared absorption, because SiH groups cannot be detected; their peak of absorption being embedded in the background noise of the network.

1.3.3.2.2. Water diffusion

Water diffuses in silica, while reacting to it [BUR 70, MOU 61]. The concentration of dissolved OH, C_{OH} is proportional to the square root of the water vapor pressure P_{H2O} [HET 62, MOU 61]. The equilibrium reaction can be written as:

$$\bullet\underset{\bullet}{\overset{\bullet}{\mathrm{Si}}}-\mathrm{O}-\underset{\bullet}{\overset{\bullet}{\mathrm{Si}}}\bullet+\mathrm{H_2O}\leftrightarrow\ 2\ \mathrm{HO}\text{-}\underset{\bullet}{\overset{\bullet}{\mathrm{Si}}}\bullet \qquad [1.9]$$

Several transport mechanisms have been considered:

a – associated diffusion of protons and OH groups [ROB 66];

b – interstitial diffusion of the water molecule. The latter is associated with the more or less quick reaction [1.9] of water with the network [DOR 69];

c – water diffusion in the ambipolar form of hydronium ions (H_3O^+) and hydroxyl OH^-, to form SiOH after the reaction with the network.

The b mechanism is proposed by Doremus [DOR 69]. It seems to better explain the experimental results. The water vapor will react in the network to form SiOH groups. Those can, in their turn, react two to two to release a water molecule, which will be able to diffuse and react in its turn. The diffusion process of the water vapor is complex.

At high temperature (T > 1,000°C), the reaction rate is much more important than the diffusion process. The concentration of SiOH, [OH] of H_2O, [H_2O], and the effective diffusion coefficient, D_{OH} will be given by the relation:

$$D_{OH} = 4D \frac{[H_2O]}{[OH]} = 4D\, k_{OH} [OH] \qquad [1.10]$$

D, being the diffusion coefficient of the water molecule, k_{OH} is the equilibrium constant of the reaction given by the relation:

$$[H_2O] = k_{OH} [OH]^2 \qquad [1.11]$$

Experimentally, we find the diffusion coefficient of D_{OH} is proportional to the concentration of hydroxyls groups.

At low temperature (T < 1,000°C), the reaction rate becomes smaller and the hypothesis of a very fast reaction is no longer valid.

The Doremus model seems to be in agreement with the experimental results obtained by isotopic labeling methods, consisting of doing an oxidation under water vapor isotopically enriched (tritium, deuterium [BUR 70, HOL 70], ^{18}O [LIG 60, MIK 81]) with a silica already saturated in natural OH.

The calculation of isotopic profiles is detailed in the reference [RIG 82].They are in harmony with the experimental results.

1.3.3.2.3. Oxygen diffusion

The solubility and diffusivity of the oxygen molecule have been deduced from permeation measurements made by Norton [NOR 61]. Several authors also point out an exchange mechanism of the oxygen, when we submit natural silica to a thermal process under an atmosphere of enriched oxygen in ^{18}O [SUC 63, WIL 65]. Nevertheless, the values of the diffusion coefficients found strongly vary from one experiment to another. They do not help us to get to a conclusion, while also raising the question of the purity of the gases used, notably their level of water vapor.

Is the oxygen exchange an intrinsic phenomenon between the interstitial O_2 molecules and the silica network, or is it the result of another transport mechanism lacunar? Several authors tried to answer this question.

D.A. Hutchingson [HUT 54] did not find an exchange with silica for temperatures going up to 1,000°C, and Rochet [ROC 84] showed that there was also no exchange between silica and dry oxygen at 930°C.

Conversely, we know that water vapor exchanges a lot with silica. This makes the following hypothesis plausible: uncontrolled water vapor traces could be responsible for the exchanges observed in the different works [SUC 63, WIL 65].

Thus, molecular oxygen diffusion without reaction in silica seems to be the dominating phenomenon, even if the following reaction has been reported, in the case of silica without hydroxyls groupings:

$$\bullet\overset{\bullet}{\underset{\bullet}{Si}}{}^{3+}\ \overset{\bullet}{\underset{\bullet}{Si}}{}^{3+}\bullet + \frac{1}{2}O_2 \rightarrow \bullet\overset{\bullet}{\underset{\bullet}{Si}}{}^{4+}\text{-O-}\overset{\bullet}{\underset{\bullet}{Si}}{}^{4+}\bullet \qquad [1.12]$$

If this reaction occurs, it is very difficult to prove, because of the difficulty in obtaining water deprived of oxygen gas.

Silica is a very stable material. In its most reduced form, it can be written SiO_{2-x}. The value of x, measured by UV absorption was found to be equal to 3×10^{-5}.

1.4. Equations of atomic transport during oxidation

In the previous sections, we presented most techniques leading to the formation of silica layers on silicon, either by oxidation or deposition. In this section, we will limit ourselves to the various physical aspects of oxidation.

The transport equations presented in this section are valid for any oxidation process; silicon oxidation being an application of the general equations.

During oxidation, the oxide growth can be done only if at least one of the oxide components is transported through the film during growth. The latter will be identified as mobile species. The subjacent principle of this transport is that mobile species move by successive hops from site to site, if they are available to host them.

In a perfect single crystal, each atom occupies a well-defined site. The transport of species through a crystal thus presupposes the existence of *defects*. They can be *interstitial*, if the species migrate in an interstitial position or *lacunar*, if the species are placed in a substitutional position compared to the atoms of the crystal lattice. The interstitial defects can be occupied by atoms, molecules or ions. These species are also called interstitial, following the example of the sites they occupy.

In single crystals, the concept of defects is clearly defined. Indeed, each atomic site has a precise position in the periodic network. The situation is more complex in a vitreous material such as silica, where atoms have a random distribution in the amorphous atomic network.

In this case, the hopping distance can enormously vary according to the position of the atom. We must keep that in mind, when we use simplified models, such as the one described below, from which the transport equations are deduced.

Nevertheless, we will use, independently of the diffusive environment, the expressions of mobile species and of mobile defects, to describe the subjacent transport mechanisms.

1.4.1. *Transport equations in the general case*

To simplify, we will use a model where transport is carried out in only one dimension of space, with two parallel planes (1) and (2), distant of a and perpendicular to the direction of displacement (a is equal to the distance between two neighboring sites). Let us choose a bench mark, so that the respective positions of the planes are $x+a/2$ and $x-a/2$ (see Figure 1.14).

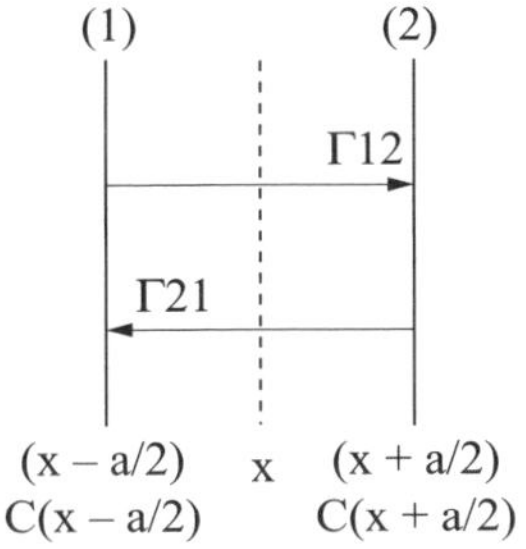

Figure 1.14. *Geometry of the 1D transport model*

Let us call J the flux, expressed in number of defects cm^{-2}.s^{-1}, due to the transport of a type of defect (or of mobile species). Let us note $C(x\text{-}a/2)$ and $C(x\text{+}a/2)$, the respective defects concentrations on the level of the planes, expressed in number per volume unit. Between the planes, we assume that there is a potential barrier that should be overcome to carry out a hop from one site towards another. Γ_{12} and Γ_{21} are the probabilities per time unit, of effective hops between plans (1) and (2) in both directions. J_{12} and J_{21} are the flux derived from it, and J is the resulting net flux expressed as follows:

$$J = J_{12} - J_{21}$$

$$J = C(x - a/2).a.\Gamma_{12} - C(x + a/2).a.\Gamma_{21}$$

$$J = a\left[C(x-a/2).\Gamma_{12} - C(x+a/2).\Gamma_{21}\right] \qquad [1.13]$$

By developing *C(x),* we obtain:

$$J = (\Gamma_{12} - \Gamma_{21})\left[a.C(x) + \sum \frac{a^{2n+1}}{2^{2n}2n!}\frac{\partial^{2n}C(x)}{\partial x^{2n}}\right]$$

$$-\frac{\Gamma_{12}+\Gamma_{21}}{2}\left[a^2.C(x) + \sum \frac{a^{2n+2}}{2^{2n+1}(2n+1)!}\frac{\partial^{2n+1}C(x)}{\partial x^{2n+1}}\right] \qquad [1.14]$$

At this stage, it is necessary to make approximations to make the expression of the transport equation more easily usable. We discuss in the next two sections the two most frequently used approximations.

1.4.2. *First approximation: C(x) varies slowly with the depth x*

If C(x) varies slowly with the depth, we can limit the development to the first term, as follows:

$$C(x \pm a/2) = C(x) \pm \frac{a}{2}\frac{\partial C(x)}{\partial x} \qquad [1.15]$$

and the flux is reduced to:

$$J = a.C(x).(\Gamma_{12} - \Gamma_{21}) - a^2 \frac{\partial C(x)}{\partial x}\frac{\Gamma_{12}+\Gamma_{21}}{2} \qquad [1.16]$$

1.4.2.1. *Application to the transport of neutral species in an isotropic environment*

In such a situation, we have $\Gamma_0 = \Gamma_{12} = \Gamma_{21}$. Thus the first term of the flux is cancelled and reduced to:

$$J = -a^2\Gamma_0 \frac{\partial C(x)}{\partial x} = -D\frac{\partial C(x)}{\partial x} \qquad [1.17]$$

with: $D = -a^2\Gamma_0$.

This is in fact the first Fick law with 1D.

1.4.2.2. *Application to the transport of species charged under electric field* $\varepsilon(x)$

In this case $\Gamma_{12} \neq \Gamma_{21}$. The periodic potential of the network is altered by the electric field, as illustrated in Figure 1.15, for $Zq\varepsilon(x) > 0$, with Zq the electric charge of the species.

We will make the following approximations:

– the distance between a minimum and a maximum of the potential barrier is half the hop distance a. The latter is not altered by the electric field $\varepsilon(x)$;

– the potential is projected in only 1D in x;

– the fluctuation of the potential barrier, due to the electric field between two *minima* is worth $a.\varepsilon(x)$ and $\varepsilon(x)$, is the mean value of the field between $x-a/2$ and $x+a/2$.

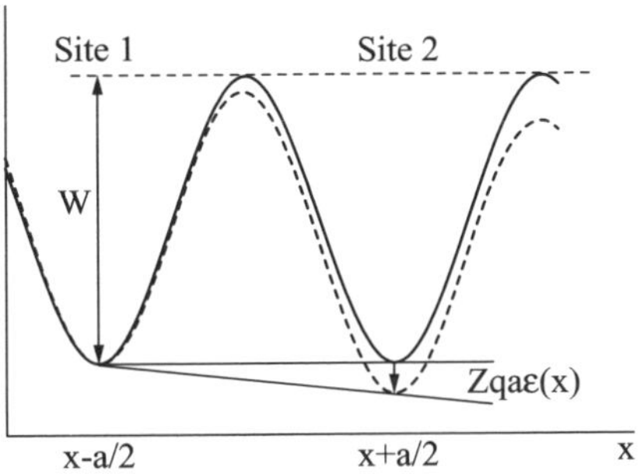

Figure 1.15. *Potential energy of the charged species in presence (----) and without electric field (-)*

When $\varepsilon(x) = 0$, the potential has a periodicity of a. To move from site to site, the mobile species must cross an energy barrier of a height W. Thus, the hopping probability from site 1 to site 2 is expressed as follows:

$$\Gamma_0 = \Gamma_{12} = \Gamma_{21} = \nu.\exp(-\frac{W}{k_B T}) \quad [1.18]$$

$\nu = \nu_0 \exp(\Delta S/k_B T)$ is the frequency coefficient. ΔS is the entropy fluctuation, ν_0 is an oscillation frequency ($\nu_0 \approx 10^{12}\,\mathrm{s}^{-1}$), k_B the Boltzmann constant ($k_B = 8.67 \times 10^{-7}$ eV.K^{-1}) and T the temperature in K.

When $\varepsilon(x) \neq 0$, the potential energy is modified (dotted line in Figure 1.15). Thus, the potential barrier decreases for a hop from 1 to 2, and increases for a hop from 2 to 1:

$$\begin{aligned} \Gamma_{12} &= \nu.\exp\left[-\frac{1}{k_B T}\left(W - \frac{Zqa\varepsilon(x)}{2}\right)\right] = \Gamma_0 \exp\left(\frac{Zqa\varepsilon(x)}{2k_B T}\right) \\ \Gamma_{21} &= \nu.\exp\left[-\frac{1}{k_B T}\left(W + \frac{Zqa\varepsilon(x)}{2}\right)\right] = \Gamma_0 \exp\left(-\frac{Zqa\varepsilon(x)}{2k_B T}\right) \end{aligned} \quad [1.19]$$

Thus:

$$J(x) = 2a.C(x).\Gamma_0.\sinh\left(\frac{Zqa\varepsilon(x)}{2k_B T}\right) - a^2\Gamma_0.\cosh\left(\frac{Zqa\varepsilon(x)}{2k_B T}\right)\frac{\partial C(x)}{\partial x}$$

Two limiting cases must be considered:

1) Low field, if $\left|Zqa\varepsilon(x)/2k_B T\right| << 1$, then by developing the hyperbolic functions at the first order:

$$J(x) = a.C(x).\Gamma_0.\frac{Zqa\varepsilon(x)}{k_B T} - a^2\Gamma_0.\frac{\partial C(x)}{\partial x}$$

$$J(x) = \mu.C(x).\varepsilon(x) - D.\frac{\partial C(x)}{\partial x} \quad [1.20a]$$

with $\mu = a\Gamma_0.\frac{Zqa}{k_B T} = D\frac{Zq}{k_B T}$ (Einstein relation).

2) Strong field, if $\left|Zqa\varepsilon(x)/2k_B T\right| >> 1$, then the flux can be written:

$$\begin{aligned} J(x) &= \beta a.C(x).\Gamma_0.\exp\left|\frac{Zqa\varepsilon(x)}{2k_B T}\right| - \frac{a^2}{2}\Gamma_0.\exp\left(\frac{Zqa\varepsilon(x)}{2k_B T}\right)\frac{\partial C(x)}{\partial x} \\ J(x) &= \beta.a.\Gamma_0.\exp\left|\frac{Zqa\varepsilon(x)}{2k_B T}\right|.\left(C(x) - \frac{a^2}{2}\Gamma_0.\frac{\partial C(x)}{\partial x}\right) \end{aligned} \quad [1.20b]$$

with $\beta = +1$ for $Z\varepsilon(x) > 0$ and $\beta = -1$ for $Z\varepsilon(x) < 0$.

If C(x) varies little from one site to another, the flux is written as follows:

$$J(x) = \beta.a.C(x).\Gamma_0.\exp\left|\frac{Zqa\varepsilon(x)}{k_BT}\right| \quad [1.21]$$

1.4.3. *Second approximation: $\varepsilon(x)$ varies slowly with the depth x*

This case was solved by Dignam [DIG 72], who formulated the following hypothesis:

a) the oxide growth is in a stationary state, i.e. the flux $J(x) = J$ is independent of the depth *x*;

b) the fluctuation of the electric field is sufficiently low to be neglected, thus $\varepsilon(x) = \varepsilon$ and thus:

$$J = -D\frac{\partial C}{\partial x} + C(x)\frac{\sinh(Zqa\varepsilon/2k_BT)}{Zqa\varepsilon/2k_BT} \quad [1.22]$$

There are two extreme cases:

1) Low field so that:

$\sinh(Zqa\varepsilon/2k_BT) \approx Zqa\varepsilon/2k_BT$ i.e. $|Zqa\varepsilon/2k_BT| << 1$

Then:

$$J = -D\frac{\partial C}{\partial x} + C(x)\mu\varepsilon \quad [1.23]$$

2) Strong field if $|\varepsilon| >> \frac{k_BT}{q}\frac{1}{C(x)}\frac{\partial C(x)}{\partial x}$ then:

$$J = -2a.\Gamma_0.C(x).\sinh(Zqa\varepsilon/2k_BT) \quad [1.24]$$

and if, moreover, $|\varepsilon| >> \frac{k_BT}{Zqa}$, then:

$$J(x) = \beta.a.C(x).\Gamma_0.\exp\left|\frac{Zqa\varepsilon}{k_BT}\right| \quad [1.25]$$

Let us notice that if equations [1.23] and [1.25] are similar to equations [1.20] and [1.21], they do not answer to the same starting hypothesis and thus cannot be applied to the same domains of variable parameters. For example, the expression J of equation [1.23] remains valid for field values corresponding to equation [1.21].

Dignam [DIG 72] discusses in detail the validity domains of the hypothesis exposed above and indicates the corrections to be made, when they no longer apply (for very strong fields or high concentrations of charged defects). This study goes beyond the account in this chapter.

1.4.4. Applications of the transport equations to thermal and anodic oxidation

The previous hypotheses do not apply in the same manner for these two types of oxidation.

1.4.4.1. *Anodic oxidation*

In this case, a net current crosses the oxide; the electric field there is intense (ε of about 10 MV.cm $^{-1}$) and the atoms are transported by means of charged atomic defects. By analogy with an ion current, the charged atomic defects represent the ion species and the net current would be the ion current J_i:

$$J_i = a.C_i.\Gamma_0.\exp\left|\frac{Zqa\varepsilon}{2k_BT}\right| \qquad [1.26]$$

with C_i, a concentration of charged atomic species.

1.4.4.2. *Thermal oxidation*

Contrary to what occurs for anodic oxidation, there is no applied external electric field. Consequently, if the charged species take part in the species transport, the sum of all the electric currents (ions and electronic charges) in oxide and at the interfaces must be null. Thus, the ion current J_i must be counterbalanced by the electronic current J_e. We will find more details on the transport mechanisms in section 1.5.

At this stage, let us note that the transport equations have been established without knowing the nature of the mobile species, nor by which mean they move. If we want to obtain the laws controlling the growth during thermal or anodic oxidation, we must answer in each case the two following questions: which are the mobile species and how do they move? The answers are given in the following section, with two different experimental approaches.

1.5. Is it possible to identify the transport mechanisms taking place during oxidation?

1.5.1. *Identification using isotopic labeling techniques*

We present here the principles of isotopic labeling techniques, using a stable ^{18}O oxygen isotope. Silicon isotopes can be the subject of a similar reasoning, if we resort to it.

Silicon oxidation is carried out in at least two stages: a first natural oxygen oxidation (H_2O, O_2, O_3, etc.) upto a thickness x_0, then a second oxidation in oxygen strongly enriched in ^{18}O up to a thickness $x_0+\Delta x$. Various oxidation mechanisms will produce specific concentration profiles of the oxygen isotopes in the oxide film. The measurement of these profiles helps us to determine the oxidation mechanisms implied in the film growth.

Let us consider the mechanisms that could occur.

1.5.1.1. *First hypothesis: silicon is the only mobile species*

Whatever the implied mechanism, the ^{18}O will be set at the external surface of the oxide, with an abrupt interface $^{18}O/^{16}O$ (see Figure 1.16a).

1.5.1.2. *Second hypothesis: oxygen is the only mobile species*

The concentration profile of the ^{18}O depends on the implied process.

Some magnitudes must be defined:

– $[O]_n$: atom concentration in the silica network.

– $\left[{}^{18}O\right]_n$: ^{18}O atom concentration in the silica network.

– $\left[{}^{18}O\right]_n / [O]_n$: enrichment in ^{18}O in the silica network.

– $[O]_c$: O atom concentration in the gas dissolved in silica.

– $\left[{}^{18}O\right]_c$: ^{18}O atom concentration in the gas dissolved in silica.

– $\left[{}^{18}O\right]_c / [O]_c$: enrichment in ^{18}O in the gas dissolved in silica.

– $[O]_n = 0.44 \times 10^{23}$ at.cm^{-3}.

– $[O]_c = S \times P$, where S is the solubility and P the oxidation pressure. In the case of O_2 at 1,100°C, solubility is $S \approx 0.5 \text{ x } 10^{23}$ at.cm-3.atm-1 [KAJ 05].

– $[O]_c/[O]_n \approx 1$ ppm for 1 atm at 1,100°C.

– L_g^{18} : enrichment in ^{18}O of the gas.

1.5.1.2.1. Oxygen diffusion without interaction with the silica network

Oxygen diffuses through the already formed oxide during the first oxidation without reacting with it and reacts with silicon to form SiO_2. The ^{18}O atoms are set at the interface oxide/silicon; the $^{16}O/^{18}O$ interface is abrupt and the enrichment in ^{18}O of the new oxide layer is equal to that of gas (see Figure 1.16b).

1.5.1.2.2. Oxygen diffusion with interaction with the silica network

Oxygen diffuses interstitially through oxide (diffusion coefficient *D),* using a chemical reaction with it. Oxygen atom exchange occurs between the silica networks formed during the first oxidation and the oxidizing gas. γ is the exchange frequency of O atoms in the molecules of the oxidizing species and in the silica network. At the same time, oxygen reacts with silicon at the Si/SiO_2 interface, in order to form a new oxide. Due to the isotopic exchange, the enrichment in ^{18}O at the interface is lower than that of gas, and the amount of ^{18}O is higher than the one corresponding to the thickness increase. The shape of the ^{18}O profile after the second oxidation depends on the oxygen exchange frequency and on the growth rate of the oxide. The complete calculation of the profile is given in the reference [RIG 82], when the oxide growth is negligible. The calculation of various profiles, in the presence of oxide growth, is given in the reference [RIG 88].

We show here the cases where the differential equations are simplified, by considering in particular that the equilibrium is reached, i.e. that the concentration of the dissolved oxidizing species in the film $[O]_c$ is independent of the depth *x* and of the oxidation time *t*.

According to the exchange frequency γ, the oxygen profiles are as follows (see Figure 1.16c).

High exchange frequency

The exchange frequency γ is important ($\gamma t \gg 1$): isotopic equilibrium is reached (enrichment in ^{18}O is the same in the gaseous phase and in the silica network). The profile of ^{18}O is very close to a complementary error function in the presence, or not, of oxide growth:

$$\left[{}^{18}\text{O}\right]_n (x,t) = L_g^{18}\,[\text{O}]_n\, erfc\left(\frac{x}{2\sqrt{D^* t}}\right)$$

$$D^* = D\left(1+\frac{[\text{O}]_n}{[\text{O}]_c}\right) \approx D\frac{[\text{O}]_n}{[\text{O}]_c}$$

Low exchange frequency

The exchange frequency γ is small (γt << 1): isotopic equilibrium is not reached and the enrichment of the silica network is lower than that of the gaseous phase:

– in the absence of oxide growth, the profiles are as follows:

$$\left[{}^{18}\text{O}\right]_c (x) = L_g^{18}\,[\text{O}]_c \frac{\cosh\left[(x-x_0)/\lambda\right]}{\cosh(x_0/\lambda)}$$

$$\left[{}^{18}\text{O}\right]_n (x,t) = L_g^{18}\,[\text{O}]_c \frac{\cosh\left[(x-x_0)/\lambda\right]}{\cosh(x_0/\lambda)} \times \gamma t$$

$\lambda = \sqrt{D/\gamma}$ is the characteristic length of exchange, i.e. the characteristic distance between two events, during which a specific molecule of the oxidizing species exchanges one of its atoms with the silica network;

– in the presence of oxide growth, the general form of the oxide profiles will be:

$$\left[{}^{18}\text{O}\right]_c = A\cdot e^{-x/\lambda} + B\cdot e^{+x/\lambda}, \text{ with } \left[{}^{18}\text{O}\right]_c (x=0) = L_g^{18}\,[\text{O}]_c$$

$$\left[{}^{18}\text{O}\right]_n = \gamma t\cdot\left[{}^{18}\text{O}\right]_c$$

Taking into account an additional boundary condition allows us to solve the equation system.

For the linear regime, this boundary condition is: $\frac{\partial}{\partial x}\left[{}^{18}\text{O}\right]_c\Big|_{x=x0} = 0$.

The profiles will thus be:

$$\left[{}^{18}\text{O}\right]_c(x) = L_g^{18}[\text{O}]_c \frac{\cosh\left[(x-x_0)/\lambda\right]}{\cosh(x_0/\lambda)}$$

$$\left[{}^{18}\text{O}\right]_n(x,t) = L_g^{18}[\text{O}]_c \frac{\cosh\left[(x-x_0)/\lambda\right]}{\cosh(x_0/\lambda)} \times \gamma t$$

For the parabolic regime, the boundary condition is:

$$\left[{}^{18}\text{O}\right]_c(x=0) = L_g^{18}[\text{O}]_c$$

and the profiles are as follows:

$$\left[{}^{18}\text{O}\right]_c(x) = L_g^{18}[\text{O}]_c \frac{\sinh\left[(x-x_0)/\lambda\right]}{\sinh(x_0/\lambda)}$$

$$\left[{}^{18}\text{O}\right]_n(x,t) = L_g^{18}[\text{O}]_c \frac{\sinh\left[(x-x_0)/\lambda\right]}{\sinh(x_0/\lambda)} \times \gamma t \qquad [1.27]$$

1.5.1.2.3. Oxygen diffusion by a gradual movement of oxygen atoms of the network

The oxygen transport gradually takes place by a movement of oxygen atoms of the silica network. The exchange of oxygen atoms of the gas with those of the silica network occurs only on the external surface of oxide (and not in the volume).This exchange is made possible by the specific configuration of this surface, which looks like a break in the existing volume structure. At the same time, the diffusion of isolated defects present in the oxide involves the diffusion in the oxide volume of the ^{18}O atoms of the surface. We postulate the transfer coefficient between gas and silica to be sufficiently high, so that all the atoms of the surface are completely exchanged in a very short time.

Movement in all directions

If the defects migrate in all directions, according to a simple diffusion process, there is an isotopic exchange, and the total amount of ^{18}O is higher than the quantity fixed by the growth. In the absence of oxide growth, the profile is a complementary error function (erfc). Without any oxide growth, the calculation [ROC 81] shows that the profile remains close to an erfc function shape, in all cases, for neutral, single or double charged defects (see Figure 1.16d):

$$\left[{}^{18}\text{O}\right]_n(x,t) = L_g^{18}\,[\text{O}]_n\, erfc\left(\frac{x}{2\sqrt{D^* t}}\right)$$

$$D^* = D_d \frac{C_d}{[\text{O}]_n}$$

where C_d and D_d are respectively the concentration and the diffusion coefficient of the defects.

Movement in only one direction

If the defects migrate in only one direction (case of a charged defect in an electric field), the ^{18}O atoms move only in the direction of the growth. In this case, no oxygen atom leaves the silica network, and there is no isotopic exchange. The boundary ^{18}O /^{16}O has a light mix, due to the statistics of hops. The quantity of set ^{18}O corresponds to the oxide growth (see Figure 1.16e).

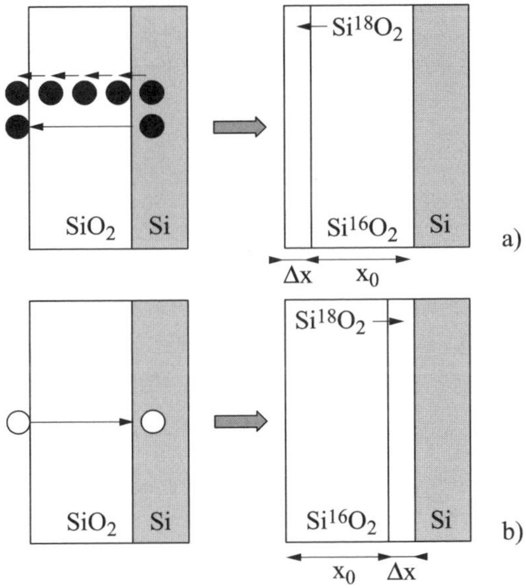

Figures 1.16. *Principle of isotopic labeling with ^{18}O applied to put the mechanisms of silicon oxidation in the spotlight. Silicon is first oxidized in $^{16}O_2$ and then in $^{18}O_2$. Five types of transport mechanisms are represented, as well as the corresponding ^{18}O profiles. The black circles represent the Si atoms, the white circles, the oxygen atoms. a) Only silicon is mobile; b) oxygen diffuses through the already formed oxide during the first oxidation without reacting with it, and reacts with silicon to form SiO_2*

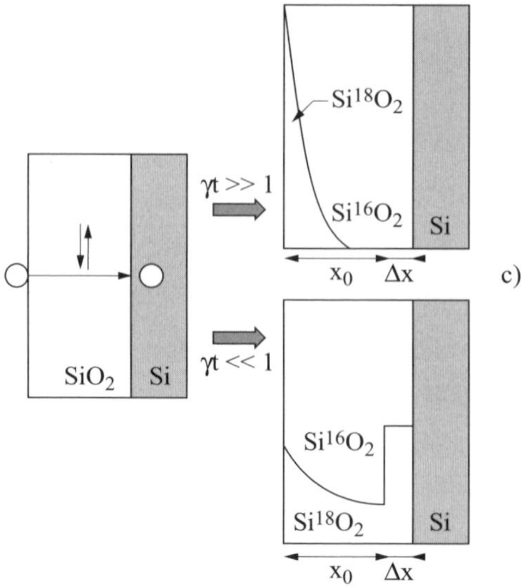

Figure 1.16c. *As previously, with c) oxygen diffuses interstitially through oxide by exchanging its oxygen atoms with it, with an exchange frequency* γ. *The form of the profile depends on the value of* γ

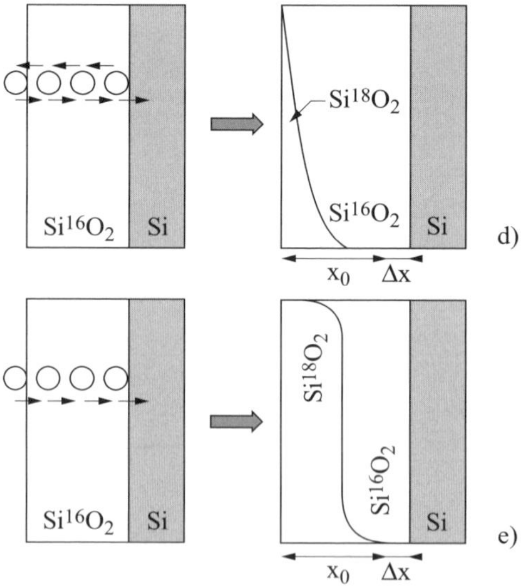

Figures 1.16d and e. *As previously, with d) and e) oxygen transport occurs by a gradual movement of oxygen atoms of the silica network, because of the presence of diffusive defects. These defects can migrate in all the directions (d) or in only one direction (e)*

1.5.2. *Important results for the thermal oxidation of silicon under dry O_2*

1.5.2.1. *Films with thickness higher than 20 nm*

Figure 1.17 shows the typical profile of ^{18}O obtained after sequential $^{16}O_2/^{18}O_2$ oxidations.

The ^{18}O profile was measured by using the narrow resonance at 151 keV of the ^{18}O (p, α) 15N reaction [BAT 92].

^{18}O oxygen is localized on the external surface of oxide and at the SiO_2/Si interface. The ^{18}O profile, close to the external surface of oxide could be adjusted with an erfc profile. The one in proximity to the SiO_2/Si interface can be adjusted by a "box" profile, with an enrichment equal to that of the oxidizing gas.

The ^{18}O concentration in the oxide volume was set at 0.2% (concentration in 18O of natural oxygen). The ^{18}O profile at the interface shows that $^{18}O_2$ oxygen diffuses through oxide, without reacting with it, and reacts with silicon at the SiO_2/Si interface. This result validates the transport model suggested by Deal and Grove [DEA 65].

The ^{18}O profile on the surface, which can be adjusted by an erfc function, is then explained by two mechanisms:

– gradual movement of oxygen atoms of the network, induced by the presence of defects in oxide;

– diffusion with exchange between oxygen and the network of an oxygenated species. This will be different from those responsible for fixing ^{18}O at the interface (that is to say enriched ^{18}O steam).

The hypothesis, according to which the water vapor would be responsible for the fixing of ^{18}O on the surface, was rejected by Rochet *et al.* [ROC 84]. Effectively, the partial pressure of H_2^{18} must be up to two orders of magnitude higher than those measured by mass spectrometry.

We can conclude that heavy oxygen fixation near the surface is due to anisotopic exchange phenomena and the ^{18}O diffusion, induced by the presence of defects in the oxide. Such a mechanism also produces a light mix at the $^{18}O/^{16}O$ interface. In Figure 1.17, where oxidation in $^{18}O_2$ is carried out at 930°C, this mix is not detectable. On the contrary, in Figure 1.18, where oxidation in $^{18}O_2$ is carried out at 1,100°C, this mix is observable.

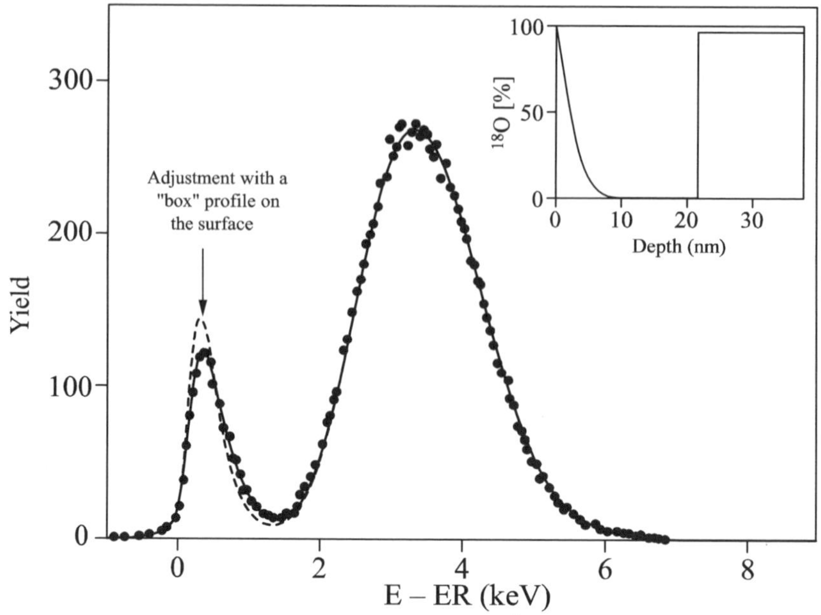

Figure 1.17. *151 keV resonance excitation curve for the $^{18}O\ (p,a)^{15}N$ reaction obtained on a 20.5 nm $Si^{16}O$, on reoxidized Si substrate under 100 mbar of $^{18}O_2$ at 930°C during 5 hours. The 18O profile used to adjust the experimental curve (unbroken line) is also presented: on the external surface, the supposed profile is an erfc function, and is constant close to the SiO_2/Si interface. The dotted line represents the adjustment when the erfc function is replaced by a rectangular profile of the same surface and surface concentration*

Isotopic labeling experiments have been carried out by Costello and Tressler [COS 84] at 1,000°C and at 1,300°C. At 1,000°C, they are qualitatively in harmony with the results mentioned above. At 1,300°C, all the initial oxygen has been exchanged by ^{18}O; the diffusion mechanism of the defects contributing to considerable oxide growth. Other experiments [HAN 88], carried out between 900 and 1,150°C, are also in qualitative harmony with the previous results.

A study using the pressure of oxidation in $^{18}O_2$ as the variable parameter was undertaken, in order to try to identify the defect responsible for the oxygen fixing on the surface [TRI 89]. The peroxyl bridge (O-O bond) has then been proposed, as the defect leading to the ^{18}O diffusion in the proximity of the surface.

It is created on the surface according to the reaction:

$$O_2 + SiO_2 \leftrightarrow 2\ Si\text{-}O\text{-}O\text{-}Si$$

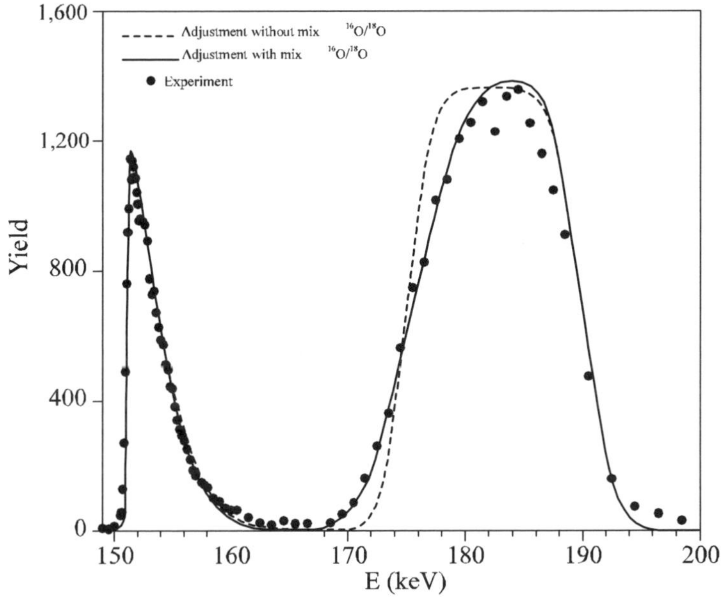

Figure 1.18. *151 keV resonance excitation curve for the reaction ^{18}O (p, α) ^{15}N obtained on a 103 nm $Si^{16}O_2$ on a reoxidized Si substrate under 200 mbar of $^{18}O_2$ at 1,100°C during 4 hours. The dotted line corresponds to the adjustment of the experimental curve with a complementary error function on the surface and a profile of type "box" at the interface. The full line corresponds to the adjustment of the experimental curve by taking into account the $^{18}O/^{16}O$ mix, produced by the presence of defects*

The quantity of ^{18}O fixed at the external surface increases when the initial oxide thickness decreases [HAN 88, ROC 84]. A study [TRI 93] could show that the increase in the fixing of ^{18}O on the surface was not related to the increase of the oxidation rate at the interface. Other mechanisms are to be considered, such as for example the modification in the structure of silica during the growth, altering the diffusion coefficient of the defects.

An oxygen exchange at the SiO_2/Si interface was highlighted by Akermark *et al.* [AKE 99], by means of sequential oxidations $^{16}O_2/^{18}O_2/^{16}O_2$ at 1,100°C (see Figure 1.19). After the first oxidation, the oxide thickness is 120 nm and the growth regime is parabolic. After the second oxidation, the quantity of ^{18}O set at the interface is equivalent to 1 nm. After the third oxidation in $^{16}O_2$, a loss in ^{18}O of 0.3 nm is observed at the interface. This result is not predicted by the Deal and Grove model, for which the oxidant concentration at the SiO_2/Si interface is null, in the regime.

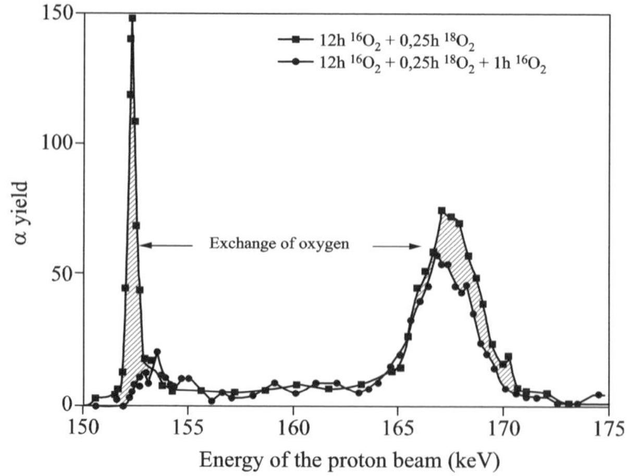

Figure 1.19. *151 keV resonance excitation curves for the reaction ^{18}O (p, α) 15N obtained after the oxidation sequences $^{16}O_2/^{18}O_2$ and $^{16}O_2/^{18}O_2/^{16}O_2$, at 1,100°C and under 210 mbar (from [AKE 99])*

1.5.2.2. *Films with thicknesses lower than 10 nm*

Rochet *et al.* [ROC 86] have grown films of thickness ≤ 6 nm, under an oxygen pressure of 13.8 mbar and in a temperature range from 600°C to 1,090°C. ^{18}O oxygen is fixed primarily on the surface. At temperatures ranging 810°C – 1,090°C, an exchange phenomenon is observed. The dominant transport mechanism is thus the interstitial transport of oxygen via defects. At 600°C, no exchange is observed, which means that the growth occurs by the movement in only one direction of the oxygen atoms of the network. This mechanism implies the transport of charged defects in the presence of an electric field.

Ganem *et al.* [GAN 92, GAN 93b] carried out sequential oxidations $^{16}O/^{18}O$ by rapid thermal oxidation. An oxide film with an initial thickness of 5 nm, reoxidized in $^{18}O_2$, up to a thickness of 10 nm presents a profile similar to films obtained by conventional thermal oxidation. For a film with an initial thickness of 3 nm, reoxidized in $^{18}O_2$, up to a thickness of 5 nm, we observe the presence of a pure $Si^{18}O_2$ layer at the interface, but also the fixing of ^{18}O in the oxide volume with a concentration of up to 30%. These results can be explained by the presence of silicon fragments in these very thin oxides. These non-oxidized silicon fragments can be the result of the roughness of the silicon surface, before oxidation and injection of silicon atoms in the SiO_2/Si interface, during oxidation. Direct observations of such fragments have been made by electron microscopy in transmission [GIB 89] and by electron diffraction [OCR 86].

In the case of the 3 nm film, the oxidation of these fragments is incomplete and continues during the second oxidation in $^{18}O_2$. This leads to the fixing of ^{18}O in the oxide volume. In the case of the 5 nm film, the fragments would be completely oxidized and the reoxidation under $^{18}O_2$ does not cause any more fixing of ^{18}O in the oxide volume. This experimental result could explain the first stages of oxidation, characterized by a regime of rapid growth. Indeed, the oxidation rate at the SiO_2/Si interface is proportional to the number of reactive sites. Therefore, as the oxidation rate at the SiO_2 interface is proportional to the number of reactive sites, the presence or the formation of silicon fragments or of a roughening of the interface, would involve an increase amongst reactive sites, and thus of the oxidation rate. When these silicon fragments are finally oxidized, the oxidation rate drops to reach its value in steady state.

1.5.3. *Important results for wet thermal oxidation*

$Si^{16}O_2$ silica layers with thicknesses ranging between 90 and 250 nm have been thermally processed in water vapor strongly enriched in ^{18}O (> 93%), under an atmosphere of 13 Torr and at temperatures from 400 to 900°C [RIG 82].

Regardless of the oxidation durations and temperatures, the fixed quantities of ^{18}O are higher than those corresponding to the increased thickness, revealing a strong isotopic exchange.

At 930°C, the ^{18}O oxygen profiles could be adjusted with an erfc function. From 400°C to 600°C, most ^{18}O profiles could be adjusted with a "cosh" function.

These results indicate that oxygen diffuses interstitially through oxide, while reacting with it.

The equilibrium reaction is as follows:

$$H_2O + Si\text{-}O\text{-}Si \leftrightarrow 2Si\text{-}OH$$

1.5.4. *Conclusions on the atomic transport mechanisms during silicon thermal oxidation*

In all the cases exposed here, oxygen is identified as the mobile species.

For this reason, the choice of oxygen as an isotopic label should be the most suitable.

1.5.5. *Experimental results and conclusions on the transport mechanisms during the anodic oxidation of silicon*

It was shown that water traces (typically 0.4%) contained in organic solvents used for anodic oxidation were the main oxygen source for the oxide growth [CRO 71a].

In the Croset *et al.* experiment [CRO 76], described below, it was intentionally added to the solvent, 0.4% of water enriched in ^{18}O (i.e. $H_2{}^{18}O$).

The oxides were initially grown in a bath of dihydroxydiethyl ether enriched in ^{18}O, to reach 150 V, i.e. approximately 90 nm of oxide thickness. These oxides were then reoxidized, but in a solvent containing 0.4% of $H_2{}^{16}O$ for voltages from 180 to 300 V (108 to 180 nm).

Figure 1.20 represents the profiles of ^{18}O obtained by SIMS (secondary ion mass spectroscopy) for various phases of the second stage of oxidation. We clearly see that the sequence $^{18}O/^{16}O$ is preserved, but a mixture 16/18 appears, and a loss of ^{18}O is highlighted.

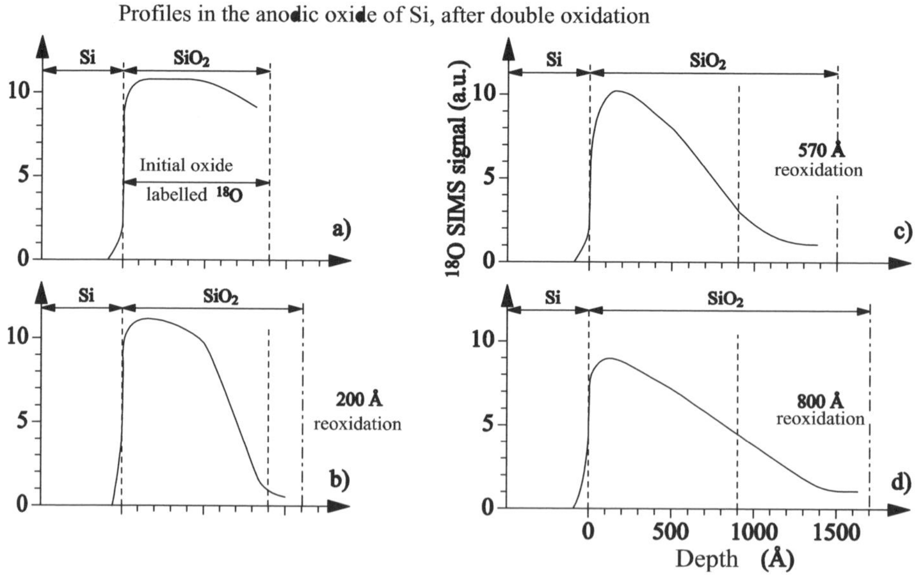

Figure 1.20. *^{18}O profiles measured by SIMS: a) after oxidation in ^{18}O; b) after a reoxidation in ^{16}O of 20 nm; c) after a reoxidation in ^{16}O of 57 nm; d) after reoxidation in ^{16}O of 80 nm*

These results show that, during anodic oxidation, the oxygen atoms move in a gradual displacement, in the direction of the electric field, as well as in the opposite direction.

Knowing that at room temperature, only the charged species move, this result requires careful study of the transport mechanisms at an atomic scale. The transport in two directions could be explained in two ways. Either the mobile oxidizing species exist under two states of charge (positive and negative), or they can switch from one state to the other during transport.

1.5.6. *Important experimental results from dry SiC thermal oxidation*

Silicon carbide is a semiconductor with wide gap, interesting for the high frequencies and high power applications. Its oxidation produces SiO_2. Various oxidation models have been proposed [HIJ 09, SON 04], based on the adjustment of the oxide growth kinetics. These models are derivative from the Deal and Grove model for silicon oxidation, and take into account the CO (or CO_2) production during the oxidation reaction of SiC under O_2. Isotopic labeling has been used for the study of the oxidation mechanisms of SiC (see [VIC 07] for a review of the results obtained by this technique).

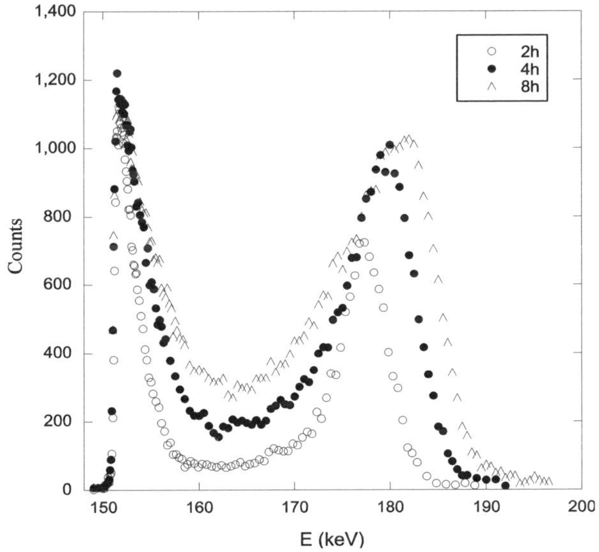

Figure 1.21. *Excitation curves of the resonance at 151 keV of the ^{18}O (p, α) ^{15}N reaction obtained on an oxide of 109 nm of $Si^{16}O_2$ on a ^6H-SiC substrate (oxidation of the carbon face), annealed under 200 mbar of $^{18}O_2$ at 1,100°C during 2, 4 and 8 hours*

Figure 1.21 represents the excitation curves of the resonance at 151 keV of the ^{18}O (p, α) ^{15}N reaction, obtained after sequential oxidations $^{16}O_2/^{18}O_2$. These curves show that the ^{18}O oxygen is located on the surface and at the interface, as in the case of silicon oxidation, but also in the oxide volume. The interaction of the CO molecule with silica was studied by isotopic labeling [CAV 09].

Results obtained during the $^{13}C^{18}O$ processing of $Si^{16}O_2$ films on silicon showed that ^{18}O oxygen has been fixed on the surface (gradual transport of oxygen atoms induced by the presence of defects) and in the oxide volume (CO diffusion through oxide with a low exchange frequency). The results of Figure 1.20 could thus be interpreted as the highlighting of CO production, during SiC oxidation.

1.6. Transport equations in the case of thermal oxidation

In section 1.4, we established the transport equations for oxidation in general and we saw how these are applied for thermal and anodic oxidation. In section 1.5, we have shown how it is possible to experimentally identify the species moving during oxidation and by which means they move, for the two types of oxidation. In this section, we will pay more attention to detailing how transport equations are applied in the case of thermal oxidation. The proceedings developed in this section could be applied in the case of silicon oxidation, as well as with other materials.

1.6.1. *General information on flux and on growth kinetics*

If Ni is the number of "defects" or of mobile species necessary to form a unit oxide volume, then the growth rate is expressed as follows:

$$\frac{\partial X}{\partial t} = \frac{|J|}{N_i} \qquad [1.28]$$

with X, the oxide thickness, t, the oxidation time and $|J|$, the flux amplitude of mobile species, where the growth occurs (SiO_2/Si interface in the case of the Si oxidation). The problem is to determine J, and two scenarios arise:

– mobile species are neutral;

– mobile species are charged.

1.6.2. *Flux calculation for neutral mobile species*

In this case, the flux is expressed by:

$$J = -D\frac{\partial C(x)}{\partial x} \qquad [1.29]$$

If the flux is conservative (there is no accumulation of mobile species in oxide), $\frac{\partial C(x)}{\partial x}$ is constant and could be calculated from the concentrations of the mobile species present at the gas/oxide interface C_s and at the oxide/silicon interface C_{in} (see following section).

1.6.3. *Flux calculation for ion mobile species*

Even though no external electric field is applied during a thermal growth for reasons of equilibrium, the chemical potential of the substrate and of the dissolved oxygen creates a potential V, through oxide.

1.6.3.1. *Case of very thin films ($X \leq 3$ nm)*

This theory has been developed by Cabrerra and Mott [CAB 49]. When the oxide thickness is lower than 3 nm, the electric field resulting from the potential can be very large. Thus, the J_i flux of "atomic defects" is given by:

$$|J_i| = a.C_i.\Gamma_0.\exp\left|\frac{Zqa\varepsilon(x)}{2k_BT}\right|$$

Cabrera and Mott showed that the interfaces (gas/oxide and oxide/substrate) determine the J_i flux. The product $a.C_i$ represents the density of mobile species injected to the interfaces, and Γ_0 the hopping probability per unit of time. The potential barrier that the species must cross at the interfaces is *a priori* different from the one existing within oxide.

If the film is sufficiently thin, an electron flux by tunneling current J_e will come to counterbalance the ion flux J_i. If the concentration of charged species is negligible in oxide, we will then be able to neglect the variations of $\varepsilon(x)$ with the depth, and we will be able to write:

$$\varepsilon(x) = \frac{V}{X}$$

X being the obtained oxide thickness, thus:

$$|J_i| = a.C_i.\Gamma_0.\exp\left(\frac{|Z|.q.a.|V|}{2k_B.T.X}\right)$$

and the growth rate is given by:

$$\frac{dX}{dt} = \frac{|J_i|}{N_i} = \frac{a.C_i.\Gamma_0}{N_i}.\exp\left(\frac{|Z|.q.a.|V|}{2k_B.T.X}\right)$$

Cabrera and Mott assume that the potential V remains constant during oxidation and that it is determined by the balance currents of the interfaces. Thus:

$$\frac{dX}{dt} = U.\exp\left(\frac{X_1}{X}\right) \qquad [1.30]$$

with $U = \frac{a.C_i.\Gamma_0}{N_i}$ and $X_1 = \frac{|Z|.q.a.|V|}{2k_B.T}$.

By integration, the processing duration t, required to have an oxide thickness X, is:

$$t = \frac{1}{U}.\int_0^X \exp\left(-\frac{X_1}{x}\right)dx \qquad [1.31]$$

This integral can only be calculated digitally [FRO 77]. We will nevertheless be able to give approximations by neglecting, for example, the higher orders in X/X_1. In this case:

$$t \approx \frac{X^2}{UX_1}.\exp\left(-\frac{X_1}{X}\right) \qquad [1.32]$$

and:

$$\frac{X_1}{X} \approx \ln\left(\frac{X^2}{UX_1 t}\right) \quad 1^{st} \text{ approximation} \qquad [1.33a]$$

$$\frac{X_1}{X} \approx A - \ln(t) \text{ 2}^{\text{nd}} \text{ approximation} \qquad [1.33b]$$

A better approximation has been proposed by Ghez [GHE 73]:

$$\frac{X_1}{X} \approx -\ln\left(\frac{t+\tau}{X^2}\right) - \ln(X_1 U) \qquad [1.33c]$$

1.6.3.2. *Case of thick films (X >> 3 nm)*

The internal electric field is now low. The atomic transport is thus now to the diffusion and the conduction of mobile species. In stationary state, the electric current must be equal to zero and the transport of the mobile charged species imposes that an electronic current must be set up, to ensure the electrical neutrality of the system, defined by:

$$\sum_i Z_i q J_i(x) + \sum \gamma q.J_e(x) = 0 \qquad [1.34]$$

where $J_e(x)$ is an electron flux (hole or electron) with $\gamma = +1$ for the holes and $\gamma = -1$ for the electrons. $J_i(x)$ is the flux of the mobile species of index i and of charge Z_i (it is an ion flux).

Thus, $J_i(x)$ and $J_e(x)$ can be expressed by:

$$J_i(x) = -D_i \frac{\partial C_i(x)}{\partial x} + C_i(x)\mu_i \varepsilon(x) \qquad [1.35]$$

$$J_e(x) = -D_e \frac{\partial C_e(x)}{\partial x} + C_e(x)\mu_e \varepsilon(x) \qquad [1.36]$$

with μ_i and μ_e the respective mobilities of ion and electronic species.

In general, the last three equations are difficult to use. However, they can be simplified if we consider that only the ion species move. Thus, the electro-neutrality will be ensured by electronic species of opposite polarity compared to the ion species, and in using the Einstein law:

$$\mu_i = D_i \frac{qZ_i}{k_B T}$$

We can write that:

$$\mu_e = D_e \frac{q}{k_B T} \frac{Z_i}{|Z_i|} \qquad [1.37]$$

Equation [1.34] becomes:

$$J_e(x) = |Z_i| . J_i(x) \qquad [1.38]$$

Thus, relations [1.35] and [1.36] become:

$$\frac{J_i(x)}{C_i(x)\mu_i} = -\frac{D_i}{\mu_i} \frac{1}{C_i(x)} \frac{\partial C_i(x)}{\partial x} + \varepsilon(x) \qquad [1.39]$$

$$\frac{|Z_i| J_i(x)}{C_e(x)\mu_e} = -\frac{D_e}{\mu_e} \frac{1}{C_e(x)} \frac{\partial C_e(x)}{\partial x} + \varepsilon(x) \qquad [1.40]$$

While combining [1.39] and [1.40], we have:

$$J_i(x)\left(\frac{1}{C_i(x)\mu_i} - \frac{|Z_i|}{C_e(x)\mu_e}\right) = -\frac{k_B T}{qZ_i} \frac{1}{C_i(x)} \frac{\partial C_i(x)}{\partial x} - \frac{k_B T}{q} \frac{|Z_i|}{Z_i} \frac{1}{C_e(x)} \frac{\partial C_e(x)}{\partial x} \qquad [1.41]$$

To satisfy the electric neutrality, we have:

$$C_e = |Z_i| C_i \qquad [1.42]$$

$$J_i(x) \frac{1}{C_i(x)} \left(\frac{1}{\mu_i} - \frac{1}{\mu_e}\right) = -\frac{k_B T}{qZ_i} \frac{1}{C_i(x)} \left(1 + |Z_i|\right) \frac{\partial C_i(x)}{\partial x}$$

$$J_i(x) = -\frac{k_B T}{qZ_i} \left(1 + |Z_i|\right) \mu_i \frac{\partial C_i(x)}{\partial x} \frac{1}{1 - \dfrac{\mu_i}{\mu_e}}$$

$$J_i(x) = -\frac{D_i\left(1+|Z_i|\right)}{1-\dfrac{\mu_i}{\mu_e}}\frac{\partial C_i(x)}{\partial x} = -D_{eff}\frac{\partial C_i(x)}{\partial x} \qquad [1.43]$$

with, $D_{eff} = \dfrac{D_i\left(1+|Z_i|\right)}{1-\dfrac{\mu_i}{\mu_e}}$ and if $\left|\dfrac{\mu_i}{\mu_e}\right| \ll 1$ and $D_{eff} \approx D_i\left(1+|Z_i|\right)$.

When several species are mobile, equations [1.34], [1.35] and [1.36] become too complex to be solved analytically. This case can be solved when electrons and holes are at the equilibrium and when their mobility is much higher than the ion mobility. The other cases are too complicated and have not been looked at for the silicon thermal oxidation.

1.7. Deal and Grove theory of thermal oxidation

The basic assumption of Deal and Grove is that oxidizing species, dissolved in oxide in interstitial position, migrate up to the SiO_2/Si interface, to react there and form the new oxide. During transport, they could also react with the silica network. Thanks to this process, the authors were able to establish the growth kinetics of silica on silicon, largely used nowadays.

This theory is in agreement with the experimental results regarding thermal oxidation under O_2 for thicknesses higher than 40 nm, as well as oxidation under water vapor for the whole range of thicknesses.

In order to establish growth kinetics, we must, first of all, determine the flux compared to the above basic assumptions, without taking into account of a possible reaction of the species in silica during their transport. This particular point will be the subject of section 1.8.

1.7.1. *Flux calculation*

Let us call F_g the flux of oxidizing species in the gaseous phase, F_1 the flux at the gas/oxide interface, F_2 the flux in oxide, and F_3 the flux at the oxide/silicon interface. Figure 1.22 shows the concentrations of species, as well as these flux.

In steady flow, all flux are equal and constant:

$$F_g = F_1 = F_2 = F_3 = F \qquad [1.44]$$

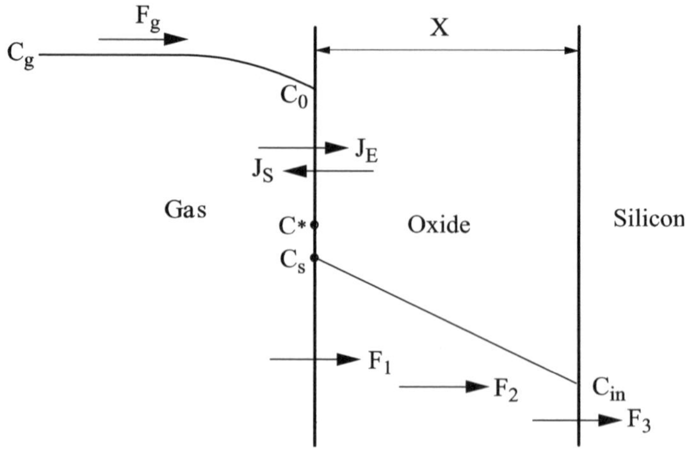

Figure 1.22. *Flux and distributions of the oxidizing species according to the Deal and Grove theoretical model*

1.7.1.1. *Flux in gas F_g and net flux at the gas/oxide interface F_1*

For a given gas pressure, Pg, we will have a concentration of oxidizing species C_g proportional to P_g (perfect gases hypothesis). Near the interface, a part of the molecules dissolves in oxide, by the way producing a slight drop of the oxidizing species concentration at the proximity of the surface in the gas C_0. It will induce a flux, $F_g = h_g (C_g - C_0)$, where h_g is a transfer speed of molecules in gaseous phase.

F_1, a net flux at the gas/oxide interface, can be expressed by the difference between the entering and outgoing flux:

$$F_1 = J_E - J_S \qquad [1.45]$$

If C_s is the concentration of oxidizing species dissolved in oxide on the surface, and if we make the hypothesis of a first order proportionality between the outgoing flux J_s and C_s, we will have:

$$J_S = h\, C_s \qquad [1.46]$$

When oxidation stops, $F_1 = 0$ and C_s reaches the limiting value C*, which depends on the external pressure of the gas.

With the free volume model, we can write:

$$C^{*} = \frac{V_g}{V_{ox}} C_g \quad [1.47]$$

where $\frac{V_g}{V_{ox}}$, is the ratio of the free volumes in the gas and in the oxide of the transported species. C* is of about a few ppm at atmospheric pressure. Expression [1.47] is equivalent to the Oswald law, stipulating that at equilibrium, the concentration of species dissolved in the solid, is proportional to the one in gas. It is also an adaptation of Henry's law, of the dissolution of a gas in a solid and not in a liquid.

NOTE.– In the case of steam oxidation, we must consider that C* is really in relation to the mobile species, and not with the OH clusters, formed in silica during oxidation.

Thus, if $F_1 = 0$ with $C_S = C^{*}$, we have $0 = J_E - h\ C_s$ and thus:

$$F_1 = h.(C^{*} - C_S) \quad [1.48]$$

1.7.1.2. *Net flux in oxide, F_2*

The first Fick law in 1D (one dimension) gives us:

$$F_2 = -D\frac{\partial C(x)}{\partial x}.$$

As the flux is constant at the equilibrium and assuming that D is a constant, we can write that:

$$F_2 = D.\frac{(C_S - C_{in})}{X} \quad [1.49]$$

if X is the thickness of the formed oxide and C_{in}, the concentration of oxidizing species present at the oxide/substrate interface.

1.7.1.3. *Flux at the oxide/silicon interface, F_3*

If we consider a first order chemical reaction at the interface (e.g. $Si + O_2 \xrightarrow{k_r} SiO_2$), then:

$$F_3 = k_r C_{in} \tag{1.50}$$

with k_r the reaction constant. By using equations [1.48], [1.49] and [1.50], and by setting that in steady state $F_1 = F_2 = F_3 = F$, we have:

$$h.(C^* - C_S) = D.\frac{(C_S - C_{in})}{X} = k_r C_{in} \tag{1.51}$$

As C* is the only variable directly connected to the gas concentration,. it is convenient to express all in functions of C^* and:

$$C_{in} = \frac{C^*}{1 + \frac{k_r}{h} + \frac{k_r X}{D}} \tag{1.52}$$

$$C_S = \frac{\frac{(1 + k_r X) C^*}{D}}{1 + \frac{k_r}{h} + \frac{k_r X}{D}} \tag{1.53}$$

Thus:

$$F = k_r C_{in} = \frac{k_r C^*}{1 + \frac{k_r}{h} + \frac{k_r X}{D}} = \frac{C^*}{\frac{1}{k_r} + \frac{1}{h} + \frac{X}{D}} \tag{1.54}$$

By writing that $\frac{1}{k_e} = \frac{1}{k_r} + \frac{1}{h}$ in [1.54]:

$$F = \frac{C^*}{\frac{1}{k_e} + \frac{X}{D}} \approx \frac{C^*}{\frac{1}{k_r} + \frac{X}{D}} \tag{1.55}$$

Bearing in mind that, generally, $h >> k_r$, even in the event of non-equilibrium, we can write that $C_S \approx C^*$ and $k_e \approx k_r$.

1.7.2. *Growth kinetics equations*

Let us remember that N_i represents the number of oxidizing species necessary to form a unitary volume of silica (N_i= 2.25 10^{22} cm^{-3} for O_2 and N_i= 4.5 × 10^{22}cm^{-3} for H_2O). Thus, in agreement with [1.54], we obtain the growth rate:

$$\frac{dX}{dt} - \frac{F}{N_i} - \frac{C^*}{N_i\left(\frac{1}{k_e} + \frac{X}{D}\right)} \qquad [1.56]$$

By integration:

$$X^2 + \frac{2D}{k_e}X = \frac{2DC^*}{N_i}t + cste \qquad [1.57]$$

By using the initial conditions that $t = 0$, $X = X_0$ (X_0 being the initial oxide thickness), then:

$$X^2 + \frac{2D}{k_e}X = \frac{2DC^*}{N_i}t + {X_0}^2 + \frac{2D}{k_e}X_0 \qquad [1.58]$$

which can be written:

$$X^2 + AX = B(t + \tau) \qquad [1.59]$$

with:

$$A = \frac{2D}{k_e},\ B = \frac{2DC^*}{N_i} \text{ and } \tau = \frac{X_0^2 + AX_0}{B}.$$

[1.59] will have as solution:

$$X = \frac{A}{2}\left[\sqrt{1+\frac{4B}{A^2}(t+\tau)}-1\right] \qquad [1.60]$$

Let us consider the two limiting cases:

– Short times, low thicknesses when: $(t+\tau) << t_c = \frac{A^2}{4B}$.

In this case, the equation [1.59] could be approximated by:

$$X \approx \frac{B}{A}(t+\tau) = k_L\,(t+\tau) \qquad [1.61]$$

with $k_L = \frac{B}{A} = \frac{k_e C^*}{N_i}$, the linear constant giving place to the linear oxidation regime.

– Long times, high thicknesses when: $(t+\tau) >> t_c = \frac{A^2}{4B}$.

In this case, equation [1.59] could be approximated by:

$$X^2 \approx B(t+\tau) = k_P\,(t+\tau)$$

That is to say:

$$X \approx \sqrt{k_P}\,\sqrt{t+\tau} \qquad [1.62]$$

with $k_P = B = \frac{2DC^*}{N_i}$, the parabolic constant giving place to the oxidation regime in $\sqrt{t}$.

In such a situation, the oxidation rate is controlled by the diffusion, but also by the concentration limit C^*, defined by the gas pressure.

The constant $A = \frac{k_P}{k_L}$, represents a critical thickness below which oxidation follows a linear law and beyond which it follows a parabolic law.

Equation [1.59] can be also written as:

$$\frac{X^2}{k_P} + \frac{X}{k_L} = t + \tau \qquad [1.63]$$

1.7.3. *Remarks on the variations of the oxidation constants k_P and k_L*

As seen previously, if we use the free volume model, C^* is proportional to 1/T; this dependence will be more complex in a statistical model.

Experimentally, it was shown that C^* slightly decreases when T increases. Thus, for a restricted temperature range, we will be able to consider as a first approximation that C^* remains constant. And if the hypothesis is made that the thin SiO_2 films behave in the same way as thicker silica, we will be able to establish that oxidation constants k_P and k_L, proportional to C^*, will be proportional to the gas pressure of C_g processing.

Moreover, $k_e \approx k_r$, then $k_L \approx \frac{k_r C^*}{N_i}$, that is to say that k_L and k_r will have the same activation energies, if we neglect the variations of C^* depending on T. In the same way, k_P will have the same activation energy as C^*, if we neglect the variations of C^* depending on T.

In addition, the critical thickness A proportional to D/k_r, will be independent of the pressure and will have an activation energy equal to the difference of the one of $D(W_o)$ and the one of $k_r(W_r)$. So, when A decreases T increases, if W_r is higher than W_o, as we will see later on.

1.7.4. *Determination of the oxidation parameters from experimental results*

By deriving equation [1.63], we have:

$$\frac{2X}{k_P} dX + \frac{1}{k_L} dX = dt \qquad [1.64]$$

which gives a linear dependence of the reverse of the speed:

$$\frac{dt}{dX} = \frac{2X}{k_P} + \frac{1}{k_L} \qquad [1.65]$$

Thus, from experimental results of X as a function of t, dt/dX is represented according to X in Figure 1.23.

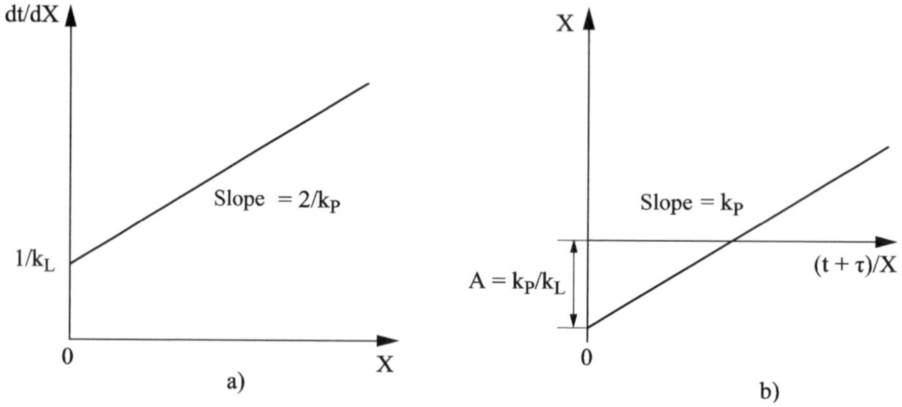

Figure 1.23. *Two curves for dt/dX functions: a) X; and of b) X according to (t+ τ)/X*

The slope of this curve will be equal to $2/k_P$ and the intersection with the axis X = 0 will be equal to $1/k_L$.

Another mean to determine the constants is to notice that equation [1.59] can be written:

$$X = \frac{B(t+\tau)}{X} - A = \frac{k_P(t+\tau)}{X} - \frac{k_P}{k_L} \qquad [1.66]$$

By delineating X according to (t+ τ)/X (Figure 1.23b), we can also determine k_P and k_L.

t can be given previously by plotting X as a function of t, by extrapolating the curve at the ordinate axis X = 0 (see Figure 1.24).

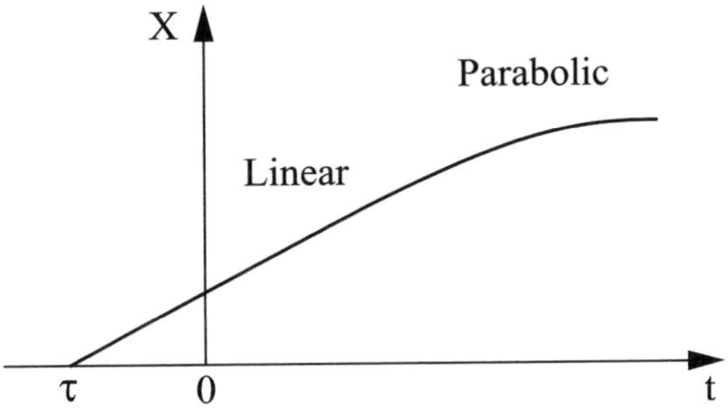

Figure 1.24. *Kinetics of growth: X functions of t curve*

1.7.5. *Confrontation of the Deal and Grove theory with experimental results*

Nevertheless, the linear parabolic law of the Deal and Grove theory is in good agreement with the experimental results in the following cases:

– in the case of wet oxidation for all thicknesses;

– in the case of dry oxidation under O_2 for thicknesses higher than 20-30 nm (for thinner thicknesses, the growth is characterized by a fast growth regime). In Figure 1.25, this restriction is illustrated for an oxidation carried out under O_2 at 700°C for variable times [DEA 65]. The value of X_0 which must be added in the Deal and Grove equation is equal to 20-30 nm for processing temperatures ranging between 700°C – 1,200°C; the processes being carried out at atmospheric pressure.

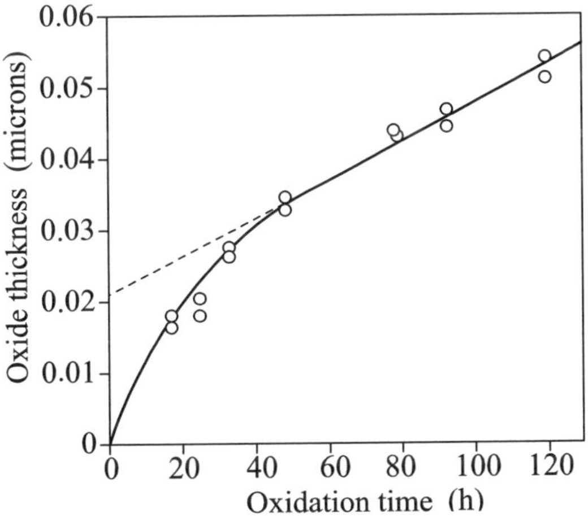

Figure 1.25. *Silicon oxidation under dry oxygen at 700°C (according to Deal and Grove [DEA 65])*

Figures 1.26 [DEA 65, WOL 69] and 1.27 [IRE 78b] show oxidation kinetics. The variations of oxidation constants B, B/A and A as functions of 1/T at atmospheric pressure are represented in Figures 1.28, 1.29 and 1.30 for wet oxidation, as well as for dry oxidation. We will further detail the experimental results. However, we can see that the constants k_P and k_L are higher for water vapor than in the case of O_2.

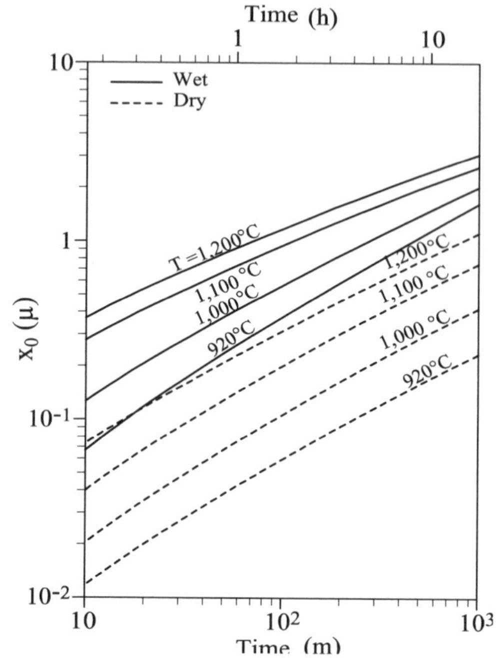

Figure 1.26. *Oxide thickness as a function of the time for various processing temperatures. The full lines represent the results for wet oxidation and the dotted lines, for dry oxidation. Silicon (111) oxidation according to Wolf [WOL 69] based on the data of Deal and Grove [DEA 65]*

The difference in the case of the parabolic constant can easily be understood by the fact that the diffusivity and solubility of the H_2O molecule in silica are higher than those of O_2. The respective values of activation energy (16.3 kcal for H_2O and 28 kcal for O_2) are in harmony with those found in the case of compact silica.

The difference observed for the linear constant can be due to the highest solubility of the water vapor in silica and/or to a more important reactivity of the latter at the SiO_2/Si interface. The corresponding value of the activation energy is independent of the environment and is worth 45 kcal, which is very close to the value necessary to break the Si-Si bond, i.e. 42.2 kcal.

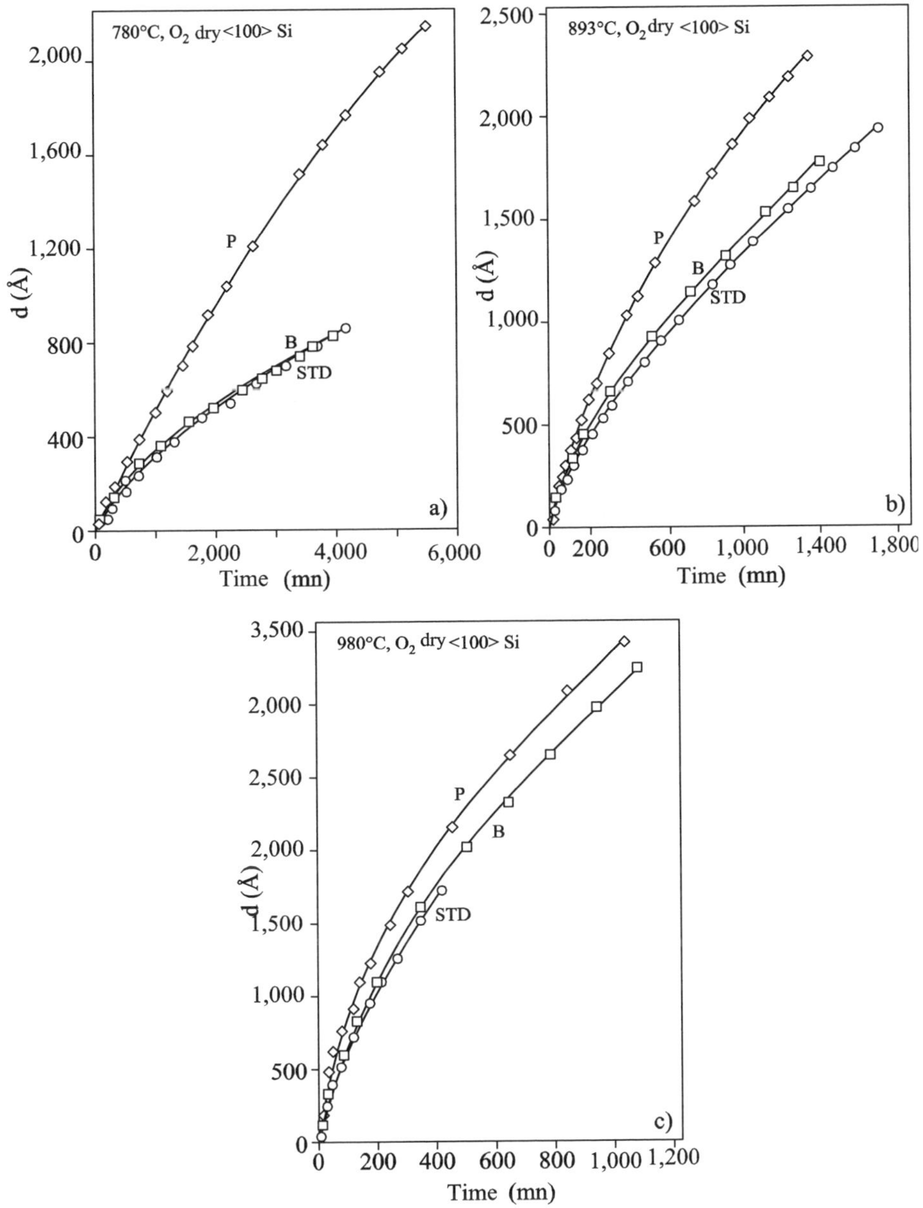

Figure 1.27a. *Oxide thicknesses obtained by silicon (100) oxidation, little doped (STD), very doped phosphorus (P), very doped boron (B), for various temperatures under dry oxygen at atmospheric pressure (according to Irene and Dong [IRE 78b])*

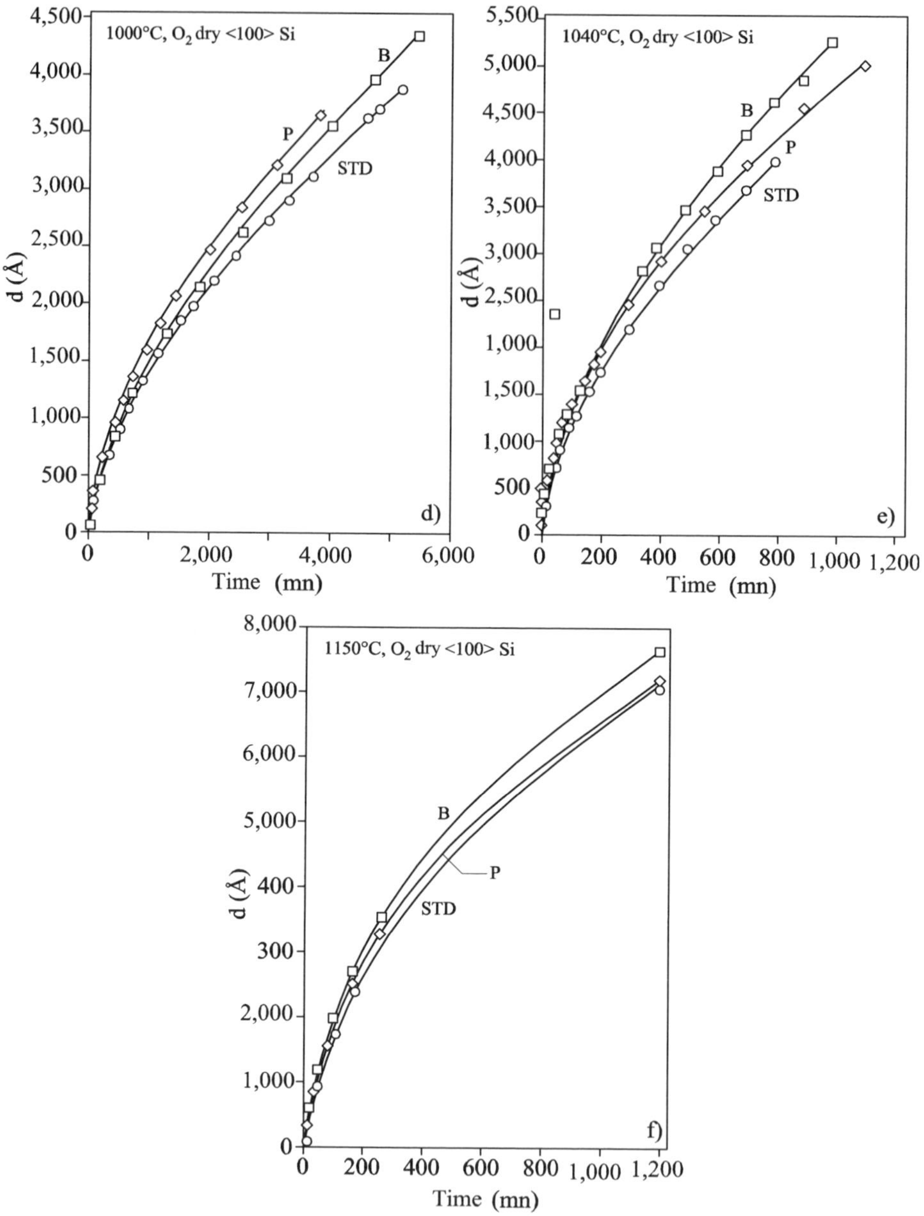

Figure 1.27b. *As in Figure 1.27a for other temperatures (according to Irene and Dong [ANGER 78b])*

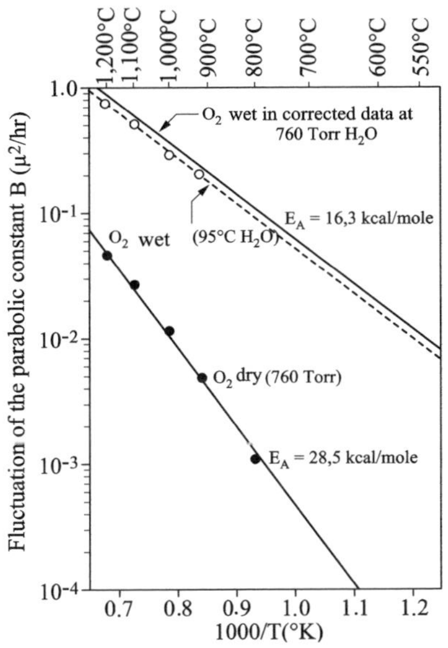

Figure 1.28. *Effect of the temperature on $k_P = B$ (according to Deal and Grove [DEA 65])*

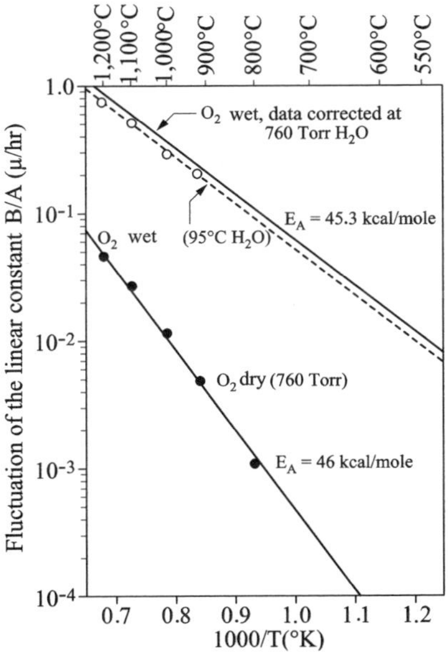

Figure 1.29. *Effect of the temperature on $k_L = B/A$ (according to Deal and Grove [DEA 65])*

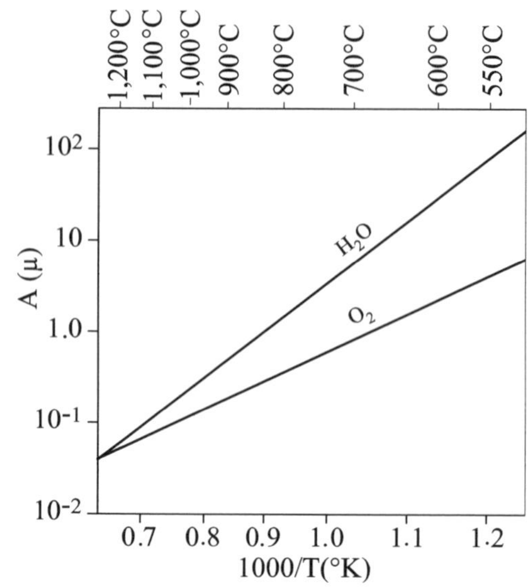

Figure 1.30. *Effect of the temperature on $k_P/k_L = A$ (according to Deal and Grove [DEA 65])*

1.7.6. *Conclusions on the Deal and Grove theory*

As seen in section 1.5, the theoretical framework leading to the equation of kinetics [1.59] does not take account of the three following phenomena:

– the exchange of oxygen on the surface;

– the oxygen loss at the interface probably due to the SiO exo-diffusion towards the gaseous phase;

– the initial phase with roughening/existence of fragments and/or presence of an electric field.

If we do not expect a specific incidence on the kinetics from the first phenomenon , on the other hand, the two other phenomena should be taken into account, in order to predict with more accuracy the growth kinetics in case of dry oxidation.

For practical reasons, we will be able, as a first approximation, to use the results obtained by the Deal and Grove theory, and more particularly the constants of this, defined in Figures 1.28, 1.29 and 1.30. However, it is necessary to pay particular attention to the processing conditions, such as the temperature, the partial pressure, the orientation, the doping, etc. This aspect is discussed in section 1.10.

1.8. Theory of thermal oxidation under water vapor of silicon

In section 1.7, the Deal and Grove theory supposes that mobile species diffuse through the oxide without reacting with it. If these species react, is it possible to produce a theory taking account of this reaction and which can satisfactorily explain the experimental results?

SiO_2 is not the only chemical species produced during the reaction of the water vapor with silica. Hydrogen is produced at the SiO_2/Si interface. Hydrogen and water react with silica and produce SiOH and SiH. The composition of the silica film is altered, which can change its physical and transport properties. Breed and Doremus [BRE 76] proposed a theory, developed below, supposing that H_2O is the mobile species.

1.8.1. *Concentration profiles expected for H_2O*

In stationary state, the flux of mobile molecules of H_2O is constant and expressed by:

$$J_{H_2O} = -D_{H_2O} \frac{\partial [H_2O]}{\partial x} = \text{constant}\,, \qquad [1.67]$$

where D_{H_2O} and $[H_2O]$ are respectively the diffusion coefficient and the concentration of water molecules.

If $[H_2O]_s$ and $[H_2O]$ are the concentrations of the water molecules on the external surface and at the SiO_2/Si interface, we have:

$$J_{H_2O} = -D_{H_2O} \frac{\partial [H_2O]}{\partial x} = D_{H_2O} \frac{[H_2O]_s - [H_2O]_{in}}{X} \qquad [1.68]$$

where X is the silica thickness. By integration, we obtain:

$$[H_2O] = \left[[H_2O]_s - [H_2O]_{in}\right]\left(1 - \frac{x}{X}\right) + [H_2O]_{in} \qquad [1.69]$$

1.8.2. *Concentration profiles expected for the OH groups*

If we postulate that the SiOH groups and H_2O molecules are in equilibrium according to the reaction:

$$Si-O-Si+H_2O \rightleftarrows 2\,SiOH$$

we have:

$$[OH]=k_{H_2O}[H_2O]^{1/2}$$

where k_{H_2O} is the equilibrium constant of the previous reaction, we obtain:

$$[OH]=k_{H_2O}\left\{\left[[H_2O]_s-[H_2O]_{in}\right]\left(1-\frac{x}{X}\right)+[H_2O]_{in}\right\}^{1/2} \quad [1.70]$$

If $[H_2O]_{in}$ is negligible, we have:

$$[OH]=[OH]_s\left(1-\frac{x}{X}\right)^{1/2} \quad [1.71]$$

1.8.3. *Concentration profiles expected for H_2*

The water molecules react at the SiO_2-Si interface, to form SiO_2 and H_2, according to the reaction:

$$H_2O+Si \rightleftarrows Si-O-Si+H_2$$

The hydrogen formed at the interface diffuse through the oxide. In stationary state, the flux of J_{H_2} of H_2 is constant, equal in absolute value to that of the water molecules:

$$\left|J_{H_2}\right|=D_{H_2}\frac{\partial[H_2]}{\partial x}=\text{constant} \quad [1.72]$$

where D_{H_2} and $[H_2]$ are respectively the diffusion coefficient and the concentration of H_2 molecules.

If $[H_2]_s$ and $[H_2]_{in}$ are the concentrations of H_2 molecules on the external surface and at the SiO_2/Si interface, we have:

$$\left|J_{H_2}\right| = D_{H_2} \frac{[H_2]_{in} - [H_2]_s}{X} \quad [1.73]$$

We thus obtain:

$$[H_2] = \left([H_2]_{in} - [H_2]_s\right)\frac{x}{X} + [H_2]_{in} \quad [1.74]$$

Since $\left|J_{H_2}\right| = \left|J_{H_2O}\right|$, we have the following relations:

$$D_{H_2}\left([H_2]_{in} - [H_2]_s\right) = D_{H_2O}\left([H_2O]_s - [H_2O]_{in}\right) \quad [1.75]$$

If $[H_2O]_{in}$ and $[H_2O]_s$ are negligible, we have:

$$D_{H_2}[H_2]_{in} = D_{H_2O}[H_2O]_s \quad [1.76]$$

$$[H_2] = [H_2]_{in}\frac{x}{X} \quad [1.77]$$

$$[H_2] = \frac{D_{H_2O}}{D_{H_2}}[H_2]_s\frac{x}{X} \quad [1.78]$$

1.8.4. *Concentration profiles expected for H*

H_2 reacts with silica. If we assume that the reaction:

$$H_2 + Si - O - Si \rightleftarrows \text{SiOH} + \text{SiH} \quad [1.79]$$

occurs at the equilibrium, we have:

$$[OH][H] = k_{H_2}[H_2] \quad [1.80]$$

where $[H]$ is the concentration of SiH groups, and k_{H_2} is the equilibrium constant of the reaction.

By using the previous expressions of $\left[H_2\right]$ and $\left[OH\right]$ in parabolic regime, we obtain:

$$\left[H\right]=\frac{\alpha\dfrac{x}{X}}{\sqrt{\left(1-\dfrac{x}{X}\right)}} \qquad [1.81]$$

α is given by the expression:

$$\alpha=k_{H_2}\frac{\left[H_2\right]_{in}}{\left[OH\right]_s}=k_{H_2}\frac{D_{H_2O}}{D_{H_2}}\left[H_2O\right]_s\frac{1}{k_{H_2O}\left[H_2O\right]_s^{1/2}}$$

$$\alpha=\frac{k_{H_2}}{k_{H_2O}}\frac{D_{H_2O}}{D_{H_2}}\left[H_2O\right]_s^{1/2}$$

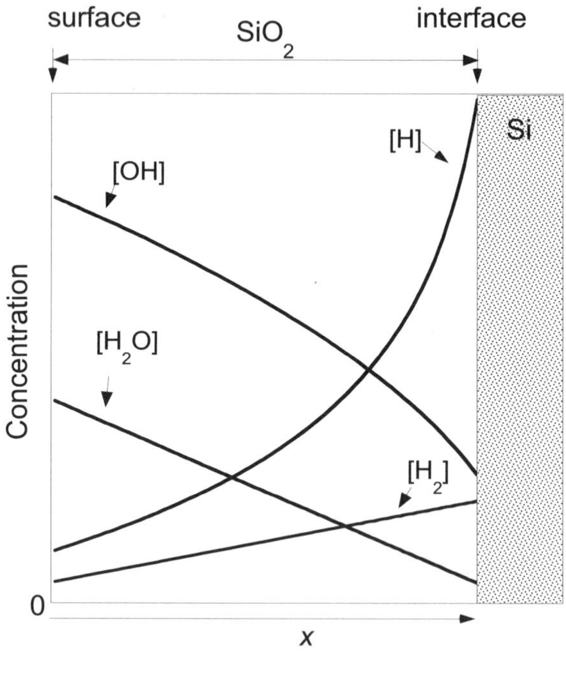

Figure 1.31. *Expected concentration profiles of the various species present during the silicon oxidation in water vapor*

1.8.5. *Comparison of the expected and the experimental profiles*

Beckmann and Harrick [BEC 71] have determined the [OH] and [SiH] concentrations using infrared analysis techniques. It was not possible to distinguish SiOH from H_2O, but it is reasonable to think that the H_2O concentration is negligible. The experimental profiles obtained are always in good agreement with the theory. The SiH group's concentration highly increases close to the interface, in agreement with the theory.

However, an increase in the concentration of the SiH groups is observed close to the external surface, contrary to the theory.

Burkard [BUR 67] grew oxides under water vapor enriched in tritium, and measured, by radioactive analysis, the total concentration of the hydrogenated groups $[H_T]$.

If we neglect $[H_2O]$ and $[H_2]$:

$$[H_T]=[OH]=[H][H_T]=[OH]_s\sqrt{1-\frac{x}{X}}+\frac{\alpha\frac{x}{X}}{\sqrt{\left(1-\frac{x}{X}\right)}} \qquad [1.82]$$

A minimum must be observed for $[H_T]$ at the following abscissa:

$$\frac{x}{X}=\frac{[OH]_s-2\alpha}{[OH]_s-\alpha}$$

We observe a good harmony between the form of the observed profile and the theory.

1.8.6. *Wolters theory*

Wolters [WOL 80] has developed a theory, according to which the diffusing species are H_3O^+ and OH^-.This theory also tallies with the experimental results of Burkard [BUR 67].

1.9. Kinetics of growth in O_2 for oxide films < 30 nm

1.9.1. *Introduction*

The Deal and Grove theory models the growth kinetics of oxide only beyond the X_0 value of the model (30 nm in the range of temperatures 700°C – 1,200°C, at atmospheric pressure). We present here alternative theories able to report a regime of faster initial growth, observed for films < 30 nm (see Figure 1.32).

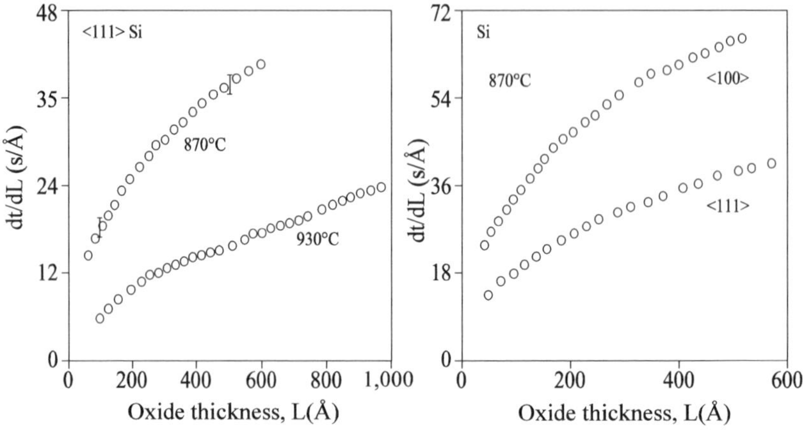

Figure 1.32. *Inverse of the growth rate from the thickness, for oxidations carried out under 1 atm of O_2 (according to [HOP 75])*

The growth of thin films is sometimes carried out at pressures lower than those of thicker films, or in inert gas/O_2 mixtures. It makes it difficult to compare between their kinetics and those of thicker films. Indeed, pressure affects the kinetics.

At low pressure, oxidation is slower on the face (111), than on the face (100). At higher pressure, oxidation is faster on the face (111), than on the face (100), the transition taking place around 10^{-1} atm [KAM 77, RAI 80]. In the case where oxygen is diluted in an inert gas, the water vapor pressure can become considerable compared to the partial O_2 pressure.

Moreover, the surface quality of the substrate (native oxide, impurities) influences the kinetics [DEL 87]. This influence will be greater, as the film thickness is low.

First, we expose below some thin film growth kinetics, which can be useful for the reader.

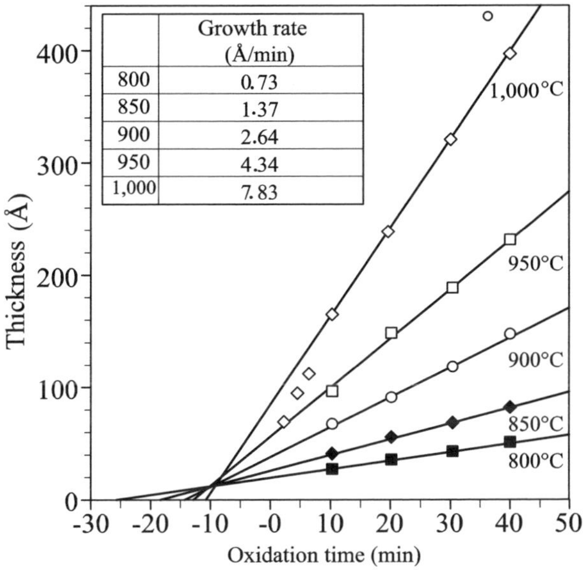

Figure 1.33. *Growth kinetics, Si (100), under O_2 atmosphere [CHA 91]*

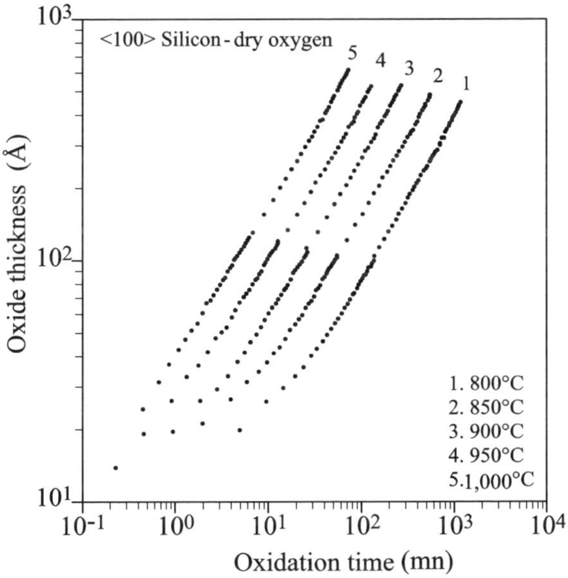

Figure 1.34. *Growth kinetics, Si (100), under O_2 atmosphere [MAS 85]*

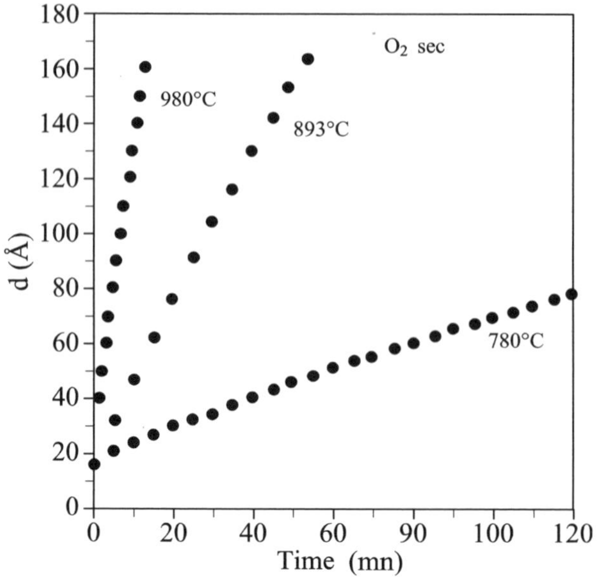

Figure 1.35. *Growth kinetics, Si (100), under O_2 atmosphere [IRE 78]*

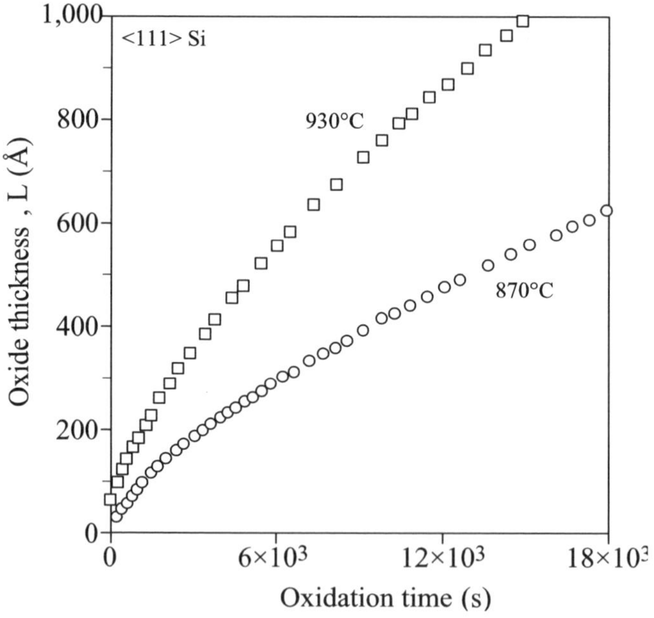

Figure 1.36. *Growth kinetics, Si (111), under O_2 atmosphere [HOP 75]*

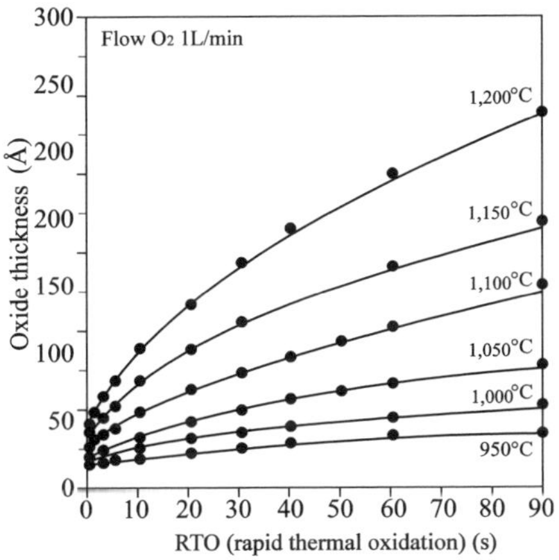

Figure 1.37. *Growth kinetics, Si (100), under O_2 atmosphere, by rapid thermal oxidation [FUK 92]*

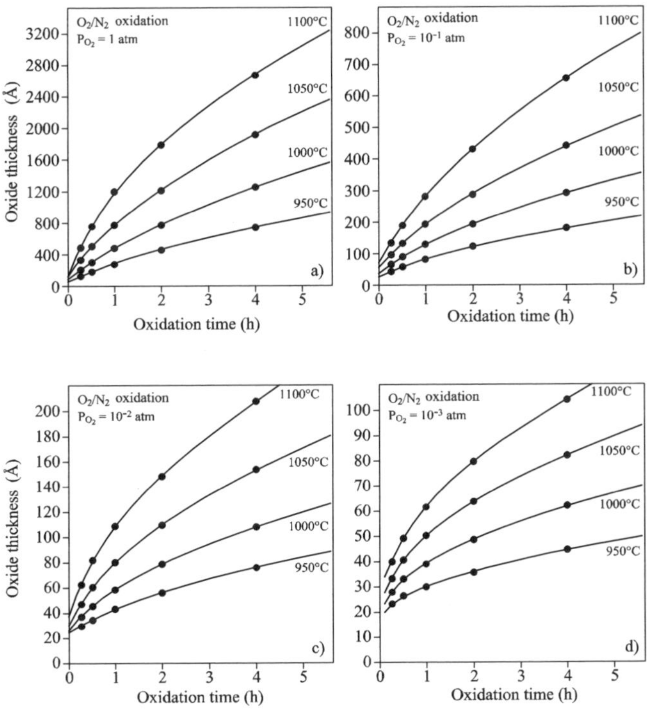

Figure 1.38. *Growth kinetics, Si (100), oxidation in a O_2/N_2 mixture for various partial O_2 pressures [KAM 77]*

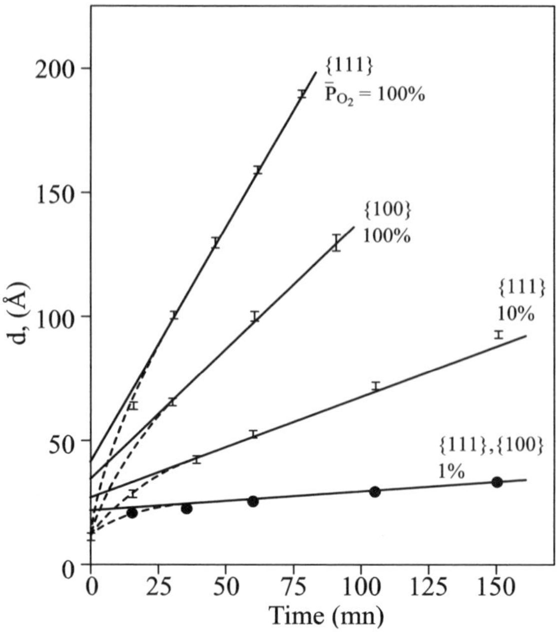

Figure 1.39. *Growth kinetics, oxidation in a O_2/N_2 mixture, for various partial O_2 pressures [VAN 72]*

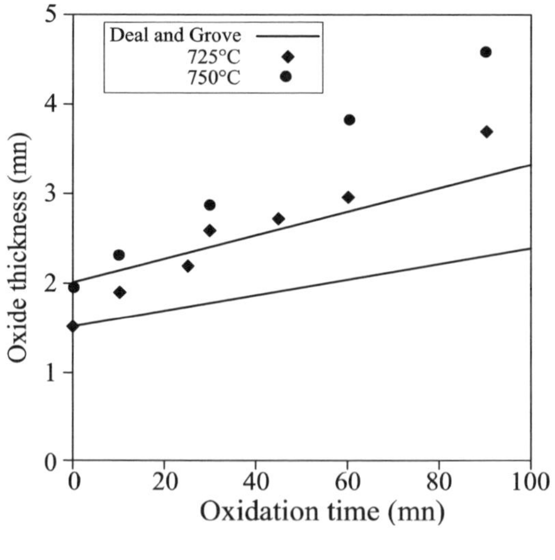

Figure 1.40. *Growth kinetics, Si (100), for ultra-thin oxide films [KRZ 07]*

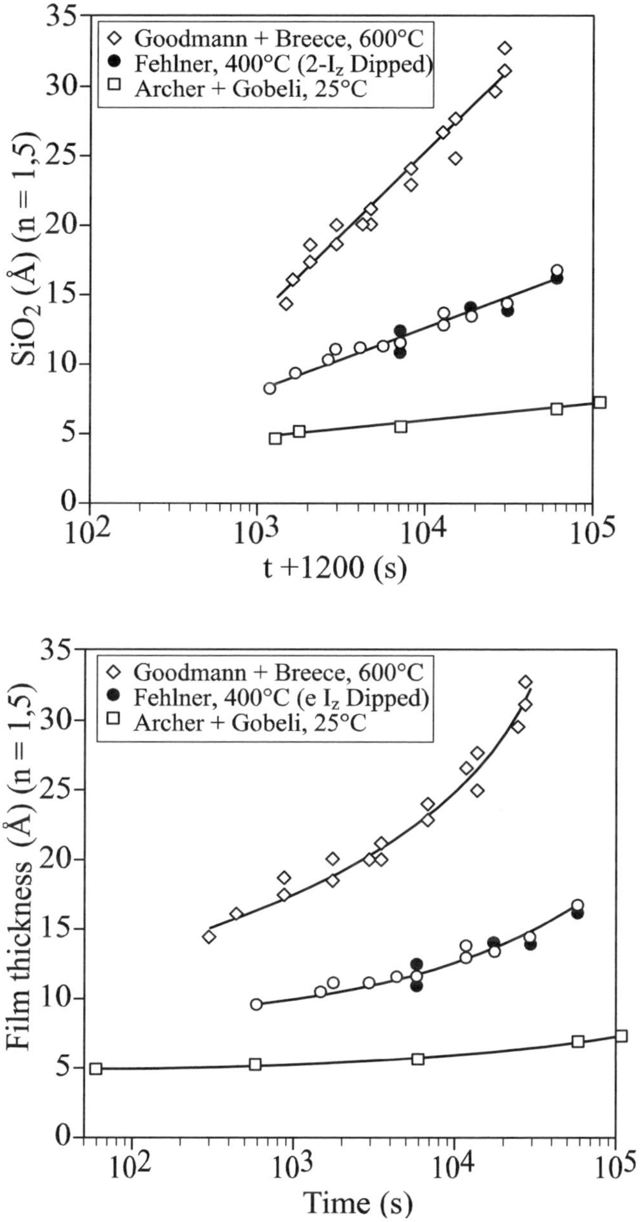

Figure 1.41. *Growth kinetics, Si (100), for ultra-thin oxide films, at low temperatures [FEH 72]*

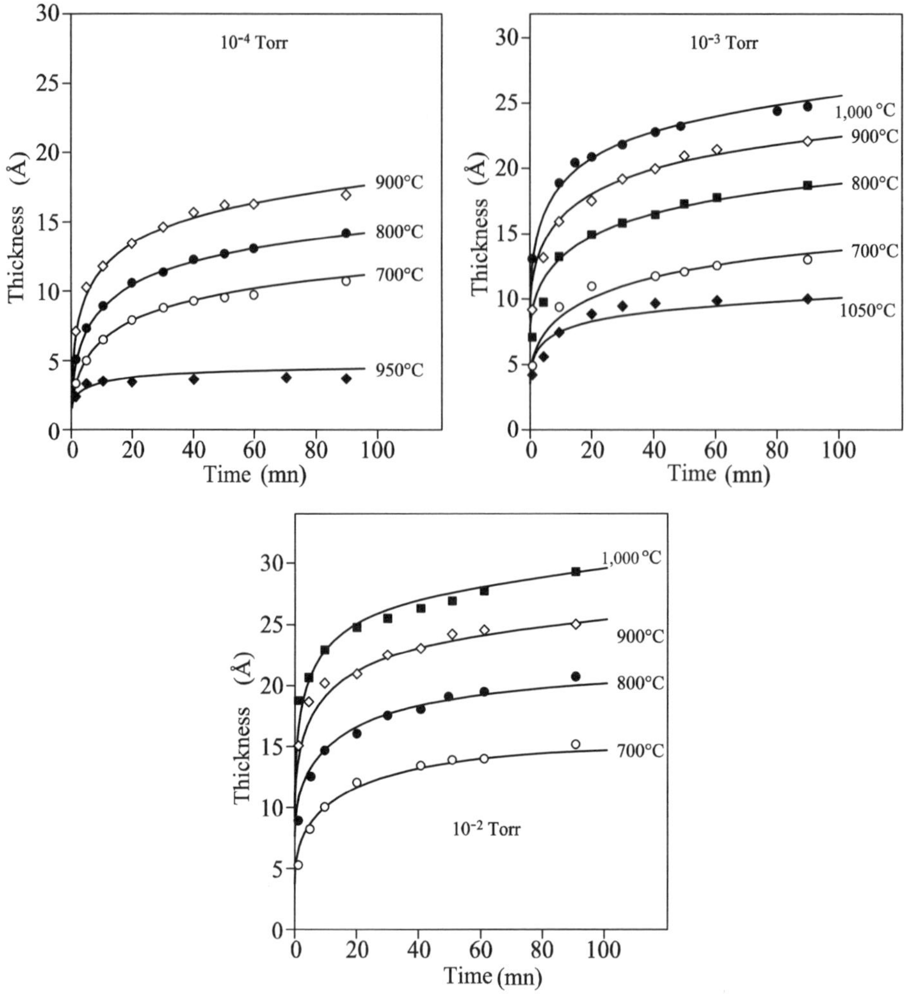

Figure 1.42. *Growth kinetics, Si (111), for ultra-thin oxide films, at low O_2 pressures [DER 82]*

1.9.2. *Oxidation models of thin films*

During the 1970s and 1980s, a very significant effort was made, in order to develop a silicon oxidation model, able to report the initial growth process. For a review of the existing models at the time, we can refer to Massoud [MAS 85] and Irene [IRE 88]. We mention here some of the models published during this period.

Hopper *et al.* [HOP 75], then Han and Helms [HAN 87], suggested that two oxidation mechanisms took place in parallel. However, isotopic labeling experiments [ROC 84] showed that the contribution of a second mechanism to the growth is negligible beyond 5 nm.

Ghez and Van der Meulen [GHE 72] have put forward the idea that three reactions were in competition at the interface. Blanc [BLA 78] simplified this model with two reactions. But none of these two models satisfactorily adjust the kinetics.

Fargeix *et al.* [FAR 83a, FAR 83b] and Tiller [TIL 83] suggested the existence of a blocking layer close to the interface, where the diffusion coefficient is lower. One of the potential explanations is a compressive stress located near the SiO_2/Si interface [FAR 83b]. With this model, we can correctly adjust the kinetics of growth, but we cannot predict the variations of the oxidation rate with the pressure. More recently, Watanabe *et al.* [WAT 07] also proposed a model based on the existence of a strained layer of about 1 nm. There again, the model does not predict the variations of oxidation rate with pressure.

In the Massoud model [MAS 85], the oxidation rate in the regime of thin films, under many variable experimental conditions (direction of the substrate, doping, partial oxygen pressure), can be empirically expressed by:

$$\frac{dX}{dt} = \frac{B}{2X + A} + C_1 e^{-X/L_1} + C_2 e^{-X/L_2}$$

The first term of the expression corresponds to the Deal and Grove equation. The two exponential terms represent the increase of growth rate in the regime of thin films. The first exponential term has a characteristic length of about 1 nm. It becomes negligible for thicknesses higher than 5 nm. The second exponential term has a characteristic length of about 7 nm and becomes negligible for thicknesses higher than 25 nm. To explain all their experimental results and the fact that the two characteristic lengths are independent of the direction of the substrate, of the doping, of the partial oxygen pressure, Massoud *et al.* suggested that the increase of the growth rate in the regime of thin films is due to an excess of reactive sites at the interface. This model leads to the correct adjustment of many kinetics, for thicknesses higher than 5 nm, and predicts rather correctly the effects of the pressure on oxidation kinetics.

Kageshima *et al.* [KAG 99] have rewritten the Deal and Grove transport equations, by adding three transport methods: i) interstitial silicon atoms are created at the interface; ii) the interstitials diffuse in the oxide layer; iii) the interstitials are absorbed in oxide to form Si-O-Si. An additional hypothesis is that the emission of

Si at the interface decreases with the interstitial density at the interface. Their equation of dX/dt as afunction of X has the same expression as that of Massoud, for $C_l=0$, if the Si emission rate is much greater than the rate of absorption in oxide. In other words, to a certain extent, their theory can explain the empirical equation of Massoud *et al.* This model, commonly called the "Si interfacial emission model", is currently used for the kinetics adjustment. It provides a sub-linear fluctuation of the oxidation rate with the pressure (m = 0.5, slightly different from the values experimentally observed: (m = 0.6-0.8). With this model we cannot correctly adjust kinetics for films grown at sub-atmospheric pressure.

The physical phenomena at the origin of this fast initial regime are not given (strained layer at the interface, silicon emission, etc.). No model has unanimous support, nor is able to universally adjust the kinetics obtained under various experimental conditions. The Massoud empirical model, with $C_l = 0$, is very largely used in simulation software, for its efficiency and its simple analytical form. The model of silicon interfacial emission is the most frequently used for the kinetics adjustment, because it is currently the most satisfactory non-empirical model, most oxidations being carried out at atmospheric pressure.

1.9.3. *Case of ultra-thin films (< 5 nm)*

We saw in section 1.5, that the gradual movement of oxygen atoms of the network, related to the presence of defects in the oxide, contributed to the growth. This contribution is more important when the film is thin and when the O_2 pressure is low [ROC 84, TRI 89].

For these ultra-thin films, Fehlner [FEH 72] obtained a satisfactory adjustment by assuming a Cabrera-Mott transport. Derrien and Commandré [DER 82] adjusted their experimental results by assuming a constant electric field. Ganem *et al.* proposed the oxidation of Si fragments present in oxide [GAN 93b] and De Almeida *et al.* [DEA 00] rewrote the transport equations without the hypothesis of a stationary state and an abrupt interface. To this day, there is no valid model for ultra-thin films that is able to fill the gaps of the models of Massoud and of the model of interfacial emission.

1.9.4. *On line simulator*

Process Lab: Oxidation [CAO 06] is an online simulator of the Deal and Grove and/or Massoud models, available at: http://www.nanohub.org/.

1.9.5. *Kinetics and models of SiC oxidation*

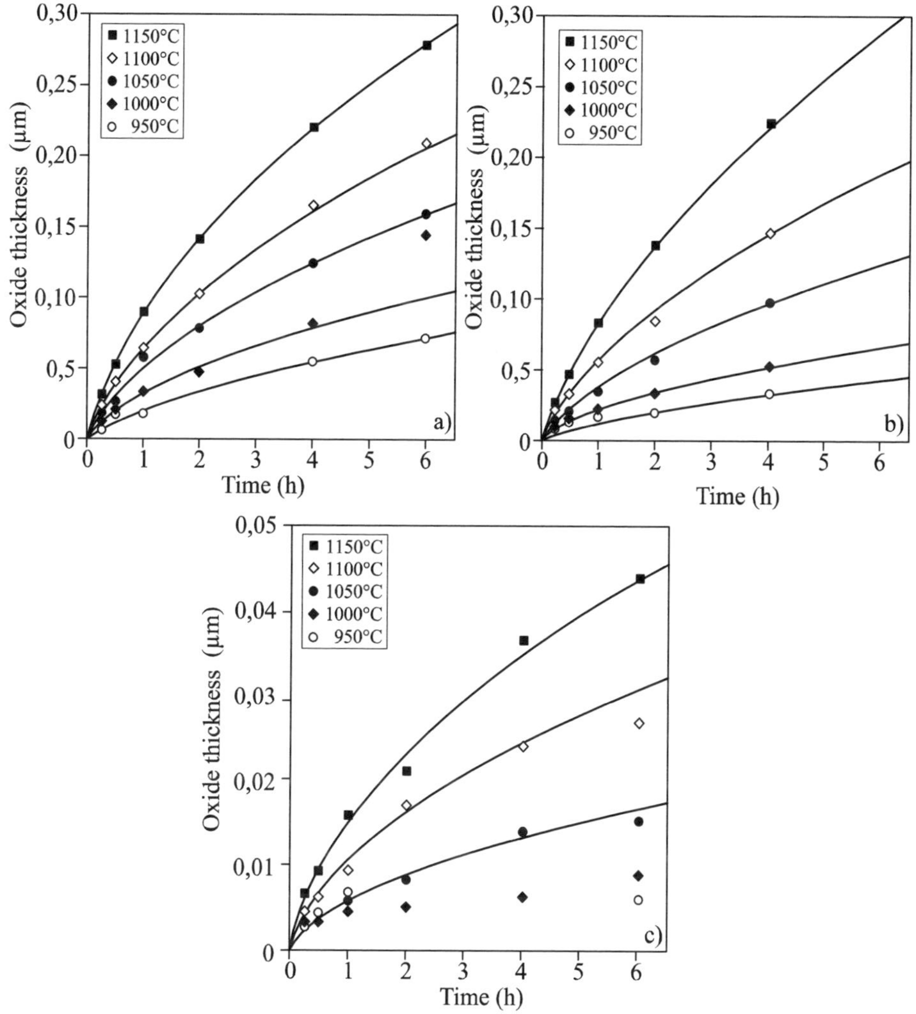

Figure 1.43. *Oxidation kinetics of the logotype 4H-SiC under an O_2 atmosphere: a) face C; b) face a; c) face Si (from [SON 04]). The lines correspond to an adjustment according to the Deal and Grove model*

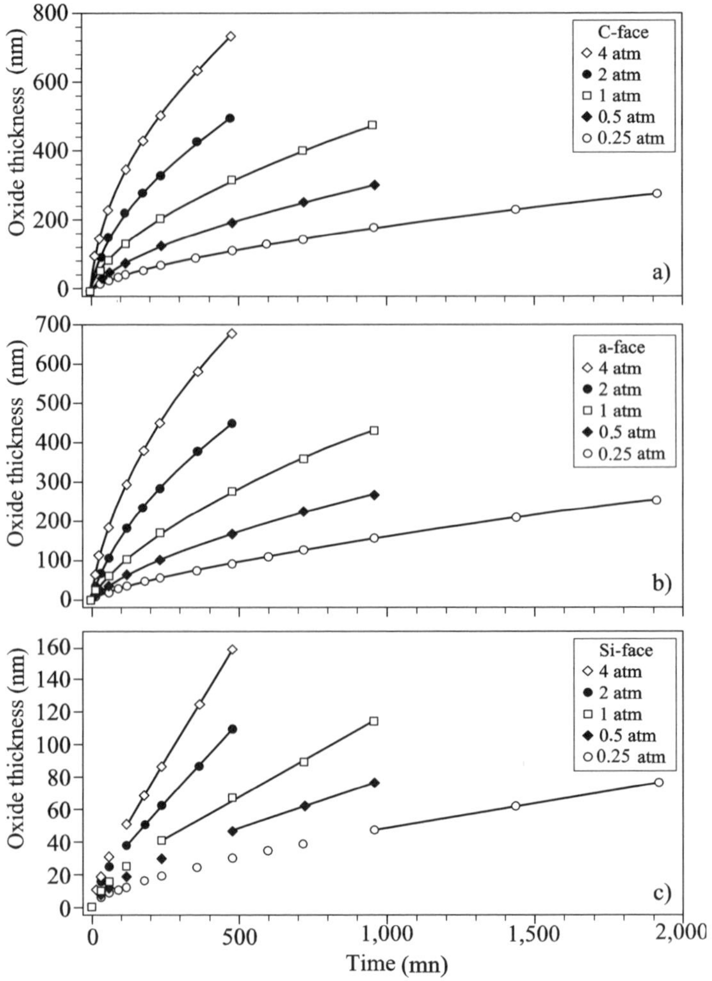

Figure 1.44. *Oxidation kinetics of the logotype 4H of SiC at 1,150°C, under atmospheres of O_2 between 0.25 atm and 4 atm: a) face C; b) face a; c) face Si (from [RAY 08]). The lines correspond to an adjustment according to the Deal and Grove model*

As seen in section 1.5, the oxidation of silicon carbide (SiC) produces SiO_2. Various oxidation models have been proposed [HIJ 09, SON 04], based on the adjustment of growth kinetics of the oxide. First of all, we present a set of kinetics of the logotype 4H, for three directions: (000-1) face C, (0001) face Si and (11-20) face a, carried out at 1 atm of O_2, for temperatures from 950°C to 1,150°C, and at 1,150°C for O_2 pressures between 0.25 and 4 atm.

We now present two models able to report SiC oxidation mechanisms.

To adjust their kinetics, Song *et al.* proposed a Deal and Grove model [SON 04], corrected in [RAY 08], taking into account the exodiffusion of the products of the oxidation reaction:

$$SiC + 1.5\, O_2 \underset{K_f}{\overset{K_r}{\leftrightarrows}} SiO_2 + CO$$

By following the Deal and Grove approach, they obtained modified expressions of *B/A* and *B:*

$$B/A \approx \frac{C^*_{O_2}}{N_i} K_f$$

$$B \approx \frac{2C^*_{O_2}}{1.5N_i} D_{O_2}\text{, if } O_2 \text{ diffusion is the limiting factor.}$$

$$B \approx \frac{2C^*_{O_2} K_f}{N_i K_r} D_{CO}\text{, if CO diffusion is the limiting factor.}$$

where K_f and K_r are respectively the constants of the direct and opposite reaction.

At 1,150°C, *B* and *B/A* are proportional to the O_2 pressure, for p = 0.25 – 2 atm [RAY 08]. Moreover, the values of *B* are similar on faces *a* and C, suggesting that the limiting factor is the O_2 diffusion through oxide.

Yamamoto *et al.* [RAY 08, YAM 07] have used the empirical equation of Massoud [MAS 85] to fit a large number of oxidation kinetics of the Si and C faces of the 4H polytype of SiC.

In the case of silicon oxidation, the empirical equation by Massoud can be justified by the model of Si interfacial emission. It was thus natural to adapt this model to the SiC oxidation. Hijikata *et al.* [HIJ 09] proposed the "model of carbon and silicon emission". This model helped them to correctly adjust curves of oxide growth rate, according to the oxide thickness, for the faces Si and C of the polytype 4H of carbide. However, demonstration of the validity of the model is difficult because of the large number of adjustable parameters that it contains.

1.10. Variations of the oxidation coefficients with experimental conditions

This section is intended for the process engineers, who must predict the oxides thicknesses once the process conditions are established. To be able to use the Deal and Grove relationship, it is necessary to know the oxidation coefficients, k_P and k_L as precisely as possible. The theory predicts how they vary with certain experimental parameters. The effective experimental conditions can give coefficients different from the theory. Moreover, some other parameters, such as crystalline direction or substrate doping, are not taken into account in the theory.

1.10.1. *Role of the pressure*

The pressure role is examined, by comparing experimental data under similar conditions.

1.10.1.1. *Wet oxidation*

In this case, k_P and k_L vary linearly with P_{H2O}, for all the temperature ranges and pressures examined (50-760 Torr for 1,000°C < T< 1,200°C [DEA 65], 1-20 atm for 800°C < T < 1,000°C [RAZ 81]). This dependence is well predicted by the theory.

1.10.1.2. *Dry oxidation*

In this case, k_P still varies linearly with P_{O2} pressure, for various conditions (50-760 Torr for T = 1,200°C [DEA 65], 15-760 Torr for 1,100°C [KAM 77], 130-1,200 Torr for 870°C < T < 930°C [HOP 75], 1-20 atm for 800°C < T < 1,000°C [LIE 82]). This result corresponds to the theory.

Contrary to k_P, k_L varies as $P^m{}_{O2}$ and m takes various values:

m = 1 for P=-760 Torr and T = 1,200°C [DEA 65];

0.7 < m < 0.8 for P=-20 atm and 800°C < T < 1,000°C [LIE 82];

m = 0.6 for P=-760 Torr and 1,100°C [KAM 77].

For films thinner than 30nm, we obtain:

$k_L \, \alpha \, P_{O_2}^{m}$, with m varying between 0.59 and 0.83 when 700°C < T < 1,000°C for substrates (111), and 0.54 and 0.66 when 850°C < T < 1,000°C for substrates (100) [GHE 72].

As seen previously:

$$k_P = \frac{2DC^*}{N_i} \text{ and } k_l = \frac{k_r C^*}{N_i} \text{ with } k_r \ \alpha \ P_{O_2} .$$

This leads us to an interfacial reaction constant, $k_r \ \alpha \ P_{O_2}^{m-1}$, which probably results from the reactive mechanisms occurring at the SiO_2 interface, the latter having various configurations.

1.10.2. *Role of the temperature*

The role of temperature is examined by comparison of various work completed for identical pressures.

1.10.2.1. *Wet oxidation*

The Arrhenius diagrams of the coefficients k_P and k_L are represented in Figure 1.45 [RAZ 81] for processes under various pressures of pyrogenic steam. For k_P the activation energy, corresponding to temperatures higher than 900°C, is 18 kcal/mole (0.78 eV). It reaches 27 kcal/mole (1.17 eV) for lower temperatures.

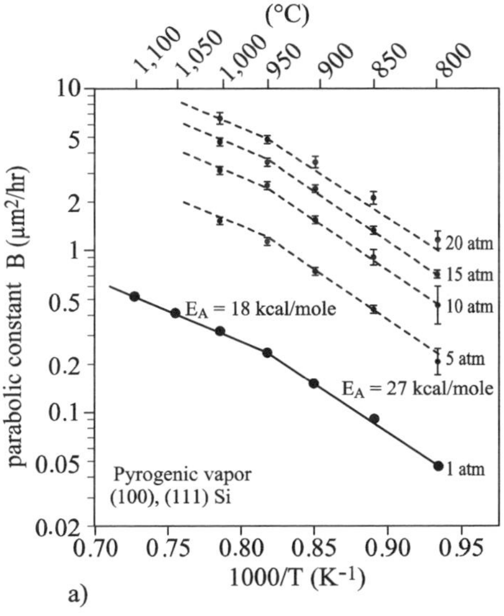

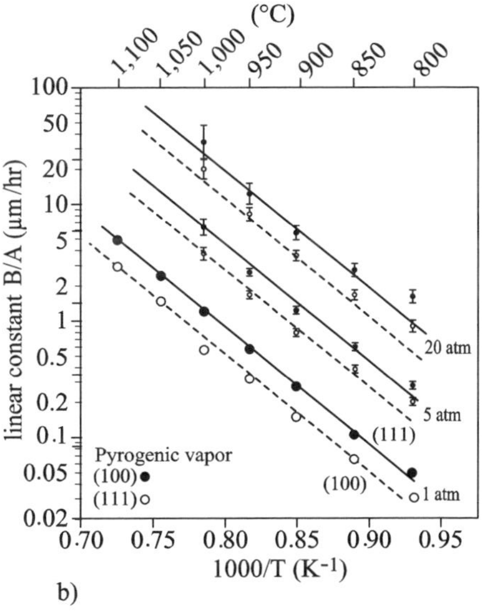

Figure 1.45. *Parabolic and linear coefficent for oxidations under water vapor of silicon (100) and (111) for various temperatures and pressures (from Razouk [RAZ 81])*

For k_L, the change of activation energy is less obvious and is around 40 kcal/mole.

The change of activation energy could be due to a change of silica structure, which would begin its viscous transition, as Eernisse suggests [EER 79], for temperatures ranging between 900 and 950°C.

1.10.2.2. *Dry oxidation*

The Arrhenius diagrams of k_P and k_L, measured by Lie, Razouk and Deal are given in Figure 1.46. They correspond to oxidations at atmospheric pressure for temperatures ranging from 750 to 1,150°C. STD represents samples of average resistivity 2 Ω .cm, B and P, bore and phosphorus doped samples of low resistivity (0.001 Ω .cm).

For higher pressures, we will refer to Figure 1.47.

The variations of the activation energies of the coefficients are enumerated and compared in Table 1.4.

The activation energy of k_P changes around 900°C as it is in the case of wet oxidation.

Above 900°C, its value is around 1.3 eV and, below, around 2.2 eV. It also refers to a change of oxide structure.

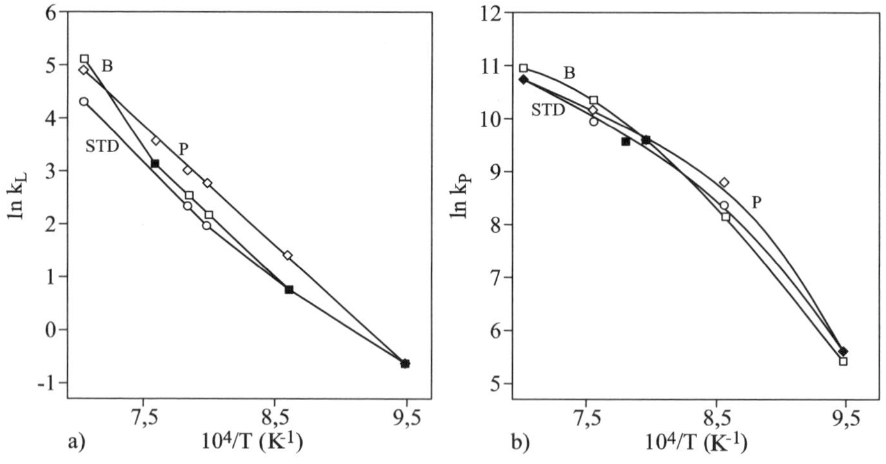

Figure 1.46. *Linear (Å/min) and parabolic (Å²/min) coefficients for oxidations under dry O_2 of silicon (100) at atmospheric pressure, for various temperatures (from Lie [LIE 82])*

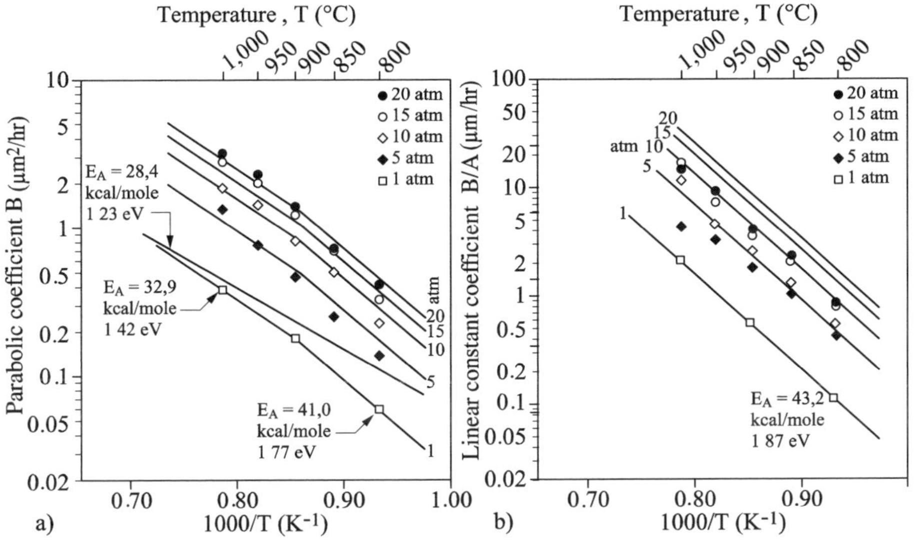

Figure 1.47. *Parabolic and linear coefficients for oxidations under dry O_2 of silicon (111) for various temperatures and pressures (from Lie [LIE 82])*

Temperature	Activation energy for k_L[eV]			Activation energy for k_L[eV]		
	STD	B	P	STD	B	P
980°C – 1,150°C	2.2	2.7	2.0	1.3	1.3	1.2
980°C – 1,150°C	1.6	1.6	1.9	2.1	2.2	2.1
980°C – 1,150°C	1.8	1.9	2.0	1.9	2.0	1.8
980°C – 1,150°C [DEA 65]	2.0			1.2		
980°C – 1,150°C [IRE 76]	1.5			2.3		

Table 1.4. *Compared activation energies for dry oxidation under O_2 (from [IRE 78b])*

Fargeix and Ghibaudo [FAR 84] point out that the activation energy of k_P, measured by the authors, is in fact only an apparent energy W_{ap}, which can take values much higher than the fundamental activation energy W_o of k_P. This can be understood, if we express:

$$k_P = k_{P_0} \exp\left(-\frac{W_O}{k_B T}\right)$$

Thus, a variation of 0.1 eV of the value of W_o can explain variations of about 1 eV for W_{ap}. 0.1 eV is compatible with a fluctuation of the oxide diffusivity, because of the unreleased stresses at low-temperature. In this case, we can express the diffusion coefficient by:

$$D = D_0 \exp\left(-\frac{\sigma \Delta V}{k_B T}\right)$$

where σ represents the stress, ΔV the consecutive variation of volume, and D_0 the diffusion coefficient in unstressed material.

For the k_L variations, Lie [LIE 82] reports values of activation energy ranging between 43.3 and 43.7 kcal/mole, for temperatures ranging between 800°C and 1,000°C, in agreement with the energy necessary to break the Si-Si bond (42.2 kcal/mole).

Irene and Dong [IRE 78b] report activation energies of 2.2 eV for temperatures higher than 980°C. However, below 900°C, these values would be 1.5 eV, which corresponds to the energy necessary to break the O-O bond.

1.10.3. *Role of the crystal orientation*

The oxidation rate also depends on the crystal orientation of the substrate. The oxidation rate is higher for the (111) orientation than for the (100) orientation.

1.10.3.1. *Wet oxidation*

For this type of oxidation, whereas the parabolic constant would not depend on the direction of the substrate, the linear constant would be strongly altered. Thus, from Figure 1.45, we find:

$$k_L(111) \approx 1,7 \times k_L(100)$$

Ligenza [LIG 61] explained why the orientation of the substrate modifies the number of reactive sites, which remains much lower than the density of the Si-Si bond on the surface and would be related to the steric hindrance of the water molecule. Thus, by a geometrical calculation, he deducted that the ratio of the densities of available sites would be 1.73 between a (111) and a (100) surface.

1.10.3.2. *Dry oxidation*

In Figure 1.48, we can clearly see the influence of the crystalline orientation of the substrate on the oxidation rate. From these results, we can express the following relation: $k_P(111) > k_P(110) > k_P(100)$. It is valid for any type of oxidation. Nevertheless, for a temperature higher than 996°C, the parabolic constant no longer depends on the orientation.

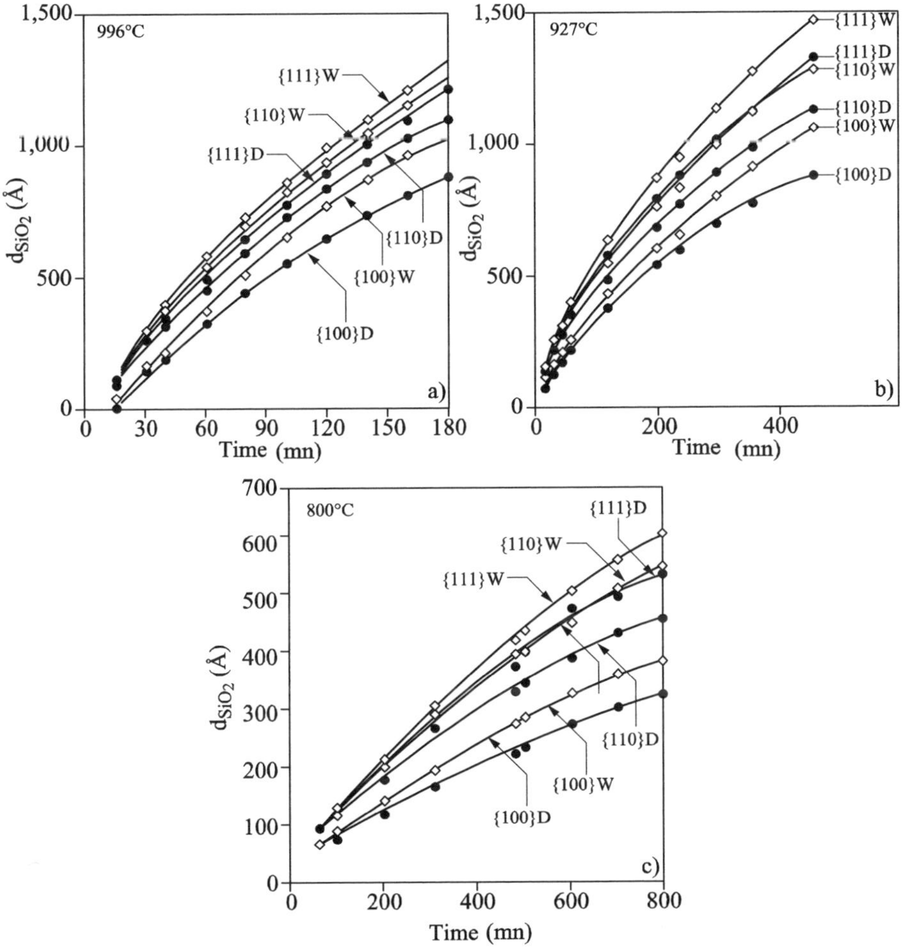

Figure 1.48. *Oxidation kinetics for various directions and temperatures. The symbol D refers to oxidations under dry atmosphere, and the symbol W refer to a wet atmosphere (from Irene [IRE 74])*

Geometrical considerations [LIG 61], lead to the following dependence: $k_P(111) \geq k_P(110) > k_P(100)$. But this calculation relies on the steric hindrance of the water molecule and not on oxygen. In practice, we will find it at high pressure, as in Table 1.5, $k_P(110) > k_P(111) > k_P(100)$.

At very low pressure, the oxide thickness follows the relation: $X(110) > X(100) > k_P(110)$ (see Figure 1.49).

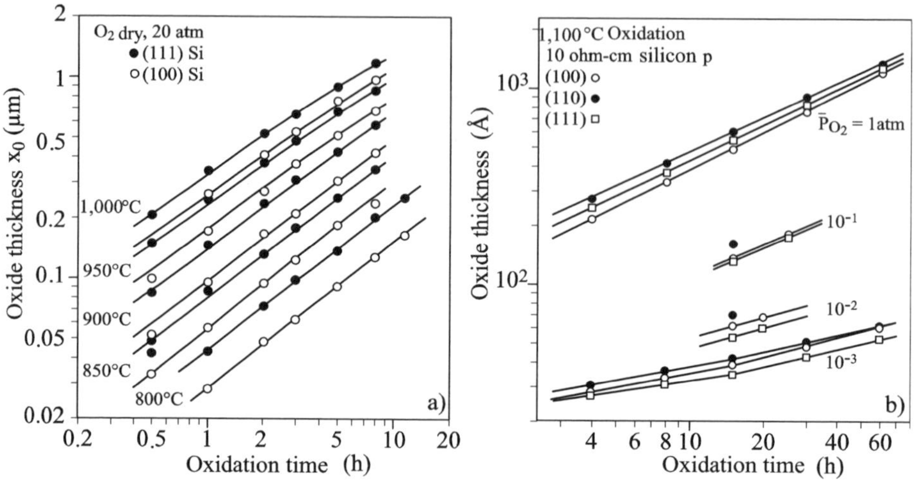

Figure 1.49. *Oxidation kinetics under dry oxygen for various crystalline orientations (On the left) P_{O2} = 20 atm and 800°C < T < 1,000°C (from Lie [LIE 82]); and (on the right), P_{O2} = 10^{-3} - 1 atm and T =1,100°C (from Kamigaki and Itoh [KAM 77])*

Coefficients	(100)	(110)	(111)	(111)[a]
A (Å)	8.47 10^2	4.02 10^2	5.87 10^2	9,0.10^2
B (Å²/min)	3.92 10^4	3.77 10^4	3.75 10^4	4,5.10^4
B/A (Å/min)	46.3	93.8	63.9	50

Deal[a] and Grove values

Table 1.5. *Values of the oxidation coefficients for T = 1,000°C under O_2 atmospheric pressure (from Kamigaki and Itoh [KAM 77])*

1.10.4. *Role of doping*

Experimentally, we observe that highly doped substrates oxidize faster than less doped substrates. To explain that phenomenon, Ho and Plummer [HO 79a, HO 79b] propose a model based on the increase of reactive sites at the SiO_2/Si interface. These would be point defects, such as vacancies or dangling bonds (Pb centers), whose concentration would increase for high doping, where the Fermi level moves towards the conduction-band (n-doping) or towards the valence band (p-doping).

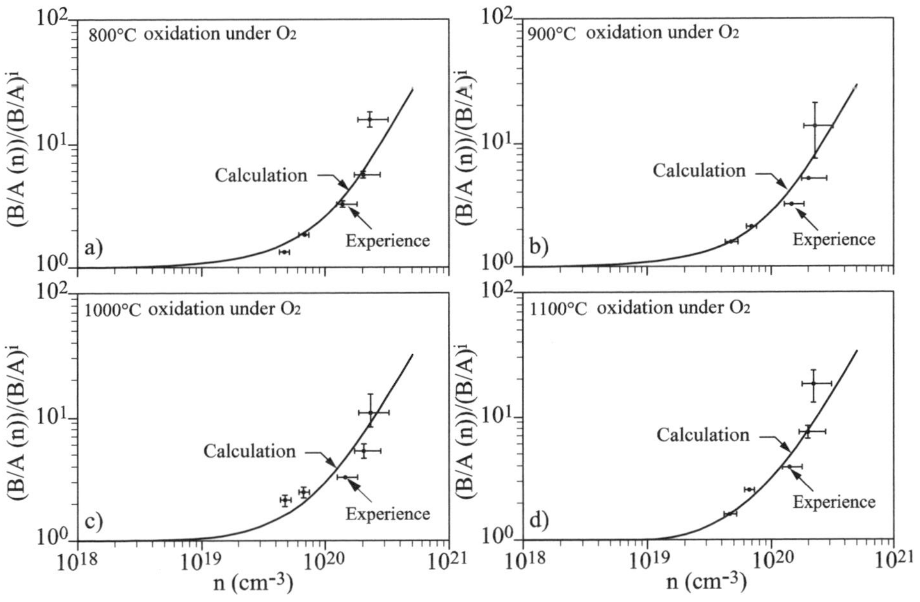

Figure 1.50. *$k_L(n)/k_L$ (i) as a function of the concentration of active dopants on a substrate (111), for various temperatures [HO 79b]*

Figures 1.50 and 1.51 present the ratio of linear coefficients of growth kinetics for doped silicon (n-doping and p-doping) to linear coefficients ofgrowth kinetics for intrinsic silicon. The theory and the experiments are in agreement.

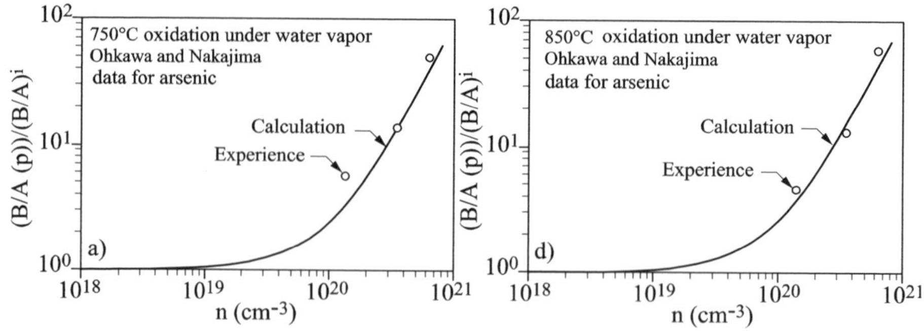

Figure 1.51. *k_L (p)/k_L (i) as a function of the concentration of active dopants (As) on a substrate (111) for various temperatures [HO 79b]*

1.11. Conclusion

Because of the many applications of silica in many domains, in particular in the electronics of integrated circuits, silicon oxidation, and more recently silicon carbide oxidation has been studied a lot in laboratories around the entire world. Still, nowadays, with the problems arising from the very high integration of transistors in circuits and in nanotechnologies, many researchers are trying to improve the understanding of the first stages of silica growth, to optimize the functional characteristics. We hope that this chapter will be a useful contribution to the development of new technologies.

1.12. Bibliography

[AKE 99] AKERMARK T., GOSSET L.G., GANEM J.-J., TRIMAILLE I., RIGO S., "Loss of oxygen at the Si-SiO_2 interface during dry oxidation of silicon", *Journal of the Electrochemical Society*, vol. 146, p. 3389-3392, 1999.

[AND 54] ANDERSON O.L., STUART D.A., "Calculation of activation energy of ion conductivity in silica glasses by classical methods", *Journal of the American Ceramic Society*, vol. 37, p. 573-580, 1954.

[BAT 92] BATTISTIG G., AMSEL G., D'ARTEMARE E., VICKRIDGEI., "A very narrow resonance in 18O(p, α)15N near 150 keV: Application to isotopic labeling. II. High resolution depth profiling of 18O", *Nuclear Instruments and Methods in Physics Research Section B*, vol. 66, p. 1-10, 1992.

[BEC 71] BECKMANN K.H., HARRICK N.J., "Hydrides and hydroxyls in thin silicon dioxide films", *Journal of the Electrochemical Society*, vol. 118, p. 614-619, 1971.

[BEL 62] BELL T., HETHERINGTON G., JACK K.H., "Water in vitreous silica. Part 2. Some aspects of hydrogen-water-silica equilibria", *Physics and Chemistry of Glasses*, vol. 3, p. 141-146, 1962.

[BLA 78] BLANC J., "A revised model for the oxidation of Si by oxygen", *Applied Physics Letters*, vol. 33, p. 424-426, 1978.

[BRE 76] BREED D.J., DOREMUS R.H., "Hydrogen profiles in water-oxidized silicon", *Journal of Physical Chemistry*, vol. 80, p. 2471-2473, 1976.

[BRU 70] BRUCKNER R., "Properties and structure of vitreous silica. I", *Journal of Non-Crystalline Solids*, vol. 5, p. 123-175, 1970.

[BUR 67] BURKARD P.J., "Tracer evaluation of hydrogen in steam-grown SiO_2 films", *Journal of the Electrochemical Society*, vol. 114, p. 196-201, 1967.

[BUR 70] BURN I., ROBERTS J.P., "Influence of hydroxyl content on the diffusion of water in silica glass", *Physics and Chemistry of Glasses*, vol. 11, p. 106-114, 1970.

[CAB 49] CABRERA N., MOTT N.F., "Theory of the oxidation of metals", *Reports on Progress in Physics*, vol. 12, p. 163-184, 1949.

[CAO 06] CAO S.V., LIU Y., GRIFFIN P., Process Lab: Oxidation, DOI: 10254/nanohub-r1879.1, 2006.

[CAV 09] CAVELLIN C.D., TRIMAILLE I., GANEM J.-J., D'ANGELO M., VICKRIDGE I., PONGRACZ A., BATTISTIG G., "An 18O study of the interaction between carbon monoxide and dry thermal SiO_2 at 1 100°C", *Journal of Applied Physics*, vol. 105, p. 033501-033501-7, 2009.

[CHA 77] CHANG R.P.H., "Multipurpose plasma reactor for materials research and processing", *Journal of Vacuum Science and Technology*, vol. 14, p. 278-280, 1977.

[CHA 80] CHANG R.P.H., CHANG C.C., DARACK S., "Fluorine-enhanced plasma growth of native layers on silicon", *Applied Physics Letters*, vol. 36, p. 999-1002, 1980.

[CHA 91] CHAO T.S., LEE C.L., LEI T.F., "Measurement of ultra-thin (<100 Å) oxide films by multiple-angle incident ellipsometry", *Journal of the Electrochemical Society*, vol. 138, p. 1756-1761, 1991.

[COP 71] COPELAND M.A., PAPPU R., "Comparative study of plasma anodization of silicon in a column of a DC glow discharge", *Applied Physics Letters*, vol. 19, p. 199-201, 1971.

[COS 84] COSTELLO J.A., TRESSLER R.E., "Isotope labeling studies of the oxidation of silicon at 1000° and 1300°C", *Journal of the Electrochemical Society*, vol. 131, p. 1944-1947, 1984.

[CRO 71a] CROSET M., PETREANU E., SAMUEL D., AMSEL G., NADAI J.P., "An O18 study of the source of oxygen in the anodic oxidation of silicon and tantalum in some organic solvents", *Journal of the Electrochemical Social*, vol. 118, p. 717-727, 1971.

[CRO 71b] CROSET M., DIEUMEGARD D., *Proceedings of the Electrochemical Society Meeting*, p. 224, Cleveland, United States, October 1971.

[CRO 76] CROSET M., DIEUMEGARD D., "Quantitative secondary ion mass spectrometry analysis of oxygen isotopes and other light elements in silicon oxide films", *Corrosion Science*, vol. 16, p. 703-715, 1976.

[DEA 00] DE ALMEIDA R. M. C., GONCALVES S., BAUMVOL I. J. R., STEDILE F. C., "Dynamics of thermal growth of silicon oxide films on Si", *Physical Review B*, vol. 61, p. 12992-12999, 2000.

[DEA 65] DEAL B.E., GROVE A.S., "General relationship for the thermal oxidation of silicon", *Journal of Applied Physics*, vol. 36 p. 3770-3778, 1965.

[DEA 78] DEAL B.E., "Thermal oxidation kinetics of silicon in pyrogenic H_2O and 5% HCl/H_2O mixtures", *Journal of the Electrochemical Society*, vol. 125, p. 576-579, 1978.

[DEL 87] DELARIOS J.M., HELMS C.R., KAO D.B., DEAL B.E., "Effect of silicon surface cleaning procedures on oxidation kinetics and surface chemistry", *Applied Surface Science*, vol. 30, p. 17-24, 1987.

[DEL 71] DELL'OCA C.J.,PULFREY D.L., YOUNG L., "Anodic oxide films", in H.A. FRANCOMBE, R.W. HOFFMANN (eds), *Physics of Thin Films*, vol. 6, Academic Press, New York, p. 1, 1971.

[DER 82] DERRIEN J., COMMANDRE M., "SiO_2 ultra-thin film growth kinetics as investigated by surface techniques", *Surface Science*, vol. 118, p. 32-46, 1982.

[DIG 72] DIGNAM M.J., "Mechanisms of ion transport through oxide films", in J.W. DIGGLE (ed.), *Oxides and oxide films*, volume 1, Marcel Dekker, New York, p. 92-286, 1972.

[DOR 62] DOREMUS R.H., in J.D. MACKENZIE (ed.), *Modern Aspects of the Vitreous State*, vol. 2, Butterworths, London, p. 12, 1962.

[DOR 69] DOREMUS R.H., in J.W. MITCHELL, R.C. DEVRIES, R.W. ROBERTS, P. CANNON (eds), *Reactivity of Solids*, John Wiley and Sons, New York, p. 667, 1969.

[DOR 94] DOREMUS R.H, *Glass Science*, (2nd edition), John Wiley and Sons, New York, 1994.

[DRE 66] DREINER R., "A-C Properties of Anodic Oxide Films on Silicon", *Journal of the Electrochemical Society*, vol. 113, p. 1210-1215, 1966.

[DRU 63] DRURY R., ROBERTS J.P., "Diffusion in silica glass following reaction with tritiated water vapour", *Physics and Chemistry of Glasses*, vol. 4, p. 79-90, 1963.

[EER 77] EERNISSE E.P., "Viscous flow of thermal SiO_2", *Applied Physical Letters*, vol. 30, p. 290-293, 1977.

[EER 79] EERNISSE E.P., "Stress in thermal SiO_2 during growth", *Applied Physics Letters*, vol. 35, p. 8-10, 1979.

[FAR 83a] FARGEIX A., GHIBAUDO G., KAMARINOS G., "A revised analysis of dry oxidation of silicon", *Journal of Applied Physics*, vol. 54, p. 2878-2880, 1983.

[FAR 83b] FARGEIX A., GHIBAUDO G., "Dry oxidation of silicon: A new model of growth including relaxation of stress by viscous flow", *Journal of Applied Physics*, vol. 54, 7153-7158, 1983.

[FAR 84] FARGEIX A., GHIBAUDO G., "Role of stress on the parabolic kinetic constant for dry silicon oxidation", *Journal of Applied Physics*, vol. 56, p. 589-591, 1984.

[FEH 72] FEHLNER F.P., "Formation of ultra-thin oxide films on silicon", *Journal of the Electrochemical Society*, vol. 119, p. 1723-1727, 1972.

[FRI 68] FRISCHAT G.H., "Sodium diffusion in SiO_2 glass", *Journal of the American Ceramic Society*, vol. 51, p. 528-530, 1968.

[FRI 69] FRISCHAT G.H., "Evidence for calcium and aluminum diffusion in SiO_2 glass", *Journal of the American Ceramic Society*, vol. 52, p. 625-625, 1969.

[FRI 75] FRISCHAT G.H., "Ion diffusion in oxide glasses", in Y. ADDA, A.D. LECLAIRE, L.M. SLIFKIN, F.H. WÖHLBIER (eds), *Diffusion and Defect Monograph Series*, 3/4, Trans Tech Publications, Aedermannsdorf, 1975.

[FRO 77] FROMHOLD A.T. JR., "Single carrier steady-state theory for formation of anodic films under conditions of high space charge in very large electric fields", *Journal of the Electrochemial Society*, vol. 124, p. 538-549, 1977.

[FUK 92] FUKUDA H., YASUDA M., IWABUCHI T., "Kinetics of rapid thermal oxidation of silicon", *Japanese Journal of Applied Physics*, vol. 31, p. 3436-3439, 1992.

[GAN 92] GANEM J.-J., Mécanismes de croissance de diélectriques par nitruration et par oxydation thermique rapide, PHD Thesis, University of Paris 7, 1992.

[GAN 93a] GANEM J.-J., BATTISTIG G., RIGO S., TRIMAILLE I., "A study of the initial stages of the oxidation of silicon using $18O_2$ and RTP", *Applied Surface Science*, vol. 65-66, p. 647-653, 1993.

[GAN 93b] GANEM J.-J., RIGO S., TRIMAILLE I., "Modellization of the silicon rapid thermal oxidation in the initial stages according to the silicon fragments model", *Microelectronic Engineering*, vol. 22, p. 35-38, 1993.

[GHE 72] GHEZ R., VAN DER MEULEN YJ., "Kinetics of thermal growth of ultra-thin layers of SiO_2 on silicon", *Journal of the Electrochemical Society*, vol. 119, p. 1100-1106, 1972.

[GHE 73] GHEZ R., "On the Mott-Cabrera oxidation rate equation and the inverse-logarithmic law", *Journal of Chemical Physics*, vol. 58, p. 1838-1843, 1973.

[GIB 89] GIBSON J.M., LANZEROTTI M.Y., "Observation of interfacial atomic steps during silicon oxidation", *Nature*, vol. 340, p. 128-131, 1989.

[GOU 80] GOURRIER S., MICEA A., BACAL M., "Oxidation of GaAs in an oxygen multipole plasma", *Thin Solid Films*, vol. 65, p. 315-330, 1980.

[GOU 81] GOURRIER S., BACAL M., "Review of oxide formation in a plasma", *Plasma Chemistry and Plasma Processing*, vol. 1, p. 217-232, 1981.

[HAN 87] HAN C.J., HELMS C.R., "Parallel oxidation mechanism for si oxidation in dry O_2", *Journal of the Electrochemical Society*, vol. 134, p. 1297-1302, 1987.

[HAN 88] HAN C.J., HELMS C.R., "18O tracer study of si oxidation in dry O_2 using SIMS", *Journal of the Electrochemical Society*, vol. 135, p. 1824-1832, 1988.

[HAU 62] HAUL R., DUEMBGEN G., "Oxygen mobility in TiO_2, SiO_2 and fused quartz by heterogeneous isotopic exchange", *Zeitschrift für Elektrochemie*, vol. 66, p. 636-641, 1962.

[HET 62] HETHERINGTON G., JACK K.H., "Water in vitreous silica. Part 1. Influence of water content on the properties of vitreous silica", *Physics and Chemistry of Glasses*, vol. 3, p. 129-133, 1962.

[HET 64] HETHERINGTON G., JACK K.H., "The oxidation of vitreous silica", *Physics and Chemistry of Glasses*, vol. 5, p. 147-149, 1964.

[HIJ 09] HIJIKATA Y., YAGUCHI H., YOSHIDA S., "A kinetic model of silicon carbide oxidation based on the interfacial silicon and carbon emission phenomenon", *Applied Physics Express*, vol. 2, 021203-1-021203-3, 2009.

[HO 79a] HO C.P., PLUMMER J.D., "Si/SiO_2 interface oxidation kinetics: a physical model for the influence of high substrate doping levels", *Journal of the Electrochemical Society*, vol. 126, p. 1516-1522, 1979.

[HO 79b] HO C.P., PLUMMER J.D., "Si/SiO_2 interface oxidation kinetics: a physical model for the influence of high substrate doping levels", *Journal of the Electrochemical Society*, vol. 126, p. 1523-1530, 1979.

[HOL 70] HOLMBERG G.L., KUPER A.B., MIRALDI F.D., "Water contamination in thermal oxide on silicon", *Journal of the Electrochemical Society*, vol. 117, p. 677-682, 1970.

[HOL 79] HOLLAHAN J.R., "Deposition of plasma silicon oxide thin films in a production planar reactor", *Journal of the Electrochemical Society*, vol. 126, p.930-934, 1979.

[HOP 75] HOPPER M.A., CLARKE R.A., YOUNG L., "Thermal oxidation of silicon", *Journal of the Electrochemical Society*, vol. 122, p. 1216-1222, 1975.

[HOR 78] HORIUCHI M., KAMIGAKI Y., HAGIWARA T., "Formation and properties of thin tunnelable SiO_2 films using a vaporized O_2 source at liquid N_2 temperature", *Journal of the Electrochemical Society*, vol. 125, p. 766-771, 1978.

[HUT 54] HUTCHISON D.A., "Isotopic exchange of oxygen in the systems water-silica and oxygen-silica", *Journal of Chemical Physics*, vol. 22, p. 758-759, 1954.

[IRE 74] IRENE E.A., "The effects of trace amounts of water on the thermal oxidation of silicon in oxygen", *Journal of the Electrochemical Society*, vol. 121, p. 1613-1616, 1974.

[IRE 76] IRENE E.A., VAN DER MEULEN Y.J., "Silicon oxidation studies: analysis of SiO_2 film growth data", *Journal of the Electrochemical Society*, vol. 123, p. 1380-1384, 1976.

[IRE 78a] IRENE E.A., "Silicon oxidation studies: some aspects of the initial oxidation regime", *Journal of the Electrochemical Society*, vol. 125, p. 1708-1714, 1978.

[IRE 78b] IRENE E.A., DONG D.W., "Silicon oxidation studies: the oxidation of heavily B- and P-doped single crystal silicon", *Journal of the Electrochemical Society*, vol. 125, p. 1146-1151, 1978.

[IRE 88] IRENE E.A., "Models for the oxidation of silicon", *CRC Critical Reviews in Solid State and Materials Sciences*, vol. 14, p. 175-223, 1988.

[KAG 99] KAGESHIMA H., SHIRAISHI K., UEMATSU M., "Universal theory of Si oxidation rate and importance of interfacial Si emission", *Japanese Journal of Applied Physics*, vol. 38, p. L971-L974, 1999.

[KAJ 05] KAJIHARA K., KAMIOKA H., HIRANO M., MIURA T., SKUJA L., HOSONO H., "Interstitial oxygen molecules in amorphous SiO_2. III. Measurements of dissolution kinetics, diffusion coefficient, and solubility by infrared photoluminescence", *Journal of Applied Physics*, vol. 98, p. 013529-013529-7, 2005.

[KAM 77] KAMIGAKI Y., ITOH Y., "Thermal oxidation of silicon in various oxygen partial pressures diluted by nitrogen", *Journal of Applied Physics*, vol. 48, p. 2891-2896, 1977.

[KAT 79] KATZ L.E., HOWELLS B.F. JR, "Low temperature, high pressure steam oxidation of silicon", *Journal of the Electrochemical Society*, vol. 126, p. 1822-1824, 1979.

[KER 77] KERN W., ROSLER R.S., "Advances in deposition processes for passivation films", *Journal of Vacuum Science and Technology*, vol. 14, p. 1082-1099, 1977.

[KRA 67] KRAITCHAMN J., "Silicon oxide films grown in a microwave discharge", *Journal of Applied Physics*, vol. 38, p. 4323-4330, 1967.

[KRZ 07] KRZEMINSKI C., LARRIEU G., PENAUD J., LAMPIN E., DUBOIS E., "Silicon dry oxidation kinetics at low temperature in the nanometric range: Modeling and experiment", *Journal of Applied Physics*, vol. 101, p. 064908-064908-8, 2007.

[LES 78] LESLIE J.D., KEITH V., KNORR K., "Auger spectroscopic evidence that plasma anodization involves mass transfer from the cathode to the anode", *Journal of the Electrochemical Society*, vol. 125, p. 44-51, 1978.

[LIE 82] LIE L.N., RAZOUK R.R., DEAL B.E., "High pressure oxidation of silicon in dry oxygen", *Journal of the Electrochemical Society*, vol. 129, p. 2828-2834, 1982.

[LIG 60] LIGENZA J.R., SPITZER W.G., "The mechanisms for silicon oxidation in steam and oxygen", *Journal of Physicsand Chemistry of Solids*, vol. 14, p. 131-136, 1960.

[LIG 61] LIGENZA J.R., "Effect of crystal orientation on oxidation rates of silicon in high pressure steam", *Journal of Physical Chemistry*, vol. 65, p. 2011-2014, 1961.

[LIG 65] LIGENZA J.R., "Silicon oxidation in an oxygen plasma excited by microwaves", *Journal of Applied Physics*, vol. 36, p. 2703-2707, 1965.

[LIG 70] LIGENZA J.R., KUHN M., *Solid State Technology*, vol. 13, p. 33, 1970.

[MAS 85] MASSOUD H.Z., PLUMMER J.D., IRENE E.A., "Thermal oxidation of silicon in dry oxygen: growth-rate enhancement in the thin regime", *Journal of the Electrochemical Society*, vol. 132, p. 2685-2700, 1985.

[MIK 81] MIKKELSEN J.C., "Oxidant transport during steam oxidation of silicon", *Applied Physics Letters*, vol. 39, p. 903-905, 1981.

[MIL 63] MILES J.L., SMITH P.H., "The formation of metal oxide films using gaseous and solid electrolytes", *Journal of the Electrochemical Society*, vol. 110, p. 1240-1245, 1963.

[MOU 61] MOULSON A.J., ROBERTS J.P., "Water in silica glass", *Transactions of the Faraday Society articles*, vol. 57, p. 1208-1216, 1961.

[MOZ 69] MOZZI R.L., WARREN B.E., "The structure of vitreous silica", *Journal of Applied Crystallography*, vol. 2, p. 164-172, 1969.

[NAN 70] NANNONI R., MUSSELIN M.J., "Anodic oxidation of silicon effects of water on oxide properties", *Thin Solid Films*, vol. 6, p. 397-406, 1970.

[NOR 61] NORTON F.J., "Permeation of gaseous oxygen through vitreous silica", *Nature*, vol. 91, p. 701-701, 1961.

[OHA 70] O'HANLON J.F., "Plasma anodization of metals and semiconductors", *Journal of Vacuum Science and Technology*, vol. 7, p. 330-338, 1970.

[OHA 73] O'HANLON J.F., SAMPOGNA M., "A new technique for high speed anodization in a DC oxygen glow discharge", *Journal of Vacuum Science and Technology*, vol. 10, p. 450-452, 1973.

[PAS 62] PASCAL P., *Nouveau traité de Chimie Minerale*, t. 8, Editions Masson, Paris, 1962.

[PER 71] PERKINS W.G., BEGEAL D.R., "Diffusion and permeation of He, Ne, Ar, Kr, and D_2 through silicon oxide thin films", *Journal of Chemical Physics*, vol. 54, p. 1683-1694, 1971.

[PFE 80] PFEFFER R., OHRING M.G., in G. LUCOVSKY, S.T. PANTELIDES, F.L. GALEENER (eds), *The Physics of MOS Insulators*, Pergamon Press, New York, p. 162-166, 1980.

[PFE 81] PFEFFER R., OHRING M., "Network oxygen exchange during water diffusion in SiO_2", *Journal of Applied Physics*, vol. 52, p. 777-784, 1981.

[PLI 77] PLISKIN W.A., "Comparison of properties of dielectric films deposited by various methods", *Journal of Vacuum Science and Technology*, vol. 14, p. 1064-1081, 1977.

[PUL 73] PULFREY D.L., HATHORN F.G.M., YOUNG L., "The anodization of Si in an RF plasma", *Journal of the Electrochemical Society*, vol. 120, p. 1529-1535, 1973.

[PUL 74] PULFREY D.L., RECHE J.J.H., "Preparation and properties of plasma anodized silicon dioxide films", *Solid-State Electronics*, vol. 17, p. 627-628, 1974.

[RAI 80] RAIDER S.I., FORGET L.E., "Reversal of relative oxidation rates of <111> and <100> oriented silicon substrates at low oxygen partial pressures", *Journal of the Electrochemical Society*, vol. 127, p. 1783-1787, 1980.

[RAY 08] RAY E.A., ROZEN J., DHAR S., FELDMAN L.C., WILLIAMS J.R., "Pressure dependence of SiO_2 growth kinetics and electrical properties on SiC", *Journal of Applied Physics*, vol. 103, p. 023522-1-023522-7, 2008.

[RAZ 81] RAZOUK R.R., LIE L.N., DEAL B.E., "Kinetics of high pressure oxidation of silicon in pyrogenic steam", *Journal of the Electrochemical Society*, vol. 128, p. 2214-2220, 1981.

[REV 69] REVESZ A.G., EVANS R.J., in J.W. MITCHELL, R.C. DE VRIES, R.W. ROBERTS, P. CANNOn (eds), *Reactivity of Solids*, John Wiley and Sons, New York, p. 425-454, 1969.

[REV 70] REVESZ A.G., "A new model for defects in non-crystalline silicon dioxide", *Journal of Non-Crystalline Solids*, vol. 4, p. 347-356, 1970.

[REV 79] REVESZ A.G., "The role of hydrogen in SiO_2 films on silicon", *Journal of the Electrochemical Society*, vol. 126, p. 122-130, 1979.

[REV 80] REVESZ A.G., "The defect structure of vitreous SiO_2 films on silicon. I. Structure of vitreous SiO_2 and the nature of the SiO_2 bond", *Physica Status Solidi (a)*, vol. 57, p. 235-243, 1980.

[RIG 80] RIGO S., ROCHET F., STRABONI A., AGIUS B., in G. LUCOVSKY, S.T. PANTELIDES, F.L. GALEENER (eds), *The Physics of MOS Insulators*, Pergamon Press, New York, p. 167-188, 1980.

[RIG 82] RIGO S., ROCHET F., AGIUS B., STRABONI A., "An 18O study of cooperative diffusion and chemical reaction during thermal treatments of silica films in water vapor", *Journal of the Electrochemical Society*, vol. 129, p. 867-876, 1982.

[RIG 86] RIGO S., "Silica films on silicon" in G. BARBOTTIN, A. VAPAILLE (eds), *Instabilities in Silicon Devices: Silicon Passivation and Related Instabilities*, p. 5-100, Elsevier, Amsterdam, 1986.

[RIG 88] RIGO S., "Si oxidation mechanisms as studied by oxygen tracer methods", in C.R. HELMS, B.E. DEAL (eds), *The Physics and Chemistry of SiO_2 and the Si-SiO_2 Interface*, p. 75-84, Plenum Press, New York, 1988.

[ROB 64] ROBERTS G.J., ROBERTS J.P., "Influence of thermal history on the solubility and diffusion of water in silica glass", *Physics and Chemistry of Glasses*, vol. 5, p. 26-31, 1964.

[ROB 66] ROBERTS G.J., ROBERTS J.P., "An oxygen tracer investigation of the diffusion of water in silica glass", *Physics and Chemistry of Glasses,* vol. 7, p. 82-88, 1966.

[ROC 81] ROCHET F., Etude par traçage isotopique à l'oxygène 18 des mécanismes de transport atomique dans les couches minces d'oxyde de silicium et d'oxyde de tantale sur silicium, Postgraduate Thesis, Pierre and Marie Curie University, 1981.

[ROC 84] ROCHET F., AGIUS B., RIGO S., "An 18O study of the oxidation mechanism of silicon in dry oxygen", *Journal of the Electrochemical Society*, vol. 131, p. 914-923, 1984.

[ROC 86] ROCHET F., RIGO S., FROMENT M., D'ANTERROCHES C., MAILLOT C., ROULET H., DUFOUR G., "The thermal oxidation of silicon the special case of the growth of very thin films", *Advances in Physics*, vol. 35, p. 237-274, 1986.

[SCH 57] SCHMIDT P.F., MICHEL W., "Anodic formation of oxide films on silicon", *Journal of the Electrochemical Society*, vol. 104, p. 230-236, 1957.

[SCH 64] SCHMIDT P.F., OWEN A.E., "Anodic oxide films for device fabrication in silicon", *Journal of the Electrochemical Society*, vol. 111, p. 682-688, 1964.

[SCH 65] SCHMIDT P.F., O'KEEFE TW., OROSHNIK J., OWEN A.E., "Doped anodic oxide films for device fabrication in silicon. II. Diffusion sources of controlled composition, and diffusion results", *Journal of the Electrochemical Society*, vol. 112, p. 800-807, 1965.

[SHA 72] SHACKELFORD J.F., STUDT R.L., FULRATH R.M., "Solubility of gases in glass. II. He, Ne, and H_2 in fused silica", *Journal of Applied Physics*, vol. 43, p. 1619-1626, 1972.

[SHE 74] SHELBY J.E., KEETON S.C., "Temperature dependence of gas diffusion in glass", *Journal of Applied Physics*, vol. 45, p. 1458-1460, 1974.

[SHE 77] SHELBY J.E., "Molecular diffusion and solubility of hydrogen isotopes in vitreous silica", *Journal of Applied Physics*, vol. 48, p. 3387-3394, 1977.

[SMI 82] SMITH F.W., GHIDINI G., "Reaction of oxygen with Si(111) and (100): critical conditions for the growth of SiO_2", *Journal of the Electrochemical Society*, vol. 129, p. 1300-1306, 1982.

[SON 04] SONG Y., DHAR S., FELDMAN L.C., CHUNG G., WILLIAMS J.R., "Modified Deal Grove model for the thermal oxidation of silicon carbide", *Journal of Applied Physics*, vol. 95, p. 4953-4957, 2004.

[SPI 61] SPITZER W.G., LIGENZA J.R., "Oxygen exchange between silica and high pressure steam", *Journal of Physics and Chemistry of Solids*, vol. 17, p. 196-202, 1961.

[STU 70] STUDT P.L., SHACKELFORD J.F., FULRATH R.M., "Solubility of gases in glass-a monatomic model", *Journal of Applied Physics*, vol. 41, p. 2777-2780, 1970.

[SUC 63] SUCOV E.W., "Diffusion of oxygen in vitreous silica", *Journal of the American Ceramic Society*, vol. 46, p. 14-20, 1963.

[SWE 61] SWEETS D.E., LEE E.W., FRANK R.C., "Diffusion coefficients of helium in fused quartz", *Journal of Chemical Physics*, vol. 34, p. 17-22, 1961.

[TIL 83] TILLER W.A., "On the kinetics of the thermal oxidation of silicon", *Journal of the Electrochemical Society*, vol. 130, p. 501-506, 1983.

[TRI 89] TRIMAILLE I., RIGO S., "Use of 18O isotopic labelling to study thermal dry oxidation of silicon as a function of temperature and pressure", *Applied Surface Science*, vol. 39, p. 65-80, 1989.

[TRI 93] TRIMAILLE I., RAIDER S.I., GANEM J.J., RIGO S., PENEBRE N.A., "Use of 18O labelling to study growth mechanisms in dry oxidation of silicon", in C.R. HELMS, B.E. DEAL (eds), *The Physics and Chemistry of SiO_2 and the Si-SiO_2 Interface 2*, Plenum Press, New York, p. 7-14, 1993.

[TSU 78] TSUCHIMOTO T., "Plasma stream transport method (2) Use of charge exchange plasma source", *Journal of Vacuum Science and Technology*, vol. 15, p. 1730-1733, 1978.

[VAN 72] VAN DER MEULEN Y.J., "Kinetics of thermal growth of ultra-thin layers of SiO_2 on silicon", *Journal of the Electrochemical Society*, vol. 119, p.530-534, 1972.

[VAN 75] VAN DER STEEN G.H.A.M., PAPANIKOLAU E., "Introduction and removal of hydroxyl groups in vitreous silica - 2. Chemical and physical solubility of hydrogen in vitreous silica", *Philips Research Report*, vol. 30, p. 192-205, 1975.

[VIC 07] VICKRIDGE I., GANEM J., HOSHINO Y., TRIMAILLE I., "Growth of SiO_2 on SiC by dry thermal oxidation: mechanisms", *Journal of Physics D.*, vol. 40, p. 6254-6263, 2007.

[VON 72] VON DER MEULEN Y.J., "Kinetics of thermal growth of ultra-thin layers of SiO_2 on silicon", *Journal of the Electrochemical Society*, vol. 119, p. 530-534, 1972.

[WAT 07] WATANABE T., OHDOMARI I., "A kinetic equation for thermal oxidation of silicon replacing the Deal–Grove equation", *Journal of the Electrochemical Society*, vol. 154, G270-G276, 2007.

[WIL 65] WILLIAMS E.L., "Diffusion of oxygen in fused silica", *Journal of the American Ceramic Society*, vol. 48, p. 190-194, 1965.

[WOL 69] WOLF H.F., *Silicon Semiconductor Data*, Pergamon Press, New York, 1969.

[WOL 80] WOLTERS D.R., "On the oxidation kinetics of silicon: the role of water", *Journal of the Electrochemical Society*, vol. 127, p. 2072-2082, 1980.

[YAM 07] YAMAMOTO T., HIJIKATAO Y., YAGUCHI H., YOSHIDA S., "Growth rate enhancement of (0001)-face silicon–carbide oxidation in thin oxide regime", *Japanese Journal of Applied Physics*, vol. 46, p. L770-L772, 2007.

[YAM 08] YAMAMOTO T., HIJIKATAO Y., YAGUCHI H., YOSHIDA S., "Oxide growth rate enhancement of silicon carbide (0001) Si-faces in thin oxide regime", *Japanese Journal of Applied Physics*, vol. 47, p. 7803-7806, 2008.

Chapter 2

Ion Implantation

2.1. Introduction

Ion implantation consists of introducing charged atoms (projectiles) into a material (the target), by communicating to them sufficient energy so that they enter beyond the surface area. The projectiles' energy is thus clearly different from those of techniques aiming for a surface process or deposition (plasma, molecular beam). Moreover, the implantation is distinguished from these other methods by the purity of the beam, selected at the isotopic level.

It is essentially thanks to the semiconductor industry that implantation technology could spread. Ion implantation is indeed the most common method of semiconductors doping[1]. Nowadays, there are thousands of ion implantation machines functioning in this industry. Most of them are devoted to the production of integrated circuits based on silicon. The method, which in the 1970s definitively led to ion implantation supplanting diffusion in the manufacturing technology of microelectronics devices, is the adjustment of the threshold voltage of MOS transistors.

Once incorporated in the process and because of its advantages, implantation quickly imposes itself into all the other doping stages, and even in processes of insulation or purification of the silicon: formation of drain and source areas and the deep n- or p-wells of CMOS circuits, the realization of burred collectors, emitter and

---

Chapter written by Jean-Jacques GROB.

1 The term "doping" is defined in detail by D. Mathiot in the introduction to Chapter 3 of this book.

base doping in bipolar technology, impurity gettering, realization of buried insulating silica layers by implantation of an important quantity of oxygen, etc. Nowadays, for example, the fabrication of the circuits of the most recent CMOS technology requires more than ten stages of ion implantation.

The main advantage of implantation is the precise control of the number and of the penetration depth of the ions. Although the word is incorrect (we should say fluence), the dose indicates the number of ions implanted per unit of surface of the target (often given in atoms/cm^2) and the energy of ions (generally in keV) is the parameter controlling the spatial distribution of the atoms. Implantation is often employed as a technique of pre-deposition, only used to introduce a specific dose in the target. It is by diffusion that the dopants are then redistributed on distances that we cannot reach with implantation.

The major disadvantage of the technique is the inevitable damaging of the target during the slowing down of the ions. The collisions of the projectiles with the atoms of the target move the latter from their initial equilibrium position. If the atom thus displaced acquires enough energy, it can cause other displacements, and be at the origin of a cascade of collisions. The damage created by the implantation must be repaired (we speak of healing), before being able to profit from the effects expected from implantation, i.e. full electrical activity of the dopants. This healing intervenes during a thermal treatment (annealing), following the implantation. During annealing, some diffusion of the implanted species is inevitable. Therefore, whatever the terms of use of the technique (pre-deposition followed by diffusion, or implantation followed by annealing), implantation and diffusion are inseparably linked.

Another advantage of the implantation, less important than the first on the technological level, but interesting in some situations, lies in the fact that it is an athermal process: the quantity of atoms that we can introduce into a target is not submitted to the laws of thermodynamics. In other words, the solubility limit can be largely exceeded; the only limit coming, possibly, from the phenomenon of surface sputtering. However, this advantage is very relative because, during annealing, the thermodynamic equilibrium reasserts itself, often causing the precipitation of a significant part of the implanted species. However, some precautions during these annealings – low temperatures and/or very short durations – help us to maintain these non-equilibrium situations. Of course, these limited thermal budgets also lead to the minimization of the dopant redistribution: it explains the appearance of the "rapid" thermal annealing techniques, using high power lasers or lamps.

All the aspects quickly mentioned in this introduction – ion path, damaging, annealing, dopant activation, application to the development of devices – will be developed in this chapter. Before this, a section will be devoted to the description of implantation machines (in short, implanters).

Abundant literature exists in this domain, notably in the proceedings of the semi-annual IBMM (*Ion Beam Modification of Materials*) and IIT (*Ion Implantation Technology*) conferences. In addition, several books have been devoted to ion implantation. Some of them are referred to: [DEA 73, FAV 93, RYS 82, WIL 84, ZIE 92].

2.2. Ion implanters

2.2.1. *General description*

Depending on their use in the various fields of microelectronics, implanters are classified into two categories: the medium current implanters (a few mA at maximum) and high current implanters (from a few mA up to several tens of mA). The range of accessible energy is the same in the two categories, typically from 10 to 200 keV. The difference between the two groups lies in their process capacity and in their respective flexibility in use. A medium current implanter processes one wafer at a time. The ions are distributed there uniformly with the help of a system of electrostatic scanning. The sample holder can often be oriented, which is useful for oblique implantations. The limitation in current is related to the difficulty in evacuating, as calories, a sufficient quantity of the energy deposited by the beam. On a high current implanter, typically several tens of targets are installed on a carousel revolving permanently in front of the beam. This target displacement constitutes one of the scanning directions, the other being of an electrostatic or mechanical nature.

In this configuration, heating problems under the beam are less severe. Figure 2.1 schematically illustrates the general disposition of a medium current implanter.

An ion implanter consists of a source of ions into which the elements to be ionized are introduced in the form of gas or vapor. With a polarized electrode, we can extract ions from the source. A magnetic analysis system separates the ions according to their mass and an acceleration column communicates to them the desired speed. Lastly, a set of polarized plates uniformly distributes the ions on the target. The loading of the wafers and their positioning in front of the beam and then their unloading are executed by a robotized system. On a medium current implanter, with the master controller of the target we can often tilt the wafers up to 60 degrees in relation to the direction of the beam. It is an interesting method for some applications in microelectronics. On a high current implanter, the flexibility is less important: the targets are simply tilted by a few degrees compared to the normal, in order to avoid the channeling phenomenon of ions (see section 2.3.3).

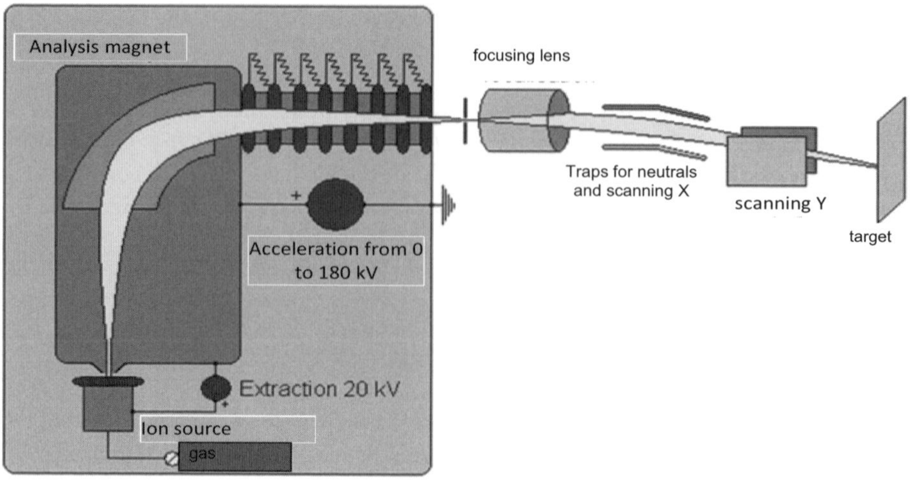

Figure 2.1. *Schematic representation of a medium current ion implanter*

There is a third category of implanters, much less common than the other two. This category comprises the high-energy machines, obviously set aside for more deep processing. These machines are often equipped with Van De Graaff voltage generators, immersed in a tank filled with gas under pressure. The ions are accelerated there, typically at energies of about a few MeV. They are more expensive and more complex to use, and thus more rare.

2.2.2. *Ion sources*

An ion source essentially consists of a closed vessel, called an arcing chamber, in which an electrical discharge is maintained in a gas or in a vapor containing the desired element. In most high current sources, the discharge is initiated and maintained by electrons emitted by a filament heated by the transit of an intense current and by a magnetic field, perpendicular to the direction of electron emission, which considerably increases their course between the cathode and anode. Thus, the probability of ionizing the vapor atoms increases by a significant factor. The source with a Freeman hot cathode is very frequently used in industrial implanters [FRE 63]. In this type of source, a voltage of about 50-100 V is applied between the cathode and anode and the intensity of the magnetic field is about 100 G. The efficiency of this source is due to the position of the cathode, parallel and very close to the extraction electrode, which produces an intense discharge at a very favorable location. In the sources with cold cathode, the discharge is maintained with the help of higher voltages and magnetic fields, respectively 1 to 2 kV and a few kG. Other methods can be used to maintain a discharge or to ionize the elements: the implementation of a field of high or very high frequency in the microwave or ECR (electron cyclotron resonance) HF (high frequency), sources, phenomena of surface sputtering and ionization in the sources with sputtering [STE 92].

Most elements are introduced into the arcing chamber in the form of a pure gas or as components of gaseous molecular compounds. Other elements, liquid or solid, with a vapor pressure of about 10^{-3} Torr in a controllable temperature range (300-700°C), are set in a furnace close to the arcing chamber. Sometimes, solids, which cannot be vaporized in the elementary form, exist in the form of compounds, which can often be more easily sublimed. Lastly, low vapor pressure materials can be vaporized using a chemical reaction. The method consists of, in the furnace mentioned above, making a highly reactive gas (fluorine, chlorine) transit on the material, pure or in oxide state, to form a more volatile fluoride or chloride.

2.2.3. *Mass analysis and beam optics*

A large amount of ions, monoatomic or molecular, with different masses or charge states, is extracted from the source. For example, when boron trifluoride is injected into the arcing chamber, the ions extracted from the source are: $^{10}B^{++}$, $^{11}B^{++}$, $^{10}B^{+}$, $^{11}B^{+}$, F^{+}, $^{10}BF^{+}$, $^{11}BF^{+}$, $^{10}BF_2^{+}$, $^{11}BF_2^{+}$, $^{10}BF_3^{+}$, $^{11}BF_3^{+}$. For each ion group, their intensities are in the isotopic abundance, namely 1 to 5, in the case of boron 10 and 11. Only one type of ion generally interests the user, for example $^{11}B^{+}$. It is thus necessary to proceed to a selection according to the mass of these ions. An analysis magnet separates the various beams. Indeed, the path of the ions, accelerated by a potential V (in volts), follows, in a magnetic field B (in Gauss), a circle trajectory of radius R (in centimeters) so that $RB \approx 144\sqrt{M_1 V / n}$, where M_1 is the mass of the ion (in atomic units) and n its charge state. The common magnets installed in the implanters have average resolutions of about $M_1/\Delta M_1 \approx 100$. This means that they are able to differentiate the ion of mass 100 from the ion of mass 101. In the range of the common beams of microelectronics technology, a good resolution in mass is essential to separate for example $^{11}BF^{+}$from$^{31}P^{+}$. In some situations, the separation is impossible and contaminations are feared, for example $^{27}Al^{+}$ and $^{11}B0^{+}$.

Most installations have a magnet deflecting the ion trajectory at 90 degrees. Under these conditions, the magnets also have a significant focusing role: the image of the extraction slot is located at the same distance from the output pole of the magnet, as the source from the input pole. It is there that a shutter is often set (a Faraday cage or a definition slot) allowing us to control the beam or to stop the undesirable ions. Beyond this point, the beam is again divergent. It is taken back a little further in quadripolar electromagnetic lenses, which refocus it before its entrance into the scanning plates.

2.2.4. *Current measurement*

Implantation stations, different on medium and high current implanters, have been broadly described in section 2.2.1. An important aspect of the design of these chambers relies in measuring the current, and thus the dose. Generally, on medium current implanters, the target is directly connected to an ammeter and the current is integrated there, in order to determine the dose ϕ, according to the relation:

$$\varphi(ions / cm^2) = \frac{1}{A} \int_0^{t_{impl}} \frac{I}{ne} dt \qquad [2.1]$$

where A (in cm^2) is the implanted surface, usually defined by a beam limiter at the entrance of the implantation chamber. I (in ampere) is the beam current, and e is the elementary charge (in coulomb). t_{impl} corresponds to the entire duration of the implantation. A certain number of phenomena possibly come to distort this measurement: (1) the ions, on their course to the target, can capture an electron and thus be found in a neutral state; (2) the impact of the ions tears off electrons from the target which, if they do not go back there, give place to a parasitic surplus current; (3) the back diffusion and the surface sputtering, by ejecting atoms, decrease the number of ions actually introduced into the target. The point (3) will be developed in section 2.3. Maintaining a low residual pressure and the reduction of the distance to cover (thanks to the interposition of a trap for neutral) are the solutions of problem (1). As for secondary electrons, they can be taken back and counted in the target current, by using a correctly conceived Faraday cage, like the one in Figure 2.2, and/or thanks to the installation, in front of the target, of a negatively polarized electrode, the role of which is to push back the electrons.

Sometimes, the main Faraday cage is replaced by four small cages, located at the four corners of the x-y scanning zone. Besides the current measurement, we can with this method check the beam uniformity.

On high current implanters, the power is measured using a Faraday cage located behind the disc supporting the samples. At each turn, the beam transits through an opening of the disc. The current is thus not measured on the samples and the dose errors coming from the secondary electrons are avoided. There is no longer an electron spinner in front of the target. On the other hand, an electron “shower” compensates for the possible accumulation of positive charges at the surface of the wafers, notably when they contain resins or oxides (see the following section).

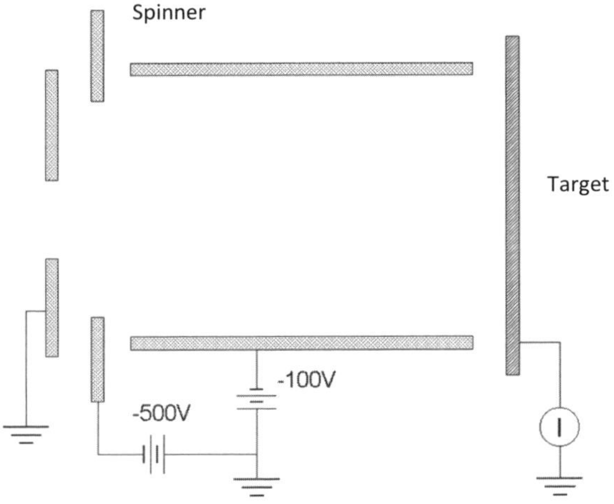

Figure 2.2. *Faraday cage for proper current measurement*

2.2.5. *Production throughput, temperature control and charge effects*

It is difficult to compare the performances of the various implantation machines, because of the large number of various possibilities used, to adapt a given system to the operations required by such or such a technology. We can however try to figure out some general trends, as for the orders of magnitude and the limits of these machines. Most of the implanters have loading systems, where samples are processed by batches of about ten wafers. The loading operations of each sample, its handling (to place it in front of the beam) and its unloading, take in general a few seconds. The maneuvers are longer between each batch (valves opening/closing, ventilation/pumping of the implantation chamber, replacement of the cassette or disc) and can generally last several minutes. For implantations of the lowest doses, the output rate is only given by these manipulations, and will thus be, at best, about 100 or 200 samples per hour.

For the strongest doses, the implantation duration starts to play a role and the output will then strongly depend on the current. However, as mentioned previously, an ill-considered increase in the current density leads to significant heating of the target. Figure 2.3 reports, for an implanter with electrostatic scanning, the theoretical value of the heating of a silicon wafer, as a function of the time, for various power densities (or irradiances, in W/m^2) [BRU 79].

Such temperatures are prohibitive: photoresists, for example, can tolerate only moderated temperatures, of about 100°C. Beyond this level, they inflate, crackle or decompose. There is thus a compromise to be found between high output and reasonable heating. With a purely electrostatic scanning, the beam remains almost permanently on the same wafer that quickly heats up. On high current implanters, this way of proceeding is thus no longer possible. Thanks to a mechanical or mixed scanning, the power density is then distributed on all of the implantation batch, the beam periodically returning on the same sample only at each turn of the disc (see Figure 2.4).

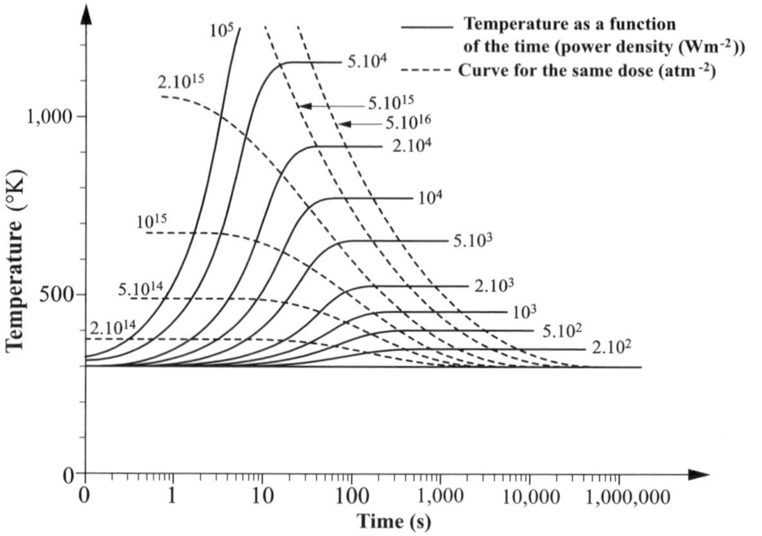

Figure 2.3. *Calculated values (full line) of the silicon temperature as a function of time and of the power density (irradiance). The curves in dotted lines correspond to the iso-dose contours reached under these conditions. According to [BRU 79]*

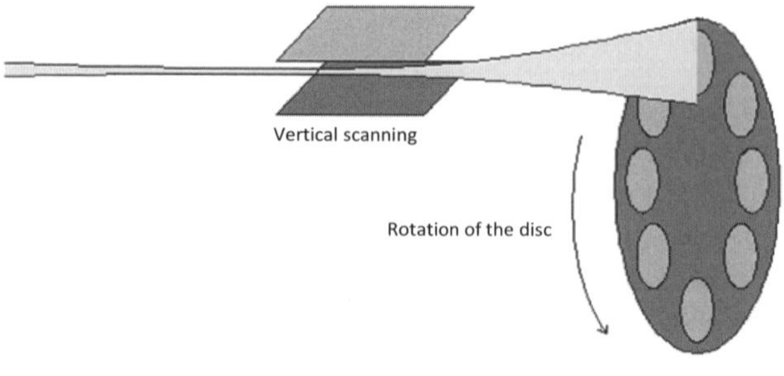

Figure 2.4. *Example of mixed scanning of the beam on a high current implanter: the vertical scanning is electrostatic and the disc rotation ensures the ions distribution in the horizontal direction*

However, for beam currents higher than 1 mA, these solutions are not sufficient and it is necessary to evacuate the calories by cooling the back face of the wafers. For that, there are two solutions: to directly cool with the help of a fluid (generally a gas) or indirectly, by thermal conductivity towards the sample holder, while trying to improve the contact.

Ion implantation in the semiconductor industry often implies insulating structures. A significant charge effect can be induced there by ions, causing a progressive deterioration of the dielectric qualities of the oxides, going, possibly, up to a disastrous breakdown. In addition, this charge effect can also disturb the behavior of the ions up to the target, producing a divergence of the beam. These problems are all the more severe, as the currents become more important. It is then necessary to maintain the target neutrality. This is generally carried out with the help of a "shower" of secondary electrons (therefore, of low energy) placed near the target, like, for example, the one in Figure 2.5.

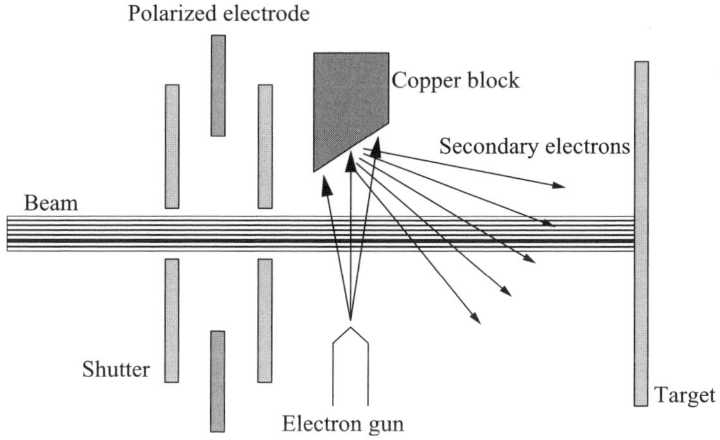

Figure 2.5. *Example of an "electronic shower" device intended to neutralize the charge induced in the target by positive ions*

2.3. Ion range

2.3.1. *Binary collision and stopping power*

The slowing down of the ions in the material is characterized by the stopping power, noted (dE/dx), corresponding to the quantity of energy dE lost per elemental distance dx. It is a function of the energy E. The interaction between the projectile and a target atom is described by assuming two distinct processes: the collision between the two nucleui and the interactions with the electrons. The first corresponds to the Coulomb repulsion, causing an important deviation of the trajectory. The second is related to excitations and ionizations, leading to significant energy loss for the ion, but which do not alter its direction. The stopping power can thus be separated into two components: nuclear $(dE/dx)_n$ and electronics (dE/dx)e:

$$\frac{dE}{dx} = (\frac{dE}{dx})_n + (\frac{dE}{dx})_e = N(S_n + S_e) \quad [2.2]$$

The stopping cross-sections for each type of interaction, S_n and S_e, are physical quantities independent of the density of the slowing environment, N is the atomic density.

In the context of classical mechanics, the elastic binary collision with an atom of the target initially immobile, following the laws of conservation of energy and of momentum, we can write the energy transfer T as:

$$T = \frac{4M_1M_2E}{(M_1 + M_2)^2}\sin^2\frac{\theta_c}{2} \quad [2.3]$$

where M_1 and M_2 are respectively masses of the ion and of the target atom. E is the energy and θ_C the scattering angle in the system of the center of mass (see Figure 2.6).

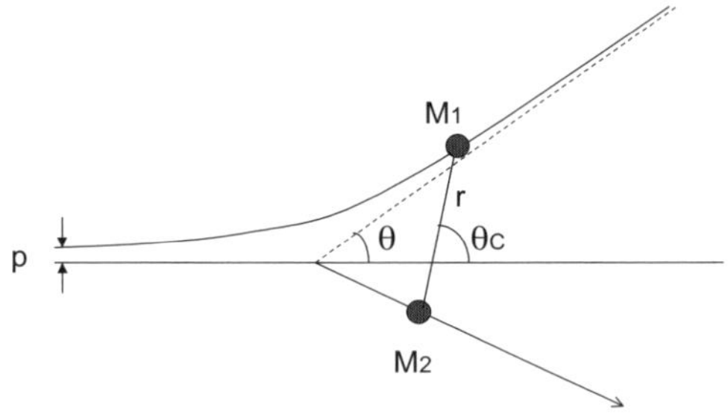

Figure 2.6. *Diagram of the collision between the projectile (M_1) and the target atom (M_2)*

The scattering angle is related to the interaction potential V(r) by the relation:

$$\theta_c = \pi - 2p\int_{r_{\min}}^{\infty}\frac{dr}{r^2\sqrt{1-\frac{V(r)}{E_c}-\frac{p^2}{r^2}}} \quad [2.4]$$

where p is the impact parameter (see Figure 2.6) and E_C is the energy in the system of the center of mass ($E_c=EM_2/(M_1+M_2)$). r_{min} is the collision radius between the two partners.

The potential used at implantation energies is of the Coulomb type, moderated by the shielding effect of the electrons. It is written:

$$V(r) = \frac{Z_1 Z_2 e^2}{4\pi\varepsilon_0 r}\varphi(\frac{r}{a}) \qquad [2.5]$$

where Z_1 and Z_2 are the atomic numbers of the projectile and of the target, and ϕ is the shielding function, tending to 1 when r decreases. a is the shield radius estimated by Bohr [BOH 48], from the Thomas-Fermi atomic model:

$$a = \frac{0.8853 a_0}{\left(Z_1^{2/3} + Z_2^{2/3}\right)^{1/2}} \qquad [2.6]$$

a_0 is the Bohr radius ($5.29.10^{-11}$m).

The cross-section (i.e. the probability that a collision, leading to the deflection θ_C, occurs) is:

$$d\sigma(\theta_c) = 2\pi p dp = -2\pi p(\frac{dp}{d\theta_c})d\theta_c \qquad [2.7]$$

And finally, the stopping cross-section is:

$$S_n = \int_{T_{\min}}^{T_{\max}} T(\theta_c) d\sigma(\theta_c) \qquad [2.8]$$

T_{max} is given by equation [2.2] with $\theta_c=180°$. T_{min} is more difficult to define. It is of the order of magnitude of the displacement energy of the target atoms. To integrate this equation, Lindhard, Scharf and Schiott [LIN 63] introduce reduced variables, ε and ρ respectively proportional to the energy and to the ion range R:

$$\varepsilon = \left[\frac{4\pi\varepsilon_0 a}{Z_1 Z_2 e^2}\frac{M_2}{M_1 + M_2}\right]E \qquad [2.9]$$

$$\rho = N\pi a^2 \frac{4M_1M_2}{(M_1+M_2)^2} R \qquad [2.10]$$

By using these notations, the reduced stopping power of Lindhard, Scharf and Schiott (LSS) is connected to the actual stopping power by:

$$S_{LSS} = \frac{\varepsilon}{\pi a^2 \gamma E} S(E) \text{ with } \gamma = \frac{4M_1M_2}{(M_1+M_2)^2} \qquad [2.11]$$

Many works compare the merits of such or such a shielding function [WIL 77, ZIE 85a]. In the traditional implantation domain ($\varepsilon<30$), the nuclear stopping cross-section takes the following universal analytical form [ZIE 92]:

$$S_n = \frac{\ln(1+1.21\varepsilon)}{2(\varepsilon + 0.0065\varepsilon^{0.154} + 0.242\varepsilon^{1/2})} \qquad [2.12]$$

The nuclear stopping power dominates at very low speed. At higher energies, the electronic stopping power takes increasing importance and, always in the domain of implantation, its value is proportional to the speed:

$$S_e = k\varepsilon^{1/2} \qquad [2.13]$$

k is a constant without dimension, which depends on M_1, M_2, Z_1, Z_2. Thus, in opposition to the case of the nuclear stopping power, the electronic deceleration cannot be represented in the form of a universal curve. Nevertheless, for common projectiles, k varies between 0.1 and 0.2 (from the heaviest to the lightest) and, in the domain of interest, the stopping powers depend on the energy, as indicated in Figure 2.7. In this figure, the electronic stopping power is represented for k = 0.15.

The nuclear stopping power is at maximum at $\varepsilon \approx 0.35$, which corresponds to 3 keV for boron, 15 keV for phosphorus or 70 keV for arsenic in silicon. The electronic stopping power becomes dominant at $\varepsilon > 2$ for the lightest ions, and at $\varepsilon > 4$, for the heaviest, which, always in silicon, corresponds to 20 keV for boron, 140 keV for phosphorus or 800 keV for arsenic.

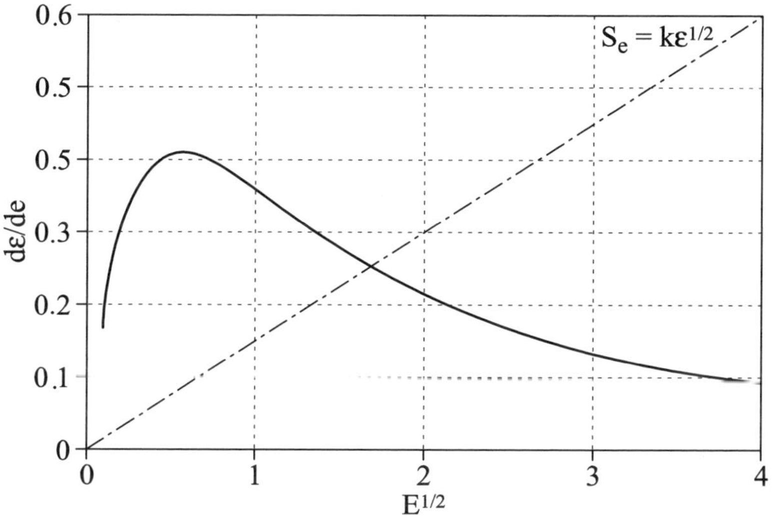

Figure 2.7. *Stopping power cross-sections as a function of the square root of energy in LSS units [FLAX 63]*

2.3.2. *Profile of the implanted ions*

The total distance covered by an ion along its trajectory is called the range (noted R), but what interests the user is the distance covered in the normal direction to the target, which we call the projected range. The slowing down process being essentially statistical, the relevant parameters are the average projected range (noted R_P) and the scattering of the values around R_P, characterized by a longitudinal standard deviation ΔR_P and a transverse standard deviation $\Delta R_\perp$ (see Figure 2.8).

To calculate these magnitudes, the concepts related to the binary collision and to the stopping powers are applied to the statistical problem of a series of collisions. For that, there are two solutions: the resolution of a transport equation [LIN 63] and the Monte Carlo approach [ROB 63]. The transport equation is written for the function probability density f(R, E) for an ion to have a range comprised between R and R +dR:

$$\frac{\partial f(R,E)}{\partial R} = N\int[f(R,E-T)-f(R,E)]d\sigma_n - \frac{\partial f(R,E)}{\partial E}(\frac{dE}{dx})_e \qquad [2.14]$$

This equation cannot be solved in this form. It is possible to develop f in moments f_n, to rewrite the equation for the f_n moment and to solve it. By definition, f_1=Rp is the moment of order 1:

$$f_1 = R_p = \int_{-\infty}^{+\infty} xf(x,E)dx \qquad [2.15]$$

and ΔRp is related to the moment of order 2 by the relation:

$$\Delta R_p^2 = \int_{-\infty}^{+\infty} (x - R_p)^2 f(x,E)dx \qquad [2.16]$$

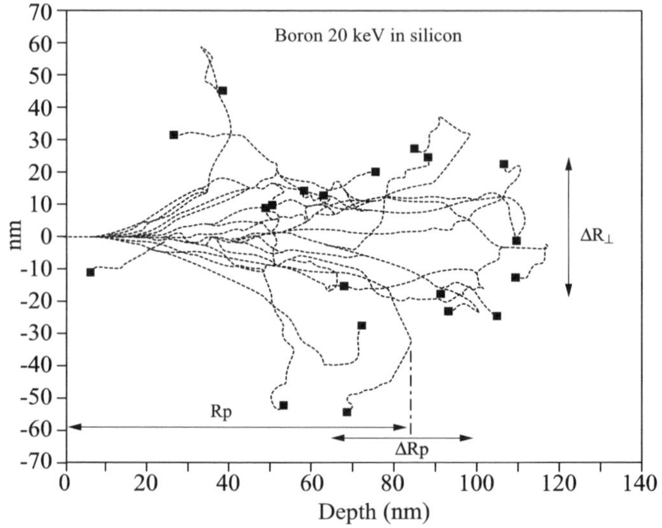

Figure 2.8. *Range of about 20 boron ions of 20 keV in silicon (the trajectories are represented by a dotted line and the stopped boron by a black square) and definitions of the average projected range and of the longitudinal and transverse scatterings*

The Monte Carlo calculation simulates the entire history of the projectile, collision after collision [BIE 80]. The final statistical assessment is carried out by summing and/or averaging on a large number of simulated trajectories. The position, direction and energy of the ion are the input data. The distance between two collisions, the impact parameter and the azimuth of each collision are the random variables. The elastic collisions deflect the trajectory. Between each collision, the ion loses energy proportionally to the electronic stopping power at the value of its instantaneous energy. The story stops when the energy of the ion drops under a certain value or when the ion leaves the target. The advantage of the Monte Carlo method compared to the transport equation relies on the 3D character of the reconstitution of the trajectories. The disadvantage is the calculation slowness.

The two methods give similar results, namely Rp and ΔRp. As a first approximation, we can then describe the profile of the implanted atoms as a Gaussian function:

$$C(x) = \frac{\varphi}{\sqrt{2\pi}\Delta R_p} \exp\left[-\frac{(x - R_p)^2}{2\Delta R_p^2} \right] \quad [2.17]$$

By comparing to the experimentally measured distributions, it is clear that, except when $M_1 = M_2$, the Gaussian approximation is not satisfactory. From the moments of order 3 and 4 of the function f(x, E), we define two dimensionless parameters, γ and β, respectively describing the profile asymmetry (*skewness*[2]) and the sharp or squashed character of the maximum concentration peak (*kurtosis*):

$$\gamma = \frac{\int_{-\infty}^{+\infty} \left(x - R_p\right)^3 f(x,E)dx}{(\Delta R_p)^3} \quad [2.18]$$

$$\beta = \frac{\int_{-\infty}^{+\infty} \left(x - R_p\right)^4 f(x,E)dx}{(\Delta R_p)^4} \quad [2.19]$$

There are various forms of distributions to rebuild the profiles from the first four moments [SEL 84]. The most usual one, regarding implantation in silicon, is the Pearson IV distribution:

$$f(x) = k\left| b_2\left(x - R_p\right)^2 + b_1\left(x - R_p\right) + b_0 \right|^{1/2b_2}$$

$$\exp\left[-\frac{b_1 / b_2 + 2b_1}{\sqrt{4b_2 b_0 - b_1^2}} \arctan\frac{2b_2\left(x - R_p\right) + b_1}{\sqrt{4b_2 b_0 - b_1^2}} \right] \quad [2.20]$$

where k is a constant obtained thanks to the normalization condition:

2 Beware of notations: do not mix up γ skewness with the γ from equation [2.11].

$$\int_{-\infty}^{+\infty} f(x)dx = 1 \qquad [2.21]$$

The constants b_0, b_1, b_2 are related to the four moments by:

$$b_1 = -\frac{\gamma \Delta R_p(\beta+3)}{A} \quad b_0 = -\frac{\Delta R_p^2(4\beta-3\gamma^2)}{A} \qquad [2.22]$$

$$b_2 = -\frac{(2\beta-3\gamma^2-6)}{A} \quad A = 10\beta - 12\gamma^2 - 18$$

The maximum of a Pearson IV distribution is at $x = R_p + b_1$. If $\gamma < 0$, the peak is beyond R_P and the concentration is more important towards the surface than towards the depth. This is the case for light projectiles in a heavy target, for example boron in silicon (see Figure 2.9.). The situation is reversed for $\gamma > 0$. We find this shape for a heavy projectile in a light target, such as arsenic in silicon (see Figure 2.10.). $\gamma = 0$ and $\beta = 3$ correspond to a Gaussian distribution.

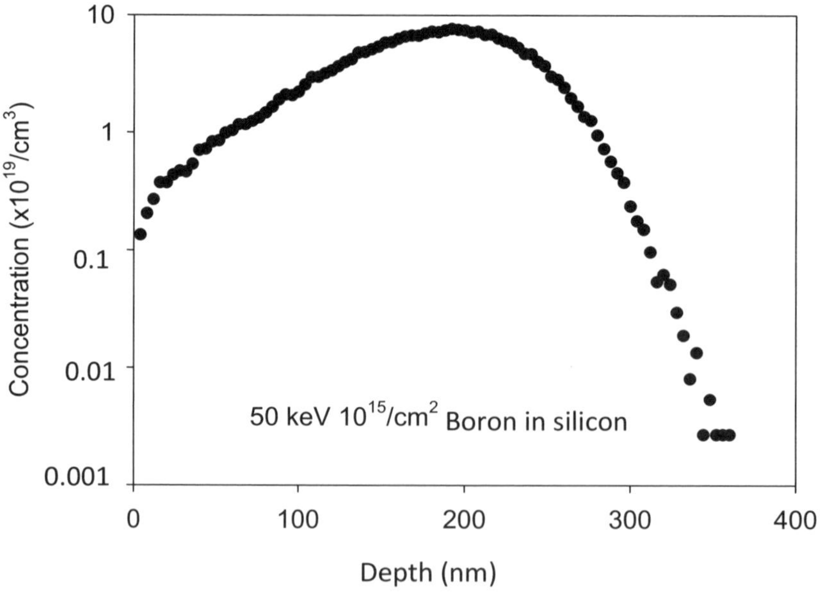

Figure 2.9. *SRIM simulation [ZIE 85a] of the profile of boron atoms ($50\ keV\ 10^{15}/cm^2$) implanted in silicon*

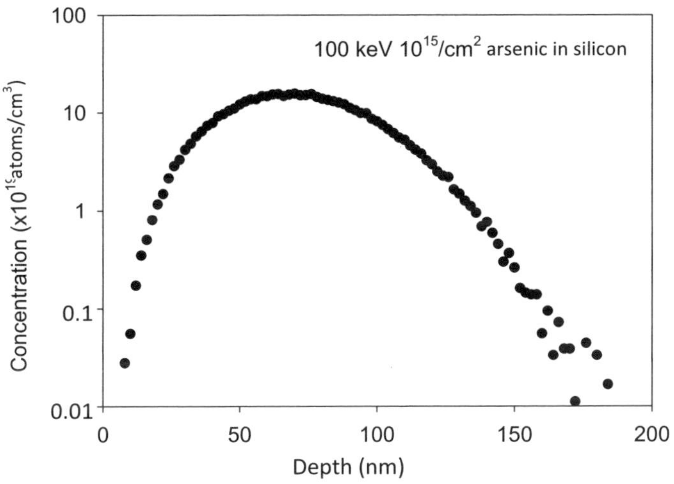

Figure 2.10. *SRIM simulation [ZIE 85a] of the profile of arsenic atoms (100 keV $10^{15}/cm^2$) implanted in silicon*

2.3.3. *Backscattering, surface sputter and channeling*

A certain number of phenomena can alter the shape of the concentration profile of implanted atoms. They appear under some specific conditions and are generally not taken into account in simulators.

Backscattering corresponds to the events of almost head-on collisions during which an energy close to $T_{max}=\gamma E$ is transferred to the recoil atom, which can lead, if $M_1<M_2$, to a reflection of the projectile on the target surface. The backscattering coefficient, i.e. the number of reflected ions divided by the total number of projectiles can go up to 30% for low energy boron ions into silicon. This phenomenon must thus be taken into account to estimate the dose actually received by the target.

Sputtering corresponds to the ejection of surface atoms of the target under the action of elastic energy transfers. It is characterized by a sputtering coefficient S, corresponding to the number of atoms ejected per incident ion[3]. It is a function of M_1, M_2 and E. S reaches its strongest value at the energy corresponding to the maximum of nuclear stopping power. In silicon, it is then of the order of 5 for an antimony beam, of 3 for arsenic, of 1.5 for phosphorus and lower than 1 for boron.

---

3 Beware of notations: do not confuse S (sputtering) with the stopping cross-section introduced previously.

The thickness of eroded material x_s is related to S and to the dose ϕ by:

$$x_s = \frac{S}{N}\varphi \qquad [2.23]$$

By assuming that the sputtering rate remains constant during implantation, and that the profile is Gaussian, for high-dose implantation, the distribution will obey:

$$C(x) = \frac{N}{2S}\left(erf\frac{x - R_p + x_s}{\sqrt{2}\Delta R_p} - erf\frac{x - R_p}{\sqrt{2}\Delta R_p}\right) \qquad [2.24]$$

and with saturation ($\phi \rightarrow \infty$), we will have:

$$C(x) = \frac{N}{2S} erfc\frac{x - R_p}{\sqrt{2}\Delta R_p} \qquad [2.25]$$

with a maximal concentration located at the surface, given by:

$$C_{\max} = \frac{N}{2S} erfc\frac{-R_p}{\sqrt{2}\Delta R_p} \approx \frac{N}{S} \qquad [2.26]$$

For high implantation doses, the profiles will evolve as shown in Figure 2.11. The surface sputtering is thus the only phenomenon actually limiting the quantity of atoms that we can introduce into a material by implantation[4].

The critical angle ψ_C is very simply calculated, by writing the channeling condition, namely that the transverse kinetic energy of the ion remains lower than the repellent potential of the lattice row or plan.

This condition, for a crystalline axis, leads to the relation [LIN 65]:

$$\psi_c = \sqrt{\frac{2Z_1Z_2e^2}{Ed}} \qquad [2.27]$$

---

4 The erf and erfc functions are defined by: $erf(x) = \frac{2}{\sqrt{\pi}}\int_0^x e^{-u^2}du$ and erfc = 1 – erf.

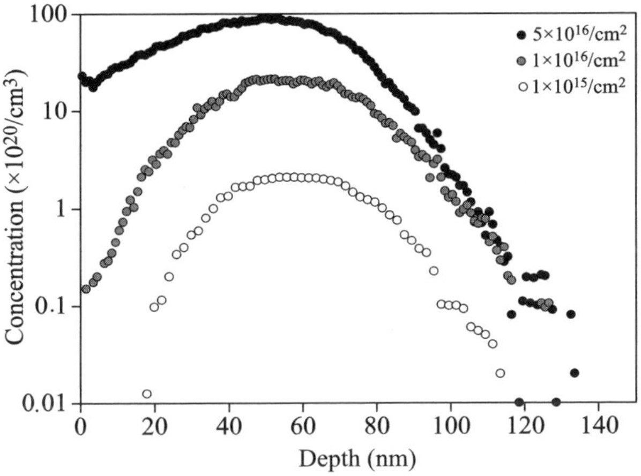

Figure 2.11. *TRIDYN simulation [MOL 84] of the profile of antimony ions of 100 keV implanted in silicon at various doses*

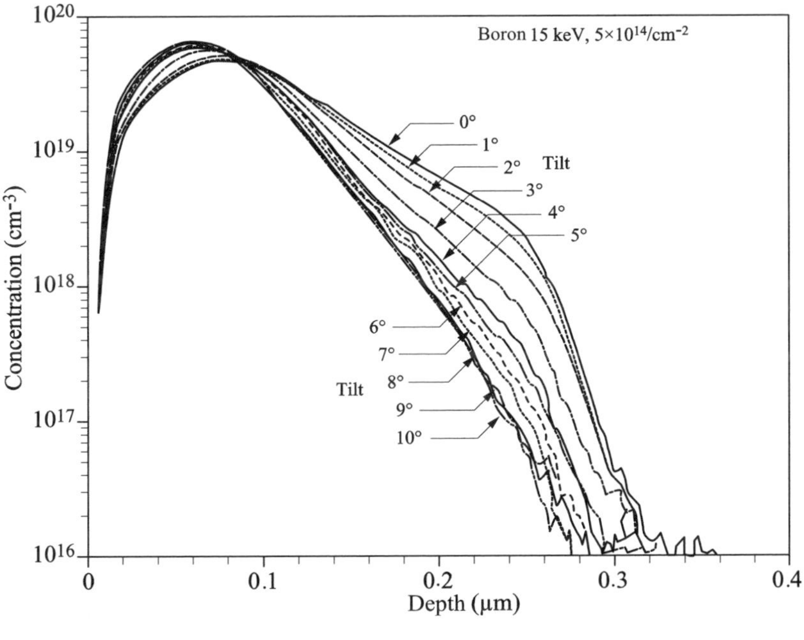

Figure 2.12. *SIMS profiles of the boron implanted at 15 keV in silicon according to the incidence angle compared to the direction <100> of the crystal (according to [KLE 91])*

The order of magnitude is of a few degrees for usual implantation energies in silicon. It is higher in axial channeling than in planar channeling. The latter constitutes a major limitation to the realization of surface junctions by boron implantation. The tails in distributions can obviously be reduced, thanks to an accurate control of the orientaion of the wafers as a function of the beam direction. But, even with these precautions, the angular divergence of the ions compared to their initial direction, after a certain number of collisions, quickly becomes very important. This means that a part of them will always end up adopting a trajectory focused along the axis or secondary planes of the crystal (see Figure 2.13). This phenomenon explains the persistence of a tail in the distributions of Figure 2.12, even for incidence directions far from any channeling axis. The only way to completely remove this effect consists of destroying the crystalline order by pre-amorphizing the target.

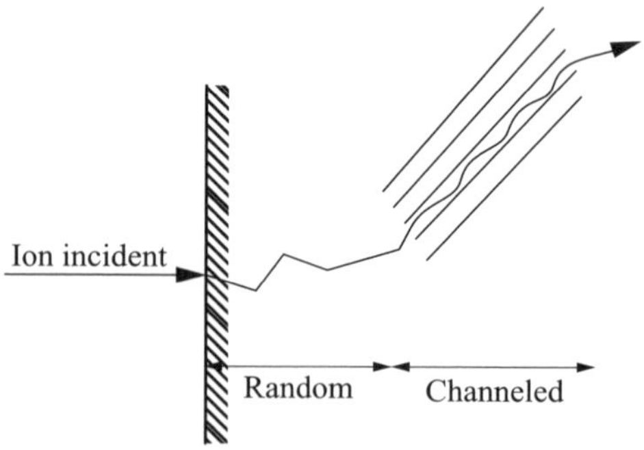

Figure 2.13. *Schematic representation of the ion channeling at the end of the course*

2.3.4. *Implantation through a mask*

In the fabrication technology of electronic circuits, the implantations are almost always carried out through a mask containing openings, corresponding to zones to dope, and protected zones, where ions must be completely stopped. Resists, oxide layers (SiO_2) or silicon nitride (Si_3N_4) are used to stop the projectiles. Because of their nature and their thickness d, the efficiency of these films is characterized by the transmission coefficient τ, i.e. the number of ions managing to cross, normalized to the total number of incident ions. By admitting a Gaussian distribution of the atoms stopped in the mask, an approximation formula of the transmission coefficient is:

$$\tau = \frac{1}{2\sqrt{\pi}} \frac{\exp\left(-z^2\right)}{z} \quad \text{with: } z = \frac{d - R_p}{\sqrt{2}\Delta R_p} \qquad [2.28]$$

For example, if the user desires less than one projectile out of 10^4 to cross the layer, it is necessary that d is of about or higher than $R_p+4\Delta R_p$, R_p and ΔR_p being the projected range and the standard deviation of the projectiles in the masking material.

Another important technological parameter is the form of the doping profile in the opening, and in particular the horizontal spread of the distributions at the edge of the mask. This spreading, caused by multiple collisions, is often as important as ΔRp. In the ideal case of an opening at the abrupt edges of 2l width, the distribution in 2D is obtained by multiplying in-depth the profile C(x) by the term:

$$\frac{1}{2}\left[erfc\frac{y-l}{\sqrt{2}\Delta R_{\perp}} - erfc\frac{y+l}{\sqrt{2}\Delta R_{\perp}} \right] \qquad [2.29]$$

Figure 2.14 shows the contours of equal concentration (from 10^{19} to $10^{12}cm^{-3}$) for boron ions of 60 keV, implanted at the dose of 10^{15} cm^{-2} through a mask with 1µm width. We can note that the doping can extend a few hundred nanometers and risk to interfer with the neighboring device, if the latter is set too close. The horizontal spread is thus a serious limitation to the integration of a growing number of devices per unit of silicon surface.

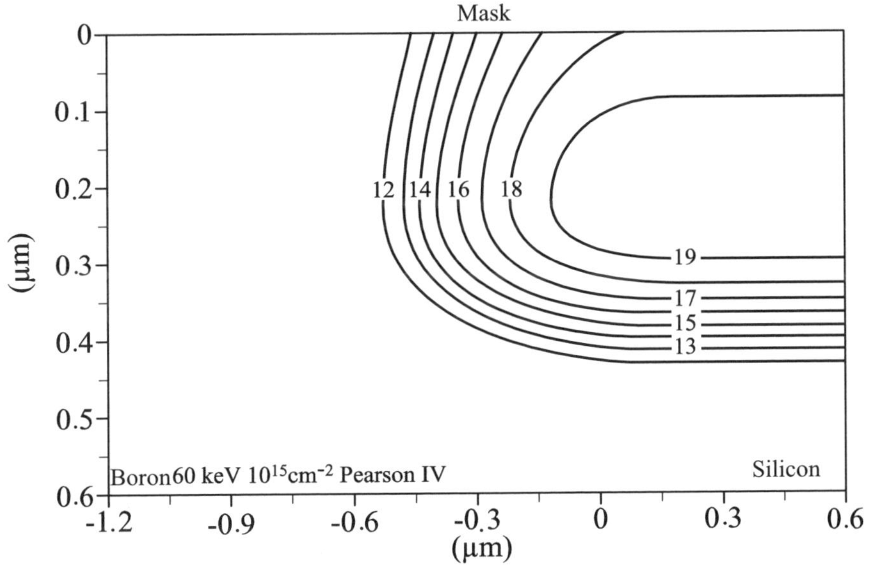

Figure 2.14. *2D profile calculated in the case of a vertical mask edge [SEL 84]*

2.4. Creation and healing of the defects

2.4.1. *Primary collision and cascade*

When an ion of energy E comes to strike an atom of the target, the transferred energy is T (see relation [2.3]). If T is higher than a threshold energy E_d, called the displacement energy (of several eV), the recoil atom thus released will be able in its turn to cover a certain distance in the target, losing its energy through nuclear and electronic collisions. Repeating this, a series of atomic displacements is created, called a cascade. This lasts until the remaining energy in each individual cascade, and in particular the one of the primary ion, is lower than E_d. The entire duration of the cascade is very short: less than one picosecond.

In addition, transfers under the threshold often represent a significant part of the energy transfered to the slowing-down environment. They sometimes constitute the majority of events, resulting in an important creation of phonons, i.e. thermal oscillations, and thus finally in a local temperature rise. This thermalization proceeds on a time scale definitely longer than the cascade duration, of about one nanosecond. In a very simple model, Kinchin and Pease [KIN 55] estimate that at each collision, the average transferred energy is divided by two and thus, at the nth collision, the average transferred energy is $T/2^n$. Finally, this model leads to a number of displaced atoms (per unit of target surface), proportional to the dose and to the total energy quantity ν lost in elastic collisions:

$$N_d = \frac{\nu}{2E_d}\varphi \qquad [2.30]$$

with:

$$E = \nu + \eta \qquad [2.31]$$

ν and η are thus respectively the total energy quantities lost in elastic collisions and electronic interactions.

However, in practice, we notice (see Figure 2.15) that the number of created defects (here in silicon at low-temperature) is often much higher than predicted by the Kinchin-Pease (KP) model. The order of magnitude is a few hundred atoms moved by incident ion for a light projectile, a few thousand for a medium mass and several tens of thousands for a heavy projectile. At higher energy, the curves again take the linear slope of the KP relation, indicating that the creation of damage at high-energy obeys this law. At low energy, the important creation of defects is due to an effect related to the high density of transferred energy, leading to the overlay of the cascades initiated by the recoil atoms. This is equivalent to a collapse of the displacement energy. This phenomenon is usually called a displacement spike.

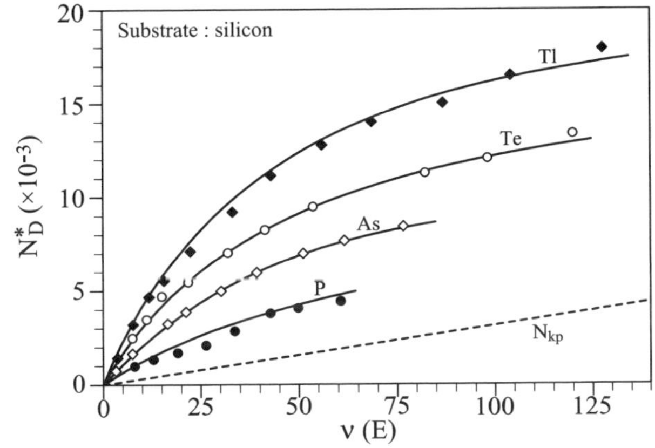

Figure 2.15. *Total number of silicon atoms moved by incident ion, according to the total quantity of energy lost in elastic collisions (according to [WAL 78]). The dashed line corresponds to the standard Kinchin and Pease model (relation [2.30])*

This phenomenon of cascade overlay is obviously even more important during the implantation of polyatomic ions, such as BF_2^+ or decaborane, commonly used in microelectronic technology. Figure 2.16 shows how the number of displaced silicon atoms gradually deviates from the linear behavior, depending on the number of atoms contained in the molecule [GRO 87].

For an in-depth calculation of the distribution of displaced atoms in a linear cascade, the two methods described in section 2.3 (the transport equation or the Monte Carlo calculation) are used. The transport equation [SIG 72] is written in a general way for the defect distribution function $F_d(r,\vec{v})d^3r$, which is the amount of deposited power (in nuclear or electronic collisions) in the element of volume (r, d^3r):

$$-\frac{\vec{v}}{v}\frac{dF_d}{d\vec{r}} = N\int d\sigma\left[F_d(\vec{r},\vec{v}) - F_d(\vec{r},\vec{v}') - F_d(\vec{r},\vec{v}'')\right] \qquad [2.32]$$

where v is the initial speed before the collision of probability dσ, v' is the speed of the ion after scattering and v" is the speed of the recoil atom.

Obviously, the standardization condition is:

– for elastic collisions:

$$\int F_{d,n}(\vec{r},\vec{v})d^3r = \nu \qquad [2.33a]$$

– for electronic collisions:

$$\int F_{d,e}(\vec{r},\vec{v})d^3r = \eta \qquad [2.33b]$$

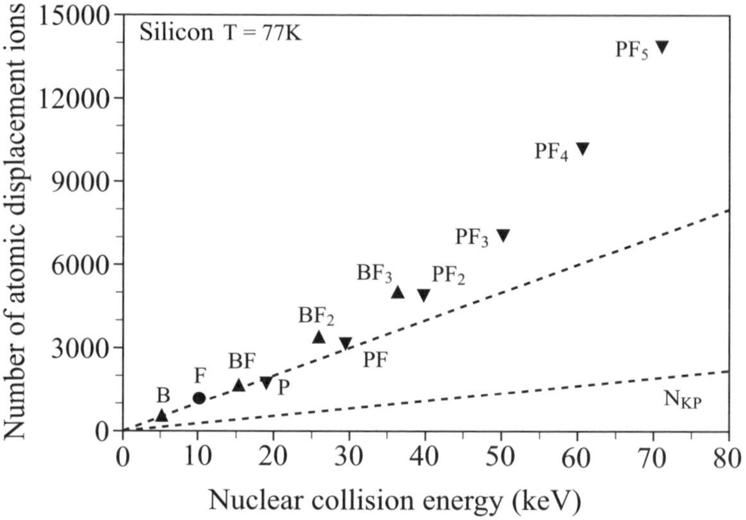

Figure 2.16. *Number of silicon atoms moved at 77 K (measured by RBS-C in situ) by the implantation of various polyatomic projectiles, according to the energy lost in elastic collisions. This number grows quicker than linearly (dshed line) and it is much more important than the Kinchin and Pease standard model predicting it (dotted line) (according to [GRO 87])*

With TRIM or SRIM [ZIE 85a], after the reconstitution of a number n of trajectories, we can obtain the profile of energy deposition, i.e. the F_d function, in the two categories of interaction. Figure 2.17 shows the profile of energy deposition in elastic collisions, for 100 keV arsenic in silicon, compared to the distribution of the implanted ions. These profiles do not coincide: a maximum of energy is deposited before the ions stopping. The defects (see the following section) are created according to the profile of energy deposition.

If F_d (x = 0) is the energy lost in elastic collisions at the surface of the target, we can then estimate the surface sputter rate S:

$$S = \Lambda \frac{F_d(x=0)}{NE_{ds}} \qquad [2.34]$$

where E_{ds} is the energy of surface displacement (close to the sublimation energy) and $\Lambda = 0.042$ Å^{-2} [SIG 72].

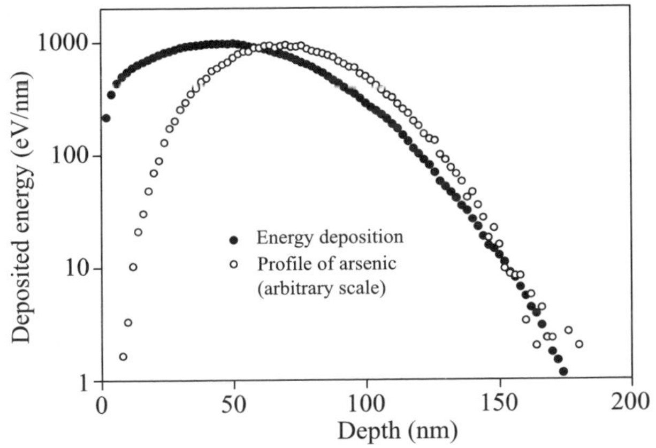

Figure 2.17. *Comparison between the profile of the deposited energy in elastic collisions (black dots and left scale) and distribution of the arsenic implanted (white dots and arbitrary scale) at 100 keV in silicon (SRIM simulation)*

2.4.2. *Point defects*

A solid is always constituted of a more or less ordered atom assembly. The concept of point defect is thus relatively easy to define, especially in the microelectronics materials, which are often perfect single crystals. Elementary point defects are thus the vacancies, i.e. vacant sites in the network, and the interstitials, which are supernumerary atoms located between the atoms of the network. After implantation, the interstitial species are the implanted atoms, but also the atoms moved from the target, called self-interstitials. Several configurations are possible for these interstitials, all of them allowing the atom to be inserted in the network. Moreover, around the vacancy, as well as around the interstitial, a local rearrangement is necessary. In silicon, we estimate that the defects' formation energy, corresponding to the difference of the free enthalpies, before and after their formation, is about 1 eV for a vacancy and 3 to 5 eV for an interstitial.

An energy transfer higher than the displacement threshold produces a Frenkel pair (vacancy + interstitial). The simple addition of the formation enthalpies makes it possible to estimate those of a Frenkel pair in silicon at approximately 4 to 6 eV. To that, we must add the energy we had to spend to break the Si-Si bonds. Moreover, if the separation distance between the two components of the Frenkel pair is not sufficient, the thermal diffusion process and sometimes the electrostatic attraction will cause their recombination. We thus need a little more important energy transfer, so that the interstitial can cover a certain distance separating it sufficiently from the vacancy. In other words, the threshold formation energy of a stable Frenkel pair is rather about 10 to 80 eV (respectively from gallium arsenide to diamond). It is of 15 eV in silicon.

It is quite obvious that, beyond this little simplistic classification in vacancies and interstitials, the defects also differ by their configuration and by the connections made in their environment, including possible hybridizations of the target atoms or of impurities. For example, an interstitial often takes a configuration of shared site with an atom of the material, in energetically favorable crystallographic directions such as <100> or <110>. This is particularly true in ionic-covalent crystals, where there is an important lattice–electron interaction, which often produces important relaxation effects around a defect.

We said that the first stage, the cascade, occurs in approximately 10^{-13} seconds, and the second one, thermalization, in 10^{-9} s. The rearrangement described above is the third stage, bringing the system towards a new equilibrium. It occurs on a much longer period of time. In the second stage and especially in the third stage, events such as diffusion, precipitation or chemical reactions can occur. We then understand that other defects can be formed. They are more complex and considered as point defects, but are already defect combinations: for example the double or triple vacancy, etc., the vacancy-impurity pairs and, in covalent compounds, the antisite defects.

2.4.3. *Accumulation of damages, amorphization*

During implantation, elementary defects (interstitial and vacancies) are generated proportionally to the projectiles' flux. They can recombine and/or destroy themselves on "sinks" (grain boundary, surfaces, interfaces, dislocations, etc.) or diffuse, proportionally to their local gradient. Without writing too complex equations, two extreme situations can arise:

– the generation rate (i.e. the projectile flux or mass) is low and/or the diffusion is important. The creation of Frenkel pairs is permanently compensated by re-combinations and annihilations. We thus reach a stationary state, with a low concentration of vacancies and isolated interstitials. This is the case during a high temperature implantation or with light ions;

– the ion flux is important (this is the case, as we saw, with modern implanters) and/or the diffusion is low, so that the generation rate of the Frenkel pairs is more important than the re-combinations or the other forms of elimination of point defects. These pairs will thus accumulate proportionally with the in-depth profile of the energy deposition, leading first to the formation of small defects aggregates, and then finally of real amorphous zones. If $n_d(x)$ is the number of atoms moved at the depth x, we can, as a first approximation, say that $a(x) = n_d/N$ is the "amorphous" fraction at this same depth. The increase (da) in the amorphization for a dose increase ($d\phi$) will be proportional to the remaining crystalline fraction (1-a), to the value of the local energy deposition $F_d(x)$ and inversely proportional to the displacement energy:

$$\frac{da(x)}{d\varphi} = [1 - a(x)]\frac{F_d(x)}{NE_d} \qquad [2.35]$$

So that:

$$a(x) = \frac{n_d(x)}{N} = 1 - \exp(\frac{-F_d(x)\varphi}{NE_d}) \qquad [2.36]$$

At low doses, we find that the defects creation is directly proportional to the energy deposited in elastic collisions:

$$n_d(x) \approx \frac{F_d(x)}{E_d}\phi \qquad [2.37]$$

This simple model describes, rather well, the accumulation of damage in most materials, without self-healing, i.e. at low temperature. In reality, we note that it becomes increasingly easier to move atoms in an already damaged material, so that the displacement cross-section F_d/NE_d varies itself with the dose. Assuming a linear variation, we find:

$$\frac{da(x)}{d\phi} = \frac{F_d(x)}{NE_d}[1 + ka][1 - a] \qquad [2.38]$$

The solution of this equation is:

$$a(x) = 1 - \frac{1+k}{k + \exp\left[\frac{F_d(x)}{NE_d}(1+k)\phi\right]} \qquad [2.39]$$

This sigmoidal variation law describes the amorphization of silicon for usual implantation energies at ordinary temperature rather well. More complex models have been proposed [WEB 00]. Lastly, let us note that above a critical damage density, corresponding to a critical dose ϕ_c, the crystal cohesion is no longer ensured and the material collapses very quickly towards the amorphous phase. This critical density is about 10 to 20% in silicon. Moreover, as previously discussed, this law of damage accumulation is valid as long as the re-combinations, healings, migrations or other rearrangements of defects are not too important. In the opposite case (at high temperature), the Frenkel pairs' generation is largely counterbalanced by re-combinations. We thus reach a stationary state, which will never lead to the material amorphization. At intermediate situations, the critical amorphization dose backs down, as the implantation temperature increases (it also depends, of course, on the projectile mass). All these behaviors are evidenced on the experimental curve in Figure 2.18, describing silicon damaging by high-energy silicon ions (linear cascades) [HEC 89].

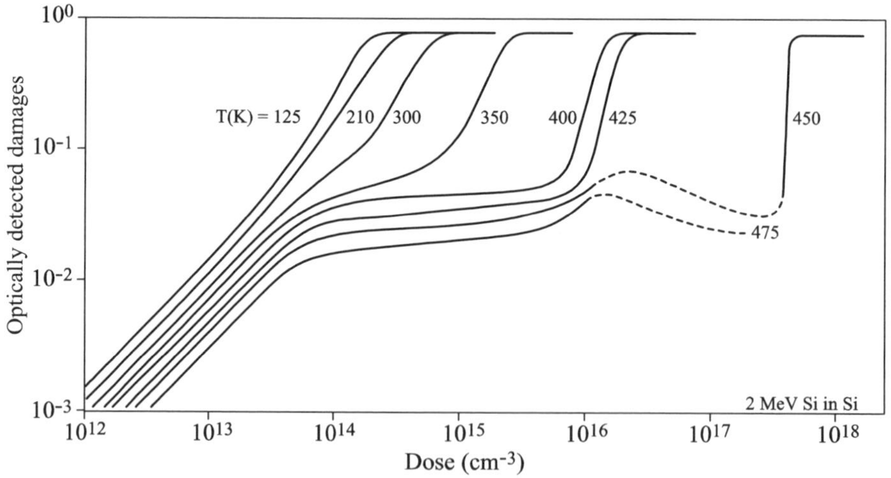

Figure 2.18. *Silicon fraction damaged by Si ions of 2 MeV as a function of the implantation dose and temperature [HEC 89]*

Using an amorphization model based on similar mechanisms, Morehead and Crowder [MOR 70] determined, for various projectiles, the critical dose for silicon amorphization as a function of the implantation temperature (see Figure 2.19).

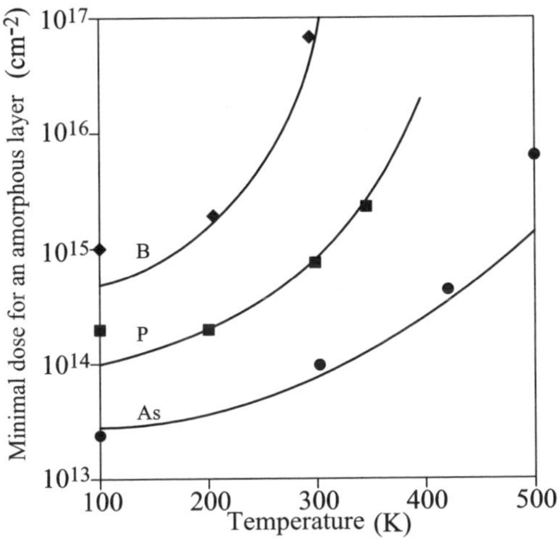

Figure 2.19. *Necessary critical dose to form an amorphous layer in silicon at various temperatures and using various projectiles [MOR 70]*

Therefore, including the temperature influence, it is better to write (for low doses):

$$n_d(x) \approx \xi(T)\frac{F_d(x)}{E_d}\varphi \qquad [2.40]$$

where $\xi(T)$ accounts for the efficiency of defect creation, the net result of the creation-healing process. By integrating $n_d(x)$ on all the implanted thickness, we obtain a relation similar to the Kinchin and Pease relation [2.30], giving the total amount N_d of displaced atoms proportional to the amount ν of power lost in elastic collisions. With the critical amorphization dose, the product $\xi(T)F_d(x)\phi_c/N$ represents a critical energy deposition per atom, at the temperature T. At ordinary temperature, in silicon, this critical deposition is about 10 to 12 eV/atom [VIE 87]. For example, boron, at the doses usually used in microelectronics, does not reach this threshold and will thus not amorphize silicon at ambient temperature.

However, even at ordinary temperature, point defects (vacancies and interstitials) are very mobile. Behaviors other than the simple accumulation in the form of amorphous agglomerates will thus appear: first of all the already quoted recombination, equivalent to a self-healing of the damages; the combination, also already mentioned, of interstitials and vacancies, between them or with impurities; the clustering of the vacancies and interstitials in a more ordered form, which leads, especially when the temperature is raised, to the formation of extended defects (for example dislocation loops[5]). A simple calculation shows, for example, that above a certain number of atoms, the insertion of a loop in the silicon lattice is less expensive in energy than the insertion of an amorphous sphere. Obviously, the transition from one to the other also requires a certain energy, which can be provided, for example, by a rise in temperature during or after the implantation (see section 2.4.4 – Damage healing). In practice, in the absence of nonlinear effects such as displacement spikes, there is thus a critical transition temperature below which the displaced atoms will evolve towards the amorphous phase, and above which extended defects will be formed, for example, in silicon, {311}[6] defects or interstitial loops. Table 2.1 gives the value of this temperature in some common semiconductors.

Material	Critical temperature	Material	Critical temperature
Si	120	SiC	350
GaAs	140	InSb	-40
GaP	220		

Table 2.1. *Critical temperature (in °C) beyond which it becomes impossible to amorphize some common semiconductors*

2.4.4. *Damage healing and dopant activation*

The aim of the implantation method in semiconductors is to set the dopants in substitutional position, where they are electrically active. The first condition is thus to restore the initial crystalline order. This recrystallization occurs during a thermal process (annealing). The conditions of this annealing will ideally be selected, so that the dopants can be positioned at the same time in substitutional sites. The first condition is related to the morphology of the damaged area, according to whether it is amorphized or not. In the first case, it is the recrystallization front (the amorphous-crystal interface), which, by passing through the implanted area, will cause dopant activation. In the second case, there is no epitaxy front. The dopant

5 Dislocation loops are planar defects formed by accumulation of self-interstitials or vacancies. They correspond to small-disks intercalated between low indexes plans, forming edge dislocations at their periphery.

6 The {311} defects are linear defects, resulting from the ordered agglomeration of self-interstitials, forming segments in the <311> directions of silicon.

itself will have to diffuse to the vacant sites. Figure 2.20 illustrates these two situations with the active phosphorus fraction in silicon, depending on the implantation temperature.

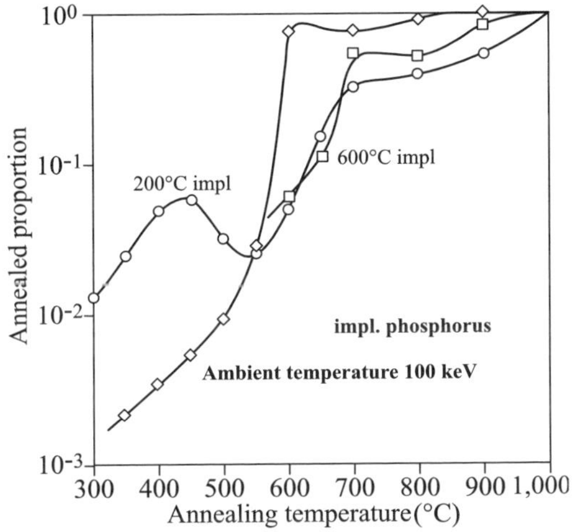

Figure 2.20. *Variations of the carriers concentration with the annealing temperature in silicon implanted with phosphorus at various temperatures (according to [TAM 71])*

At ambient temperature, i.e. below the critical temperature reported in Table 2.1, an almost total phosphorus activation is reached with the annealing at 600°C. Above this value (implantations at 200 or 600°C in this example), the full dopant activation requires a much higher annealing temperature. Moreover, for annealing between 300 and 600°C, the activity increases and then decreases again. This effect of reverse annealing is well-known in the case of boron implantations which, at traditional microelectronics doses, do not amorphize silicon at room temperature. It is related to the formation, then to the dissolution, of the intermediate extended {311} defects [SEI 71]. Beyond this temperature range, the setting in substitution of the dopants is controlled by a process, the activation energy of which is about 5 eV. The common interpretation assigns this value to the generation and the migration of silicon vacancies, which, while reacting with boron or interstitial phosphorus, lead to the full activation [HOF 75].

A very different situation arises when the implanted layer is amorphized. Indeed, the latter recrystallizes at rather low temperatures (approximately 500°C in silicon) by solid phase epitaxy (SPE), from the underlying crystalline substrate [CSE 78]. The amorphous-crystal interface moves towards the surface with a constant speed v_{EPS} given by:

$$\nu_{EPS} = \nu_0 \exp(-\frac{E_{EPS}}{kT}) \qquad [2.41]$$

with an activation energy of 2.68 eV, measured up to 1,300°C in silicon [OLS 88] and of 1.55 eV in InP or AsGa [LIC 91]. v_0 depends on the crystal orientation and on the presence of impurities. The most favorable orientation is <100> and v_0 is the lowest for the <111> orientation. [CSE 76]. A high dopant concentration increases the epitaxy rate, by a factor of 2 or 3 for approximately 10^{20} cm^{-3} of phosphorus for example. An equivalent oxygen concentration slows down the recrystallization in the same proportions [LAU 80]. However, all impurities, even the dopants, as soon as their concentration exceeds the solubility limit, slow down the epitaxy and, thus, low concentrations of not very soluble impurities can potentially completely inhibit it, leading then to the formation of a polycrystalline phase. The same mechanisms intervene in other semiconductor materials, more or less shifted in the temperatures scale: for example the typical temperature of recrystallization of germanium is 300°C, and that of SiC is about 1,500°C.

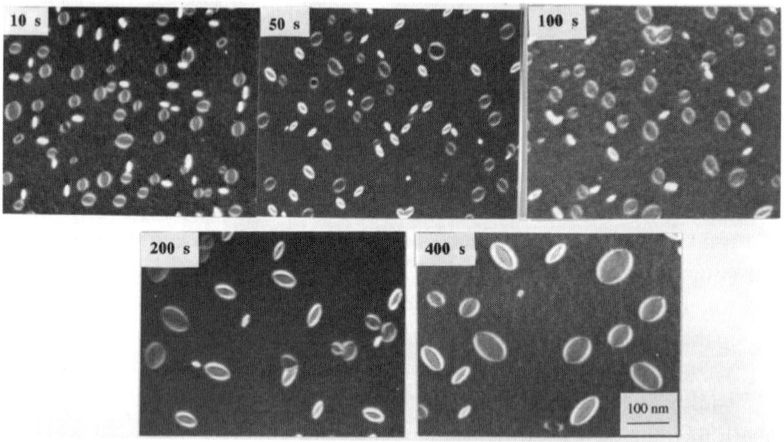

Figure 2.21. *Images of transmission electron microscopy (plane view) of the "end of range" defects formed in amorphized silicon (Ge 150 keV, 2×10^15/cm^2) and annealed at 1,000°C for increasing times (10, 50,100,200 and 400 s) [BON 96]*

Some residual defects remain after the solid phase epitaxy of an amorphous layer. They result from the presence, beyond the original amorphous-crystal interface, of an excess of recoil self-interstitials, compared to the local concentration of vacancies. In silicon, during annealing, a significant part of them will condense in {311} defects in the temperature range 600-750°C or in dislocation loops in the range 800-1 100°C. These defects are called EOR (end of range defects). They are set in the (111) planes. The series of photomicrographs in Figure 2.21 show the morphology and development in size and population of these defects during an annealing at 1,000°C. In fact, the {311} are intermediate defects dissolving in their turn for more important thermal budgets, releasing thus interstitials, which then evolve towards loops. They are not really healing any more, since recent studies [BON 96], [CLA 95] have shown that loops obey a conservative competitive growth mechanism (see Figure 2.22). Their radius increases, while their density decreases, but the total number of interstitials included in the loops is constant. However, the proximity of the surface can modify the conservative character of this equilibrium.

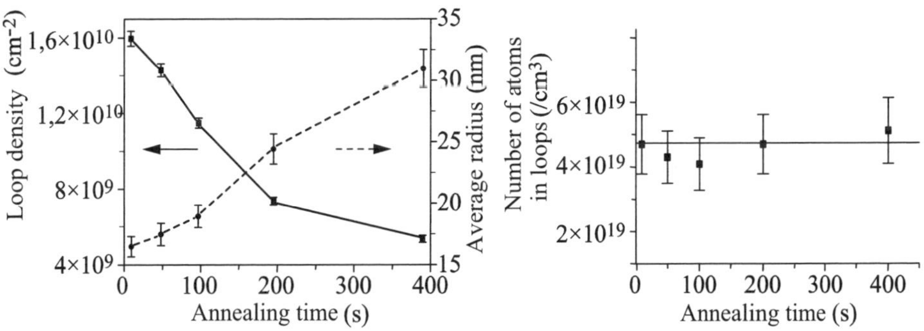

Figure 2.22. *Left: development during the annealing time (at 1,000°C) of the density and of the average radius of EOR defects observed in Figure 2.21. Right: development of the total number of atoms (per cm^2) taken in the loops functions of the annealing time at 1,000°C [BON 96]*

These defects can be traps for impurities or for charge carriers, altering the electric behavior of the devices. Moreover, excess interstitials can also, transitorily, take part in the diffusion of the dopants located in this zone. Generally, since the dopants diffuse in semiconductors via mechanisms assisted by vacancies and/or self-interstitials, any local and/or temporal supersaturation (or undersaturation) of one or the others, created by implantation, will cause an "abnormal" diffusion. This aspect of implantation is sometimes used to adjust a doping profile, in particular to slow down diffusion [NIS 92].

Regarding dopant activation, we need to recall that the quantity of impurities that can be set in substitutional sites is limited. Beyond this limit, they precipitate in the form of immobile (in terms of diffusion) and inactive (from an electric point of view) clusters. However, for some dopants, such as arsenic or phosphorus in silicon, there is an "absolute" solubility limit and another one concerning their electrical activity. This characteristic is due to the fact that these elements, implanted at high-dose and placed in substitution during annealing, have a high probability of finding another identical atom (or several) as first (or second) neighbors in the crystal. This modifies their energy level in the band gap, and thus their probability of ionization. Table 2.2 gives the maximum solubility value (at high temperature) of the main silicon dopants. Except at high temperature, where it saturates and sometimes decreases, solubility is the result of a thermal activated process. It is thus much lower at low-temperatures and the dopants can thus precipitate during annealing at these temperatures. The forced recrystallization around these precipitates also leads to the formation of dislocations. However, the fast motion of an solid phase epitaxy front through the dopants profile can result in largely exceeding the solubility values and thus in incorporating, in a metastable state, a large amount of impurities [NAR 82].

Dopant	**Solubility (cm^{-3})**	**Dopant**	**Solubility (cm^{-3})**
B	5×10^{20}	As	1.5×10^{21} (5×10^{20})
P	10^{21} (4×10^{20})	Sb	7×10^{19}
Ga	4.5×10^{19}	In	8×10^{17}

Table 2.2. *Maximum value of the solubility of various dopants in silicon. The value between brackets is the solubility of the electrically active atoms*

2.5. Applications in traditional technologies and new tendencies

The implantation is used in most doping operations of MOS and bipolar technologies. We will describe the common operations in these sectors and the new tendencies constituting advances towards new technological lines [HIL 91, KUS 91]. The first are schematized in Figure 2.23, in the case of CMOS technology:

– doping of the p- and n-wells;

– adjustment of the threshold voltage by a light over doping of the channel under the gate oxide;

– doping of the source and drain extensions (LDD);

– pockets or “anti-punch through” halos;

– source and drain doping;

– gate polycrystalline silicon doping.

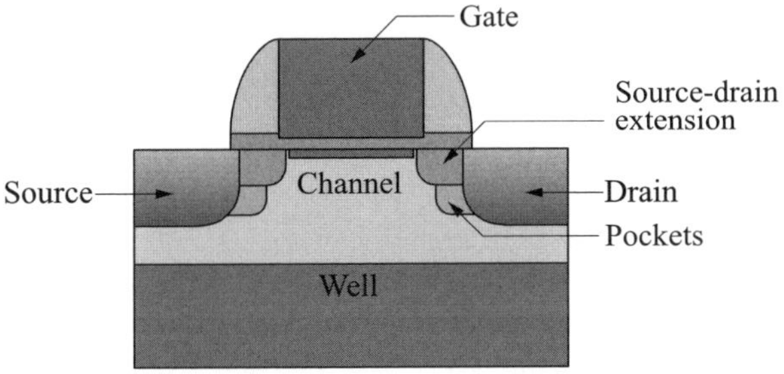

Figure 2.23. *Areas implanted in traditional CMOS technology*

2.5.1. *Common implantations*

The N- or P-MOS transistors are placed in "wells" of opposite type in order to form a junction insulating the transistor. The surface concentration in this well, just like the nature of the gate electrode or the oxide thickness, affects the threshold voltage at which the device lets the current pass. It also conditions the importance of the leakage currents between neighboring circuits. The doping precision, necessary for well, makes the implantation essential in this case. Typically, p and n doping are respectively carried out with boron and phosphorus ions, with energies in the 100 - 200 keV range and doses between 10^{12} and 10^{13} cm^{-2}. These implantations are followed by a long annealing at high temperature, in order to form a deep junction. The tendency in this sector is to replace these diffusions with high thermal budget by deep implantations, therefore at higher energy. This also has the advantage of forming a "retrograde" doping profile, advantageous to decrease the gain of parasitic transistors.

The threshold voltage to which the channel of the MOS transistor starts to conduct is a parameter with a strong importance. The adjustment of this voltage, in the n and p parts of the transistor, is carried out by introducing a small quantity of boron or phosphorus (about 10^{11} cm^{-2}) right under the gate oxide. The energy and dose are selected according to the thickness of this oxide.

The performances of the submicron transistors depend on the impurity profile close to the drains. A gradual profile, called LDD (lightly doped drain) can reduce the electric field in this zone and thus avoid the injection of "hot" electrons, which could deteriorate the N-transistor. The effect also exists, but in lower proportions, for the holes in the P-transistor. This doping of the channel-drain transition zone is carried out by a self-aligned implantation, possibly at an oblique incidence (up to 60 degrees) of phosphorus or arsenic (for n-type dopant) and boron (for p-doping). Doses are about several 10^{13} ions/cm^2 (see Figure 2.24).

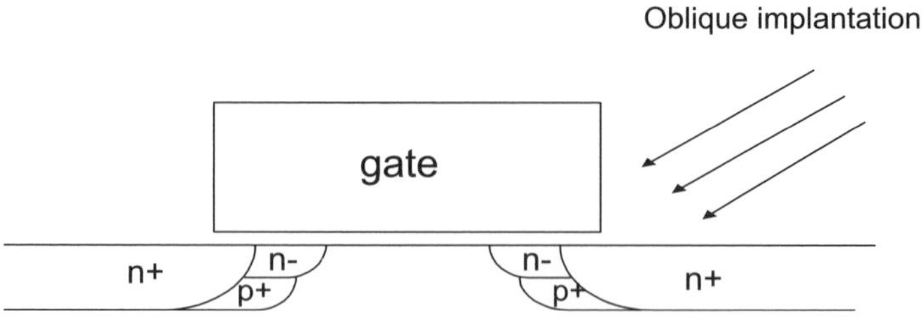

Figure 2.24. *Schematic representation of the realization of LDD and anti-"punch through" doping*

In this same zone, when the channel is very short, the transistor behavior can be alter by the "punch through" effect, if, under functioning (under polarization), the depletion zone of the drain-channel junction reaches the source-channel junction. This punch through effect is avoided by a phosphorus or boron implantation, possibly again in oblique incidence, with doses of about several 10^{12} ions/cm^2, right under the active channel, in the area adjacent to the drain and source. Accurate control of the zone to be doped requires slightly higher energies: about 100 keV for boron and 150 keV for phosphorus (see Figure 2.24).

An implantation, not mentioned in Figure 2.23, relates to the recent lateral insulation technology by the formation of trenches (shallow trench isolation STI). An over-doping, at the bottom of the trench and close to the walls can be carried out by boron implantation (typically 100 keV, 10^{13}/cm^2), by varying the incidence angle by a few degrees around the trench axis.

The "source" and "drain" zones are doped at high concentrations, about 10^{20} to 10^{21}/cm^3, with the opposite type of the channel zone, itself doped at 10^{16} or 10^{17}/cm^3. With the common circuit dimensions, these junctions must be very thin.

Arsenic (a heavy ion) is thus preferred to phosphorus, to build the n-junction. Boron being the only p-type dopant sufficiently soluble to reach the required level of concentration, there is hardly any choice but to implant it at very low energy (several keV) or to implant the BF_2^+ molecule. The doses are about 5×10^{15}/cm^2. We will discuss further the prospects regarding the realization of even more superficial junctions.

The gate electrode of the CMOS transistors, for both the N- and P-type, is formed by a pattern of deposited polycrystalline silicon. But this material, in its as-deposited form, is too resistive. It must thus be doped to increase its conductivity. This doping can possibly be made through chemistry during the deposition step, but, more often, the Si-poly is doped by arsenic or BF_2^+ implantation, with doses of about a few 10^{15} with $10^{16}/cm^2$. This doping is often carried out at the same time as the doping of the source and drain.

Figure 2.25 summarizes all these operations in the form of a dose-energy graph.

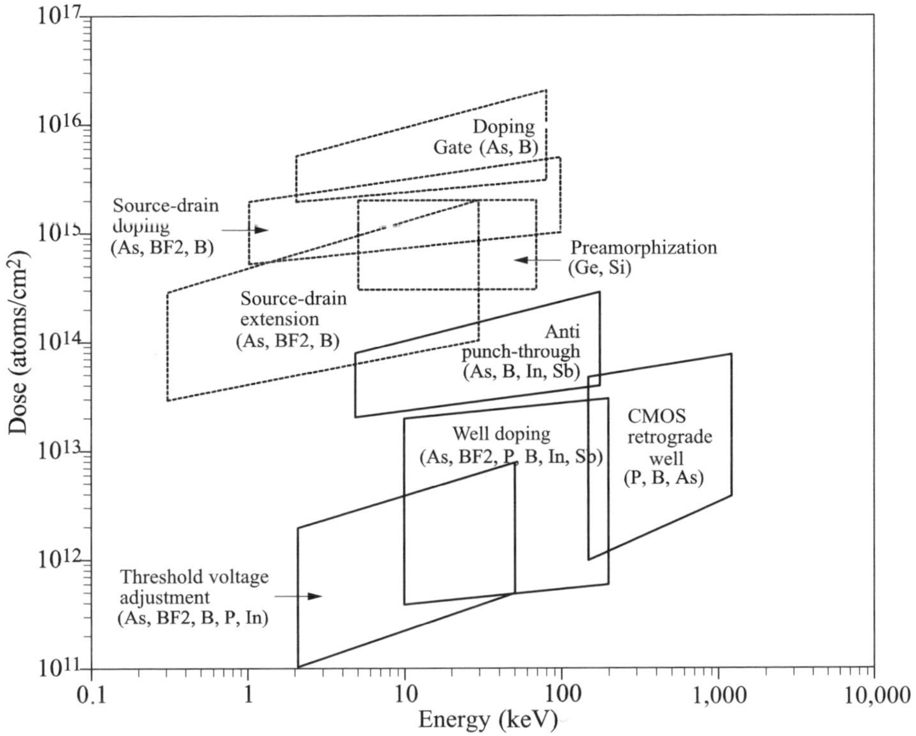

Figure 2.25. *Domains of energies and doses corresponding to the various operations of doping or amorphization of silicon in CMOS technology*

Implantation was also essential in bipolar technology when, for reasons of speed of the circuits and of more advanced integration, the need was felt to precisely control the vertical spacing between the emitter, the base and the collector. In this context, the parasitic "emitter push effect"[7] should absolutely be avoided. This was achieved by using an arsenic implantation (several tens of keV, at doses higher than $10^{15}/cm^2$) instead of a phosphorus diffusion.

---

7 For a detailed description of this effect, see section 3.2.2 of Chapter 3 of this book, written by D. Mathiot.

There again, once installed in the technological lines, the implantation was spread to all the other doping steps: creation of the buried emitters (As, Sb, several tens of keV, $>10^{15}/cm^2$) and of the base (B, several keV, $<10^{14}/cm^2$), channel stop (B, $10^{13}/cm^2$), contact doping, etc.

2.5.2. *Other applications and new tendencies*

2.5.2.1. *Gettering*

Metal contamination is a recurrent problem for the microelectronics industry. Even at very low concentration levels, it brings various harmful effects, in particular an increase in the leakage current of the reverse-biased junctions and a deterioration of the carriers lifetime. The most common impurities (Fe, Cr, Cu, Ni) diffuse very quickly and are relatively soluble at high temperature in the common semiconductors. Their introduction occurs in all the stages of the manufacturing process: growth, cleaning, etching, ion implantation and, naturally, high temperature thermal treatments (oxidation, diffusion, annealing). Gettering processes of these impurities are commonly used. They consist of attracting and maintaining these elements far from the active areas of the circuits. We often separates the trapping mechanisms into two categories: gettering by segregation, which consists of depositing or forming on the contaminated substrate a material in which the impurity is very soluble; and gettering by relaxation, which consists of causing precipitation at the desired location, using an annealing at a temperature lower than that of contamination. Gettering by ion implantation is part of the second category. Indeed, any site of heterogeneous nucleation is a gettering center and the implantation can thus be used to intentionally create these sites. In fact, most impurities and defects introduced by implantation are gettering centers: boron, which forms Fe-B complexes [KIM 83], carbon capturing self-interstitials [ROH 95, WON 89], helium forming cavities [MYE 96, ROQ 00], argon gathering in bubbles [GRO 96, ROH 95] or either dislocations [SUM 99], etc. It seems that the stability of the cavities formed after implantation and desorption of helium makes them the most efficient gettering traps in silicon. In addition, cavities also introduce levels near the center of the band gap. This offers the possibility to locally control the carrier lifetime. This method was applied to power devices [RAI 98, ROQ 00].

2.5.2.2. *High energy implantation*

In the manufacturing process of integrated circuits, very deeply implanted layers have some advantages, and offer new technological possibilities. First, a high energy implantation instead of a pre-deposition, followed by a long diffusion, considerably reduces the thermal budget. This also has the advantage of a significantly lower lateral spread of the doping. Moreover, the implantation doping profile, contrary to a diffusion profile, has a higher concentration in-depth than on the surface. This can be used for better control of the interactions between neighboring devices (latch-up) or to avoid the punch through effect due to the wells [ZAP 88]. More generally, a series of implantations at various energies and doses can be used to accurately adjust a doping profile with a very weak thermal budget [CAP 95].

In addition, high-energy implantation can also be used to prevent radiation effects, while forming a buried "gate", whose role will be to reduce the spread of the depletion region of surface junctions. Thus, a reduced number of possible parasitic charges will be able to go up towards the active devices [ZIE 85b].

Lastly, a deep implantation is able to form gettering centers for metal impurities close to the active areas [CHE 89].

2.5.2.3. *Ultra-thin junctions*

The realization of very shallow P+/N junctions remains a crucial manufacturing step of the current and upcoming CMOS structures, in particular for the control of short channel effects. Reducing the transient enhanced diffusion (TED)[8]) of boron and achieving its full activation is a challenge, insofar as strong doping implies a high implantation dose and thus an important quantity of defects, which will favor the dopant migration.

Let us recall that boron diffuses in silicon mainly via a mechanism where self-interstitials play a predominant role. Regarding implantation, the first precaution is obviously to reduce ion penetration. There are two solutions to implement this precaution: either, in a traditional implanter, decrease their energy up to values lower than 1 keV [CRI 04, LAL 05] and/or implant polyatomic ions such as decaborane [MAT 97, SMI 98], or immerse the sample in a plasma containing the doping element [QIA 91, SKO 04, YAN 98]. On the other hand, as seen in section 2.3.3, it is important to abolish channeling tails by pre-amorphizing silicon, for example by the implantation of a moderate dose of silicon or germanium [MIN 94]. To activate the totality of boron, temperatures of about or higher than 900°C are necessary. In this temperature range, EOR dissolve by releasing interstitials, contributing to the boron diffusion. To minimize it, we can reduce the annealing duration, by using, for example, a rapid annealing (RTA), of about one second up to 1,000 or 1,100°C [CLA 95].

We can also try to trap these self-interstitials, for example by carbon or fluorine implantation [GRA 05, VAN 05]. Carbon forms with interstitials stable C-Si complexes [MIR 02].

---

8 For a more detailed explanation of TED, the reader can refer to section 3.4.1 of Chapter 3 written by D. Mathiot.

We can then note a decrease in the quantity of the formed EOR [CAC 96, CHE 99] and a significant reduction of the boron diffusion [GRA 05, PAW 06]. Figure 2.26 shows that by combining pre-amorphization, low energy boron implantation, fluorine or carbon implantation and rapid annealing, the doping can be maintained on a depth of about 20 or 30 nm [GRA 05].

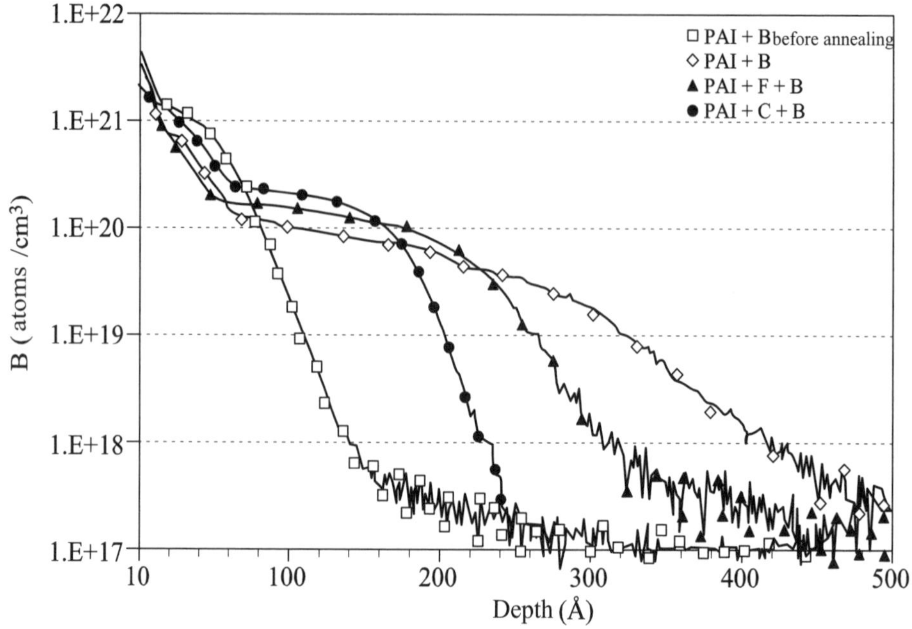

Figure 2.26. *Boron profiles obtained by boron implantation at 0.5 keV ($10^{15}/cm^2$) in pre-amorphized silicon (PAI), before (squares) and after rapid annealing at 1050°C, without (rhombuses) or with preliminary addition of fluorine (triangles) or carbon (circles) (according to [GRA 05])*

2.5.2.4. *SIMOX and Smart-Cut™*

Silicon on insulator (SOI) substrate opened up a new branch of the silicon industry, because there are many possibilities for new applications, for sensors but also for low consumption and low voltage electronics. Electronic circuits on SOI [COL 04] have many advantages; most of them related to the fact that, on such a structure, devices are completely isolated from one another, as well as from the substrate. This results in reduced parasitic capacitances, in an almost total immunity towards the interactions between the neighboring circuits or towards the radiation effects and, generally, in the performance improvement of most of the active devices. However nowadays, there are hardly more than two industrially viable ways to prepare a SOI structure. They both use ion implantation: the SIMOX (separation by implantation of oxygen) technique and, more recently, the Smart-Cut™ method (developed by the French company, SOITEC).

The SIMOX technique [AUB 95, IZU 78] consists of synthesizing the SiO_2 oxide by the implantation of a very high oxygen dose (about $10^{18}/cm^2$) in silicon. The energy of oxygen ions must be sufficient (approximately 200 keV) to form a buried layer, leaving on the surface an intact silicon film. To preserve the single-crystal character of this film, the implantation is carried out at high temperature (about 600°C).

Lastly, to cause the growth and then the coalescence of SiO_2 precipitates until the formation of a homogeneous layer with well-defined interfaces, an annealing of about one hour at very high temperature (≥1,300°C) is necessary. This is a very expensive technique, because of the very important dose, requiring a dedicated implanter (very high current) and because of the very high thermal budget. Moreover, implantation and annealing, under these conditions, require particular care, as for possible metal contaminations. Finally, the formed material often has a strong dislocations density.

The Smart-Cut™ method is done in four steps (see Figure 2.27) [BRU 95]. First, a hydrogen dose of about $10^{17}/cm^2$ is implanted in a wafer (A) of oxidized silicon. The energy of H^+ ions must be sufficient, so that they cross oxide and are implanted in silicon.

During the second stage, after cleaning, wafer A is pasted by molecular adhesion with a silicon wafer Bv (wafer bonding). Cleaning and polishing are crucial processes in this stage [MOR 96]. Then, a heat treatment is applied to the whole. Thank to the fact that wafer B acts as s stiffener and to the driving force linked to the gaseous pressure inside the cavities formed by hydrogen implantation (see Figure 2.28), stresses spread parallel with the surface, ending with the fracture occuringe at the maximum H concentration (see Figure 2.29).

The thin SiO_2/Si film of wafer A is thus "transferred" on wafer B, which then has a SOI structure. The last stage is a polishing of the surface, which is relatively rough after the transfer.

There are many advantages to this method. The most important are a low thermal budget and a limited damage, given the projectile mass. This results in an excellent material quality. Moreover, the method can be applied to the transfer of other materials, opening multiple possibilities for applications based on heterostructures (for example GaAs/Si) or on semiconductors on insulator (for example SiC/SiO_2/Si) or even silicon on quartz (see Figure 2.30) [BRU 06].

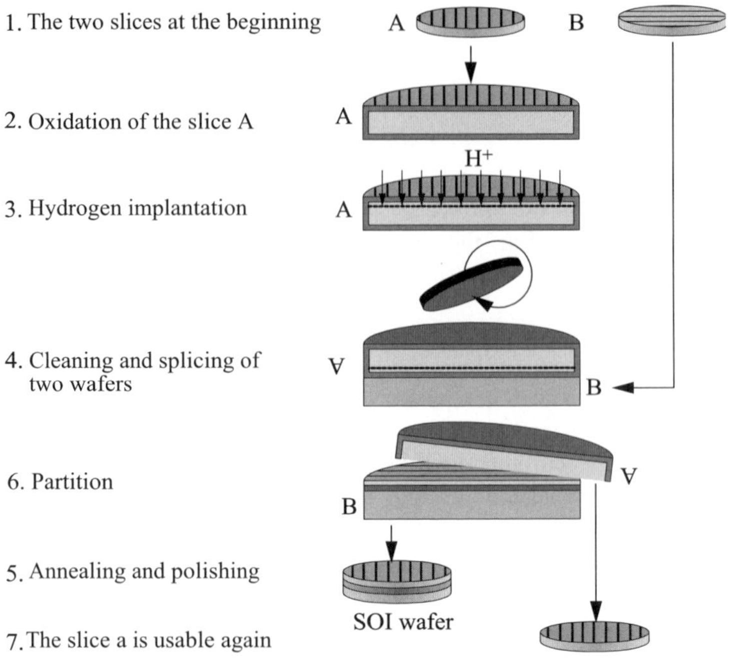

Figure 2.27. *The Smart-Cut™ method*

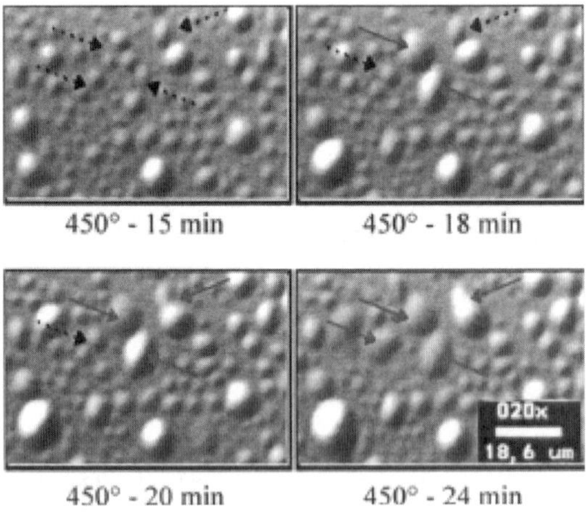

Figure 2.28. *Transmission electron microscopy image (planar view) showing the coalescence of the blisters as a function of the annealing time [ASP 01]*

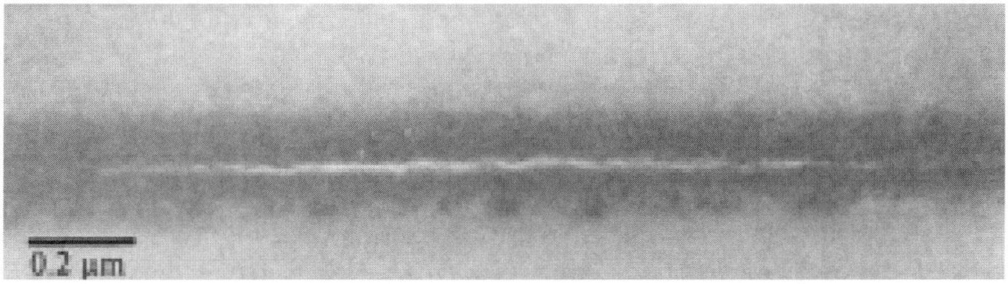

Figure 2.29. *Transmission electron microscopy image (transverse section) of a micro crack located around Rp [ASP 01]*

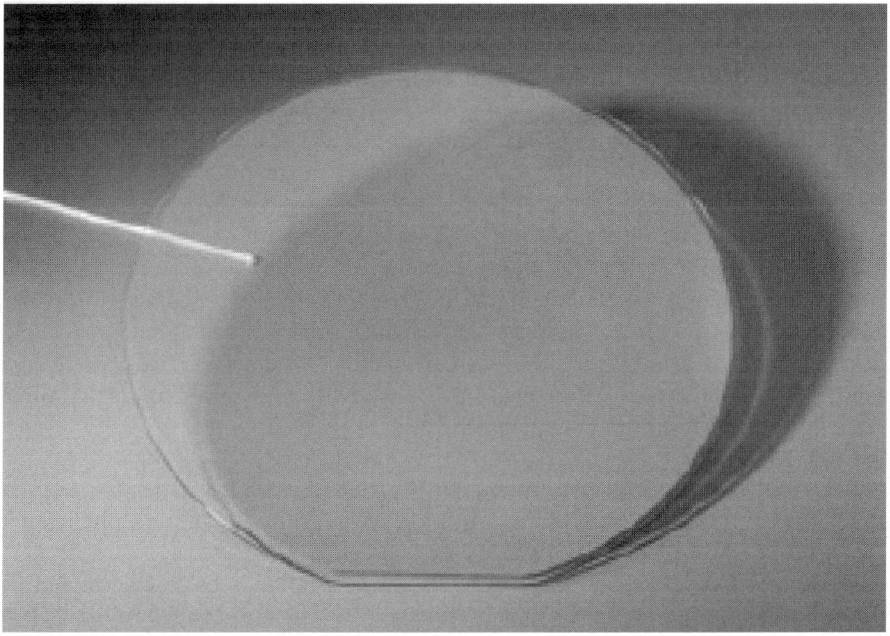

Figure 2.30. *Silicon on quartz wafer*

2.5.2.5. *Towards nanoelectronics?*

The recent development of nanotechnologies does not spare, obviously, the electronics domain. Current microelectronics is reaching its ultimate limits, regarding the reduction of the device dimensions. New approaches, involving nanoscale sizes objects, are seriously considered for the electronics of the future. The manufacturing methods of nano-objects are traditionally classified into three strategies: the first uses a very high-resolution lithography; the second consists of building these objects atom by atom, for example under the tip of a microscope with the tunnel effect; the third relies on mechanisms of a certain self-organization, concerning of course, a larger number of atoms. Ion implantation could be carried out with the first and third categories.

High-resolution lithography is indeed possible using a focused ion beam [HAR 91]. The current technology of focused beams uses sources with liquid metal (Ga), which have the characteristic of a virtual source point of very small dimensions. The object point can then be reduced to dimensions of about 15 nm, and lines 20 nm wide could be written in resist. It is thus a direct writing process (without mask), with a beam driving and/or a displacement of the target similar to electron beam lithography. However, the lateral energy deposition is, far less important, which makes it possible to reach these ultimate resolutions.

The activity concerning the surface self-organization is dominated by various growth methods (MBE, CVD). But volume self-organization (in other words the germination and the growth of an array of nano-precipitates) is governed by a certain number of non-equilibrium thermodynamics mechanisms, accessible, or at least adjustable, with the help of implantations and annealing, i.e. with tools available in the usual microelectronic technology. Nano-precipitates are organized according to the competitive growth process: the smallest are unstable and dissolve themselves during annealing for the benefit of the largest [PHI 85]. Depending on the local concentration, a so-called balance is established, forming precipitates of similar sizes, relatively regularly spaced from one another. Thanks to this phenomenon, nano-scale precipitates of all natures have already been synthesized in traditional microelectronics substrates, namely silicon and SiO_2 oxide.

Two main tendencies are prominent in these studies, depending on the applications. In the first one, optoelectronics, we seek to use these semiconductor nano-crystals to carry out light emitting devices on a silicon-based substrate [CHE 08, GON 00, LUT 00] (the emission wavelength is a function of the nature and size of the object). The second one, in nano-electronics itself, the capacity of these nano-crystal to store and restore charges seems very promising [CAR 03, DUG 06, THE 00]. An identical approach is applied to the metal nano-particles: researchers, in both cases, attempt to connect their size and density to their electronic, optical or magnetic properties. Devices, able to control the electrons one by one, up to temperatures close to ambient temperature, were recently created, with a single contact plug, as well as networks of contact plugs, necessarily disordered, prepared by surface techniques [COR 98]. However, their characteristics are still not really reproducible nowadays. Similar devices could be made by implantation and annealing.

2.6. Conclusion

Ion implantation is an essential method of micro-electronics technology. Implanters are now common clean room instruments. They are entirely automated and highly reliable. The physicochemical mechanisms determining the doping profiles and the damage are known in detail. Few innovations are thus to be expected in the field of implantation applied to conventional technology. However, new needs appear, aiming, for example, at extending the energy field, to carry out surface junctions, as well as very deep wells, or even to reduce the doping thermal budget. It raises the problem of influence of residual defects, transitorily present, on the dopants diffusion. Moreover, new methods are proposed, which are less conventional than simple doping, such as proximity gettering or the realization of semiconductor structures on an insulator. In this field, a new branch of the silicon industry has been recently developed, thanks to hydrogen implantation and to the Smart-Cut™ method. Finally, the arrival in force of nanotechnologies seems to open up new research tracks for ion implantation.

2.7. Bibliography

[ASP 01] ASPAR B. *et al.*, "The generic nature of Smart Cut® process for thin film transfer", *Journal of Electronic Materials*, vol. 30, no. 7, 2001.

[AUB 95] AUBERTON-HERVE A., WITTKOWER A., ASPAR B., "Characterization by atomic force microscopy of SDOI layer", *Nuclear Instrument and Methods in Physics Research*, section B96, p. 420, 1995.

[BIE 80] BIERSACK J.P., HAGGMARK L., "The mean projected range and range straggling of Xe ions implanted", *Nuclear Instrument and Methods in Physics Research*, vol. 174, p. 257, 1980.

[BOH 48] BOHR N., *Matematisk-Fysiske Meddelelser – Danske Videnskabernes Selskab*, Royal Danish Academy of Sciences and Letters XVIII, vol. 8, 1948.

[BON 96] BONAFOS C., Rôle des défauts end-of-range dans la diffusion anormale du bore, Thesis, Institut des sciences appliqués, Toulouse, October 1996.

[BRU 79] BRUEL M., BERTHET B., FLOCCARI M., MICHAUD J.F., *Radiation Effects and Defects in Solids Incorporating Plasma Science and Plasma Technology*, vol. 44, 173, 1979.

[BRU 95] BRUEL M., "Silicon on insulator material technology", *Electronics Letters*, vol. 31, no. 14, p. 1201, 1995.

[BRU 06] BRUEL M., "16th International conference on ion implantation technology", *AIP Conference Proceedings*, vol. 866, p. 76, Marseille, 2006.

[CAC 96] CACCIATO A.E., KLAPPE J.G., COWERN N.E.B., VANDERVOST W., BIRO L.P., CUSTER J.S., SARIS F.W., "Dislocation formation and B transient diffusion in C co-implanted Si", *Journal of Applied Physics*, vol. 79, no. 5, p. 2314, 1996.

[CAP 95] CAPPELLETTI P., FRATIN L., RAVAZZI L., "Nuclear instruments and methods in physics research section B: Beam interactions with materials and atoms", *Nuclear Instrument and Methods in Physics Research*, section B96, p. 405, 1995.

[CAR 03] CARRADA M., Manipulation de nanoparticules de Si élaborées par implantation ionique à basse énergie dans des couches minces de SiO_2 pour la fabrication de mémoires MOS non volatiles, Thesis, Paul Sabatier University, Toulouse, 2003.

[CHE 08] CHEMAM R., GROB J.J., BOUABELLOU A., "Luminescence properties of ZnS:Mn nanocrystals embedded in SiO_2 by ion implantation", *Materials Science and Engineering B 150*, vol. 26, 2008.

[CHE 99] CHEN P.S., HSIEH T.E., CHU C.H., "Removal of end-of-range defects in Ge+ pre-amorphized Si by carbon ion implantation", *Journal of Applied Physics*, vol. 85, no. 6, p. 3114, 1999.

[CHE 89] CHEUNG N.W., LIANG C.L., LIEW B.K., MUTIKAINEN R.H., WONG H., "Buried dopant and defect layers for device structures with high-energy ion implantation", *Nuclear Instrument and Methods in Physics Research*, section B37/38, p. 941, 1989.

[CLA 95] CLAVERIE A., LAANAB L., BONAFOS C., BERGAUD C., MARTINEZ A., MATHIOT D., "On the relation between dopant anomalous diffusion in Si and end-of-range defects", *Nuclear Instrument and Methods in Physics Research*, section B96, vol. 202, 1995.

[COL 04] COLINGE J.-P., *Silicon-on-Insulator Technology: Materials to VLSI*, 3rd edition, Springer, New York, 2004.

[COR 98] CORDAN A.S., GOLTZENE A., HERVE Y., MEJIAS M., VIEUC., LAUNOIS H., "Electron transport in metallic dot arrays effect of a broad dispersion in the tunnel junction dimensions", *Journal of Applied Physics*, vol. 84, no. 7, p. 3756, 1998.

[CRI 04] CRISTIANO F., CHERKASHIN N., CALVO P., LAMRANI Y., HEBRAS X., CLAVERIE A., LERCH W., PAUL S., "Thermal stability of boron electrical activation in preamorphised ultra-shallow junctions", *Materials Science and Engineering B 114–115*, vol. 174, 2004.

[CSE 76] CSEPREGI L., MAYER J.W., SIGMON T.W., "Regrowth behavior of ion-implanted amorphous layers on (111) silicon", *Applied Physics Letters*, vol. 29, p. 92, 1976.

[CSE 78] CSEPREGI L., KENNEDYE.F., MAYERJ.W., SIGMON T.W., "Substrate-orientation dependence of the epitaxial regrowth rate from Si-implanted amorphous Si", *Journal of Applied Physics*, vol. 49, p. 3096, 1978.

[DEA 73] DEARNALEY G., FREEMAN J.H., STEPHEN J., *Ion Implantation*, North Holland, Amsterdam, 1973.

[DUG 06] DUGUAY S., Propriétés de stockage de charges de nanocristaux de germanium incorporés dans des couches de silice par implantation ionique, Thesis, Louis Pasteur University, Strasbourg, November 2007.

[FAV 93] FAVENNEC P.N., *L'implantation Ionique pour la microélectronique et l'optique*, Collection technique et scientifique des télécommunications, Editions Masson, Paris, 1993.

[FRE 63] FREEMAN J.H., "A new ion source for electromagnetic isotope separators", *Nuclear Instrument and Methods in Physics Research*, vol. 22, p. 306, 1963.

[GON 00] GONZALEZ-VARONA O., PEREZ-RODRIGUEZ A., GARRIDO B., BONAFOS C., LOPEZ M., MORANTE J.R., MONTSERRAT J., RODRIGUEZ R., "Ion beam synthesis of semiconductor nanoparticles for Si based optoelectronics devices", *Nuclear Instrument and Methods in Physics Research*, section B161-163, vol. 904, 2000.

[GRA 05] GRAOUI H., FOAD M.A., "A comparative study on ultra-shallow junction formation using co-implantation with fluorine or carbon in pre-amorphized silicon", *Materials Science and Engineering*, B 124–125, vol. 188, 2005.

[GRO 87] GROB A., GROB J.J., GOLANSKI A., "Damage created in silicon by BFN+(1 less-than-or-equal-to N less-than-or*equal*to 3) and PFN++(1 less-than-or-equal-to N less-than-or*equal*to 5)", *Nuclear Instrument and Methods in Physics Research*, section B19/20, vol. 55, 1987.

[GRO 96] GROB A., ROHR P., MARIANI G., SEVELY J., GROB J.J., "Kinetics of impurity gettering on buried defects created by MeV argon implantation", *Nuclear Instrument and Methods in Physics Research section B112*, vol. 169, 1996.

[HAR 91] HARRIOTT L.R., "The technology of finely focused ion beams", *Nuclear Instrument and Methods in Physics Research section B55*, vol. 802, 1991.

[HEC 89] HECKING N., TEKAAT E.H., "2 Me V nitrogen ion implantation of Si 11 5", *Applied Surface Sciences*, vol. 43, p. 87, 1989.

[HIL 91] HILL C., HUNT P., "Ion implantation in bipolar technology", *Nuclear Instrument and Methods in Physics Research*, section B55, vol. 1, 1991.

[HOF 75] HOFKER W.K., "Implantation of boron in silicon", *Phillips Research Report Supplement*, vol. 8, 1975.

[IZU 78] IZUMI K., DOKEN M., ARIYOSHI H., "CMOS devices fabricated on buried SiO_2 layers formed by oxygen implantation into silicon", *Electronics Letters*, vol. 14, no. 18, p. 593, 1978.

[KIM 83] KIMRLING L.C., BENTON J.L., "Diffusivity of oxygen in silicon at the donor formation temperature", *Physica B (116)*, vol. 297, 1983.

[KIN 55] KINCHIN G.H., PEASE R.S., "The displacement of atoms in solids by radiation", *Report on Progress in Physics*, vol. 18, p. 1, 1955.

[KLE 91] KLEIN K.M., PARK C., TASCH A.F., SIMONTON R.B., NOVAK S, "Analysis of the tilt and rotation angle dependence of boron distributions implanted into less than 100 greater-than silicon", *Journal of the Electrochemical Society*, vol. 138, no. 7, p. 2102, 1991.

[KUS 91] KUSTERS K.H., MUHLHOFF H.M., CERVA H., "Application of ion implantation in submicron CMOS processes", *Nuclear Instrument and Methods in Physics Research*, section B55, vol. 9, 1991.

[LAL 05] LALLEMENT F., LENOBLE D., "Investigation on boron transient enhanced diffusion induced by the advanced P+/N ultra-shallow junction fabrication processes", *Nuclear Instrument and Methods in Physics Research*, section B237, vol. 113, 2005.

[LAU 80] LAU S.S., TSENG W.F., MAYER J.W., *Handbook of Semiconductors*, Chapter 7, vol. 3, Editions S.P. Keller, Amsterdam, 1980.

[LIC 91] LICOPPE C., NISSIM Y.I., KRAUZ P., HENOC P., "Solid-phase epitaxial regrowth of ion-implanted amorphized INP", *Applied Physics Letters*, vol. 49, p. 316, 1991.

[LIN 63] LINDHARD J., SCHARFF M., SCHIOTT H.E., "Range concepts and heavy ion ranges: (Notes on natomic collisions, II)", *Matematisk-Fysiske Meddelelser – Danske Videnskabernes Selskab*, vol. 33, no. 14, p. 1, 1963.

[LIN 65] LINDHARD J., "Influence of crystal lattice on motion of energetic charged particles", *Matematisk-Fysiske Meddelelser – Danske Videnskabernes Selskab*, vol. 34, no.14, 1965.

[LUT 00] LUTEROVA K., PELANT I., VALENTA J., REHSPRINGER J.L., MULLER D., GROB J.J., DIAN J., HONERLAGE B., "Red electroluminescence in Si+-implanted sol-gel-derived SiO_2 films", *Applied Physics Letters*, vol. 77, no. 19, p. 2952, 2000.

[MAT 97] MATSUO J., TAKEUSHI D., AOKI T., YAMADA I., "Cluster ion implantation for shallow junction formation", *Proceedings of the 11th Conference on Ion Implantation Technology (IEEE New York)*, p. 768, 1997.

[MIN 94] MINONDO M., Préamorphisation du silicium par l'ion germanium et formation de junctionsultra-fines P+/N, Thesis, Institut national polytechnique, Grenoble, November 1994.

[MIR 02] MIRABELLA S. *et al.*, "Interaction between self-interstitials and substitutional C in silicon: Interstitial trapping and C clustering mechanism", *Physical Review B65*, 045209, 2002.

[MOL 84] MOLLER W., ECKSTEIN W., "Tridyn- A trim simulation code including dynamic composition", *Nuclear Instrument and Methods in Physics Research*, section B2, p. 814, 1984.

[MOR 70] MOREHEAD F.F. Jr., Crowder B.L., "Formation of amorphous Si by ion bombardment as a function of ion temperature and dose", *Radiation Effects*, vol. 6, p. 27, 1970.

[MOR 96] MORICEAU H., MALEVILLE C., CARTIER A.M., ASPAR B., SOUBIE A., BRUEL M., POUMEYROL T., "Cleaning and polishing as key steps for Smart-cut® SOI process", *Proceedings IEEE International SOI Conference*, p. 152, 1996.

[MYE 96] MYERS S.M., PETERSEN G.A., SEAGER C.H., "Binding of cobalt and iron to cavities in silicon", *Journal of Applied Physics*, vol. 80, p. 3717, 1996.

[NAR 82] NARAYAN J., HOLLAND O.W., "Solid phase epitaxial growth in ion-implanted silicon", *Physica Status Solidi (a)*, vol. 73, p. 225, 1982.

[NIS 92] NISHIKAWA S., TANAKA A., YAMAJI T., "Reduction of transient boron diffusion in preamorphized Si by carbon implantation", *Applied Physics Letters*, vol. 60, no. 18, p. 2270, 1992.

[OLS 88] OLSON G.L., ROTH JA., "Kinetics of solid phase crystallization in amorphous silicon", *Material Science Reports*, vol. 3, p. 1, 1988.

[PAW 06] PAWLAK B.J., JANSSENS T., BRIJS B., VANDERVORST W., COLLART E.J.H., FELCH S.B., COWERN N.E.B., "Effect of amorphization and carbon co-doping on activation and diffusion of boron in silicon", *Applied Physics Letters*, vol. 89, 062110, 2006.

[PHI 85] PHILIBERT J., *Diffusion et transport de matière dans les solides*, Les Editions de Physique, Les Ulis, 1985.

[QIA 91] QIAN X.Y., CHEUNG N.W., LIEBERMAN M.A., CURRENT M.I., CHU P.K., HARRINGTON W.L., MAGEE C.W., BOTNICK E.M., "Sub 100 nm P+/N junction formation using plasma immersion ion-implantation", *Nuclear Instrument and Methods in Physics Research*, section B55, p. 821, 1991.

[RAI 98] RAINERI V., SAGGIO M., FRISINA F., RIMINI E., "Voids in silicon power devices", *Solid State Electronics*, vol. 42, p. 2295, 1998.

[ROB 63] ROBINSON M.T., OEN O.S., "The channelling of energetic atoms in crystal lattices", *Applied Physics Letters*, vol. 2, p. 30, 1963.

[ROH 95] ROHR P., Piégeage d'impuretés métalliques dans le silicium par implantation ionique haute énergie, Thesis, Louis Pasteur University, Strasbourg, 1995.

[ROQ 00] ROQUETA F., Piégeage d'impuretés métalliques par implantation d'hélium: applications aux composants de puissance, Thesis, François Rabelais University, Tours, 2000.

[RYS 82] RYSSEL H., GLAWISCHNIG H, *Ion Implantation Techniques*, Springer Verlag, Berlin, 1982.

[SEI 71] SEIDEL T.E., MAC REA A.U., *First International Conference on Ion Implantation* F.H. EISEN AND C.S. CHADDERTON, (Eds.), Gordon and Breach, London, 1971.

[SEL 84] SELBERHERR S., *Analysis and Simulation of Semiconductor Devices*, Springer Verlag, Berlin, p. 62, 1984.

[SIG 72] SIGMUND P., "Collision theory of displacement dalage.4. ion range and sputtering", *Revue Roumaine de Physique*, vol. 17, p. 823, 969 and 1079, 1972.

[SKO 04] SKORUPA W., YANKOV R.A., ANWAND W., VOELSKOW M., GEBEL T., DOWNEY D.F., AREVALO E.A., "Ultra-shallow junctions produced by plasma doping and flash lamp annealing", *Materials Science and Engineering*, B 114–115, p. 358, 2004.

[SMI 98] SMITH R., SHAW M., WEBB R.P., FOAD M.A., "Ultra shallow junctions in Si using decaborane? A molecular dynamics simulation study", *Journal of Applied Physics*, vol. 83, no. 6, p. 3148, 1998.

[STE 92] STEPHENS K.G., *Handbook of Ion Implantation Technology*, North Holland, Amsterdam, p. 455, 1992.

[SUM 99] SUMINO K., "Impurity reaction with dislocations in semiconductors", *Physica Status Solidi (a)*, vol. 171, p. 111, 1999.

[TAM 71] TAMURA M., IKEDA T., TOKUYAMA T., *Ion Implantation in Semiconductors*, I. RUGE AND J. GRAUL (Eds.), Springer Verlag, Berlin, p. 96, 1971.

[THE 00] THEES H.-J., WITTMAACK M., STEGEMANN K.-H., VON BORANY J., HEINIG K.-H., GEBEL T., "Microstructure and electrical properties of gale SiO_2 containing Genanoclusters for memory applications", *Microelectronics Reliability*, vol. 40, p. 867, 2000.

[VAN 05] VANDERPOOL A., TAYLOR M., "Reducing ultra-shallow boron diffusion using carbon and fluorine co-implantation", *Nuclear Instrument and Methods in Physics Research*, section B237, p. 142, 2005.

[VIE 87] VIEU C., Phénomènes de dégradation et d'amorphisation induits par implantation ionique dans du silicium monocristallin, Thesis, Paul Sabatier University, Toulouse, December 1987.

[WAL 78] WALKER R.S., THOMPSON D.A., "Computer simulation of ion–bombardment collision cascades", *Radiation Effects 37*, vol. 113, 1978.

[WEB 00] WEBER W.J., "Models and mechanisms of irradiation-induced amorphization in ceramics", *Nuclear Instrument and Methods in Physics Research*, section B166-167, p. 98, 2000.

[WIL 84] WILLIAMS J.S., POATE J.M., *Ion Implantation and Beam Processing*, Academic Press, New York, 1984.

[WIL 77] WILSON W.D., HAGGMARK L.G., BIERSACK J.P., "Calculations of nuclear stopping, ranges and straggling in low energy region", *Physical Review B*, vol. 15, p. 2458, 1977.

[WON 89] WONG H., CHEUNG N.W., YU K.M., CHU P.K., LIU J., "Impurity gettering by implanted carbon in silicon", *Material Research Society Symposium Proceedings*, vol. 147, 97, 1989.

[YAN 98] YANG B.L., JONES E.C., CHEUNG N.W., SHAO J., WONG H.,CHENG Y.C., "n(+)/p ultra-shallow junction formation with plasma immersion ion implantation", *Microelectronics Reliability*, vol. 38, p. 1489, 1998.

[ZAP 88] ZAPPE H.P., HU C., "Characterization of CMOS devices in high energy boron-implanted substrates", *IEEE Transactions Electron Devices*, vol. 35, no. 7, p. 1029, 1988.

[ZIE 85a] ZIEGLER J.F., BIERSACK J.P., LITTMARK U., "The stopping and range of ions in solids", in J.F. ZIEGLER (Ed.), *The Stopping and Ranges of Ions in Matter*, vol. 1, Pergamon Press, New York, 1984, available at: http://www.srim.org.

[ZIE 85b] ZIEGLER J.F., "High energy ion implantation", *Nuclear Instrument and Methods in Physics Research*, section B6, p. 270, 1985.

[ZIE 92] ZIEGLER J.F., *Handbook of Ion Implantation Technology*, North Holland, Amsterdam, 1992.

Chapter 3

Dopant Diffusion: Modeling and Technological Challenges

3.1. Introduction

Semiconductor doping is a key element for manufacturing electronic devices. Indeed, a pure (intrinsic) semiconductor is almost an insulator at room temperature, with a valence band almost entirely full, and a conduction band almost entirely empty. As an example, the intrinsic density of mobile carriers in silicon is only about 10^{10} cm^{-3} at 300 K. The interesting electronic properties are obtained thanks to the possibility of "doping" the material by introducing adequate impurities (dopants), making it possible to introduce mobile carriers. In fact, the functioning of all the active electronic devices mainly relies, if not entirely, on the possibility of forming P/N junctions, i.e. to put in contact areas doped with an acceptor (impurity having an electron deficit compared to the matrix, involving a conduction by "holes" in the valence band) and areas doped with a donor (impurity having an electron excess compared to the matrix, involving "free electron" conduction in the conduction band). In silicon and germanium (elements of column IV of the periodic table), donors are elements of column V (the most frequently used being phosphorus and arsenic), while acceptors are elements of column III (the most frequently used being boron).

The detail of the electric behavior of a semiconductor device thus strongly depends on the dopant distribution inside the structure. For evidence, we need to compare, for example, the equivalent electric circuit of an N-channel MOS transistor

---

Chapter written by Daniel MATHIOT.

(see Figure 3.1a) with the schematic structure of the same device realized in standard ULSI technology (see Figure 3.1b). As shown in Figure 3.1 (see the numbers indicating the correspondences on the figures), all the elements controlling the functioning of the device (access resistances, conductance and transconductance, capacitances, etc.) are directly associated with the presence of dopants.

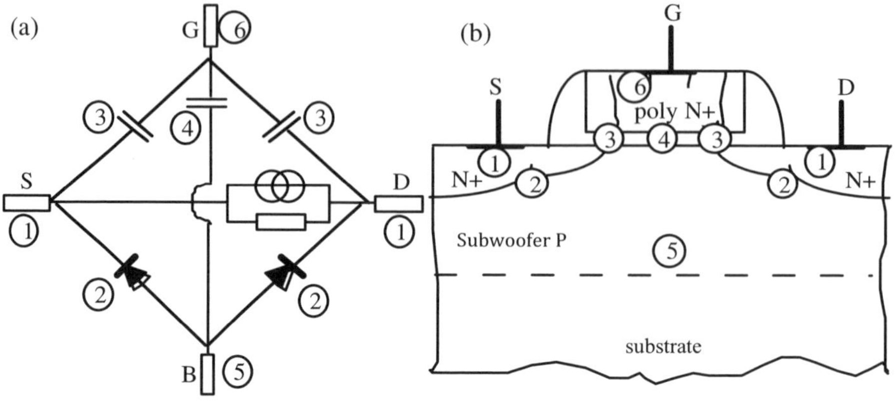

Figure 3.1. *N-MOS transistor: equivalent electric circuit (a) and corresponding technological structure (b)*

Although in modern technologies, dopants are introduced by ion implantation (see Chapter 2 of this book by J.J. Grob), high temperature annealing is necessary, to restore for example the crystalline quality of the material damaged by ion implantation and "to activate" the dopants (i.e. to position them in the substitutional site, where they are electrically active). During these high temperature processes, dopant profiles (i.e. their distribution in the material) are distorted because of the completely general phenomenon of thermal diffusion. Among other things, these diffusion phenomena are particularly critical for the control of the following key parameters:

– Channel doping: the dopant profile in the channel zone, under the gate oxide, determines the importance of the band bending in functions of the various applied voltages. Dopant profiles in this zone thus directly affect the value of the threshold voltage of the device.

– Channel length: the horizontal distance separating the drain from the source. Geometrically defined by photo-lithography steps, the effective length of the finished device is controlled by the dopant lateral diffusion from the drain and source zones.

– Depth of the drain and source junctions: the decrease of lateral dimensions also require us to use increasingly thin junctions, and thus to control as well as possible the penetration of dopant profiles.

– Access resistance: the concentration of electrically active dopants in drain and source zones and their precise distribution also determine the parasitic resistance in series with the transistor channel. This crucial parameter is one of the limiting points for the development of ultimate devices.

With the above examples, we can see that a good control of the electric properties of an electronic semiconductor device requires perfect control of the dopant profiles in the structure, and thus control as precise as possible of the diffusion steps (high temperature annealing) undergone by the circuit throughout its manufacture. This goes by the understanding and modeling of the elementary mechanisms governing the dopant diffusion (the subject of the following development). First, (section 3.2) we will present the general theory of diffusion in solids, applied to the case of semiconductors. Then, (section 3.3) we will take a closer look at the case of dopants in silicon, from experimental (sections 3.3.1 and 3.3.2) and theoretical (modeling, section 3.3.3) points of view. We will describe some practical problems for the device realization (section 3.4). Before concluding, we will briefly approach the case of germanium, a semiconductor which will probably be used in the upcoming technologies (section 3.5).

3.2. Diffusion in solids

3.2.1. *General information*

3.2.1.1. *Fick's first law*

Let us consider an element A dissolved inside a crystalline matrix. At low temperature, the system is frozen and the impurity is immobile. On the other hand, if the temperature is sufficiently raised, under the effect of thermal agitation, the atoms of A are mobile and can move inside the matrix. They are then subjected to the diffusion phenomenon: a non-uniformity of thermodynamic variables (a concentration variation of A, everything otherwise equal, being the most frequent case) induces a transport of matter (a displacement of atoms A) inside the solid. Diffusion is thus a non-reversible process. Its "driving force" is the gradient of chemical potential and it aims to homogenize this potential. The general phenomenological description of this process is only possible with the use of the thermodynamics theory of irreversible processes, which is beyond the subject of this section.

However, in the simplest case of only one diffusing species subjected only to its concentration gradient, the phenomenon is very simply described, using Fick's first law. This law stipulates that the impurity flux J (number of atoms crossing the surface unit per time unit) is proportional (in absolute value) to the concentration gradient C of the impurity:

$$\vec{J} = -D\vec{\nabla}(C) \qquad [3.1]$$

The sign "—" of the equation above simply shows the fact that diffusion occurs from high concentration zones towards low concentrations zones. This is required to be able to homogenize the concentrations.

By definition, the proportionality factor D is the diffusion coefficient, or diffusivity. The flux is generally expressed in atoms per square centimeter and per second ($cm^{-2}s^{-1}$) and concentrations in the number of atoms per cubic centimeter (cm^{-3}). D is then practically expressed in cm^2s^{-1}. We can notice the analogy of this law with the heat transfer, Fourrier's law ($\vec{J_Q} = -\kappa\vec{\nabla}(T)$) or with the elementary electric current Ohms law ($\vec{J} = \sigma\vec{E} = -\sigma\vec{\nabla}(V)$).

It is important to realize that the flux is a vector quantity and that thus, in the most general case (anisotropic environment), D is a tensor. Therefore, the flux is not necessarily parallel to the concentration gradient. This can occur in non-cubic or distorted materials, under a stress for example. Nevertheless, the case interesting to us here is silicon, a cubic material and thus isotropic in the absence of strain. In the silicon case, the diffusion coefficient is a scalar. The flux, given by the first Fick equation is then parallel to the concentration gradient.

Therefore, we will be able to consider in the following, without losing the sense of general information, only the 1D case for which Fick's first law is reduced to:

$$J = -D\frac{\partial C}{\partial x} \qquad [3.2]$$

3.2.1.2. *Generalized flow: drift terms*

Previously, we have considered that the diffusing impurity was only subjected to its concentration gradient. Let us consider now the most general case, where the impurity is submitted to driving forces of various origins (concentration gradient, stress field, electric field, etc.) and which, under the effect of these forces (and of its inevitable interactions with the crystal lattice) acquires an average stationary speed <v>. It is then easy to show that the impurity flux is given by the equation:

$$J = \langle v \rangle C \qquad [3.3]$$

The general problem thus reduces to identifying the nature of the various forces and to establishing the relation between force and speed. On the assumption that the driving force is only due to the concentration gradient, the thermodynamics of the irreversible processes shows that the driving force is the gradient of the partial mixing entropy. The calculations then lead to equation [3.2]. In the general case of a set of driving forces, while intuitively admitting that the average speed of the impurity is the sum of the elementary speeds induced by each driving force taken separately, it is convenient to write the impurity flux as:

$$J = -D\frac{\partial C}{\partial x} + \langle v \rangle C \qquad [3.4]$$

In the previous equation, the first term ("Fickian" term) shows the contribution of the lone concentration gradient and the second term ("Nernstian" or "drift" term) is the contribution of all the other driving forces. An important case is that of a charged species in the presence of an electric field. Under these conditions, the drift velocity is given by:

$$\langle v \rangle = \pm \mu E \qquad [3.5]$$

where μ is the mobility, E is the electric field and the sign $\pm$ is dependent on the sign of the charge carried by the impurity. Under these conditions, the flux is simply written:

$$J = -D\frac{\partial C}{\partial x} \pm \mu E C \qquad [3.6]$$

As we will see further, this expression of the flux, well-known in electric charge transport theory, also has its utility for the problem of dopant diffusion in a semiconductor.

3.2.1.3. *Fick's second law and simple profiles calculation*

Established for a conservative system, i.e. where there is neither loss nor creation of diffusing atoms, Fick's second law describes the temporal evolution of the concentration of the diffusing species at a given point. A simple mass balance leads to the following conservation equation (or continuity equation):

$$\frac{\partial C}{\partial t} = -\frac{\partial J}{\partial x} \qquad [3.7]$$

From where we take:

$$\frac{\partial C}{\partial t} = \frac{\partial}{\partial x}(D\frac{\partial C}{\partial x}) \qquad [3.8]$$

If D can be regarded as constant, i.e. independent of the concentration, and thus independent of x and t, in the hypothesis of a diffusion under the only effect of a concentration gradient, equation [3.2] gives:

$$\frac{\partial C}{\partial t} = D\frac{\partial^2 C}{\partial x^2} \qquad [3.9]$$

For a species submitted to additional driving forces, equation [3.4] with constant D leads to:

$$\frac{\partial C}{\partial t} = D\frac{\partial^2 C}{\partial x^2} - \langle v \rangle \frac{\partial C}{\partial x} - C\frac{\partial \langle v \rangle}{\partial x} \qquad [3.10]$$

When the system is not conservative, for example in the event of chemical reaction inducing the formation or the elimination of impurities, or of the trapping of impurities, etc., these equations are modified by the addition of generation or recombination terms, which depend on the mechanisms involved. We will see further an essential example to describe some non-equilibrium situations.

Equation [3.9] (Fick's second equation) is thus only valid for very restrictive conditions which, as we will see later on, are seldom met for the diffusion problems in semiconductors. Equations [3.9] or [3.10] are the basis for the calculation of $C(x, t)$, but regarding partial differential equations, of the first order in t and of the second order in x, their solutions are not unique. The appropriate solution for a given situation is determined by the knowledge of an initial condition and of two limit conditions describing the experimental reality. In general, the solution of these equations is not analytically possible. It is thus necessary to resort to numerical resolutions by computer.

Nevertheless, equation [3.9] has simple analytical solutions for particular initial and limit conditions. These conditions are generally imposed from an experimental point of view, in order to be able to determine the values of diffusion coefficients by comparison between the measured profiles and these simple laws. We will give here these solutions for three practical cases. The reader can refer to the reference books by Carslaw and Jaeger [CAR 59] or Cranck [CRA 75] for a more comprehensive catalogue and for the demonstration of the results below:

– *Diffusion from a thin layer*: this is the practical case, where the diffusing species is deposited at the initial moment, in the form of a thin film on the surface ($x=0$) of a very thick sample. If there is no impurity evaporation, as soon as the penetration depth is very large compared to the initial film thickness, the concentration profile follows a Gaussian law inside the sample:

$$C(x,t) = \frac{Q}{\sqrt{\pi Dt}} \exp(-\frac{x^2}{4Dt}) \qquad [3.11]$$

where Q is the total quantity of the deposited impurity (and thus dissolved in the material):

$$Q = \int_0^\infty C(x) dx \qquad [3.12]$$

– *Diffusion with a constant surface concentration*: this is the case where an infinite external source maintains on the surface of the sample a constant concentration C_0. The solution then calls upon a conventional function of the statistics, the complementary error function:

$$C(x,t) = C_0 erfc(\frac{x}{2\sqrt{Dt}}) \qquad [3.13]$$

with:

$$erfc(u) = 1 - \frac{2}{\sqrt{\pi}} \int_0^u \exp(-\eta^2) d\eta \qquad [3.14]$$

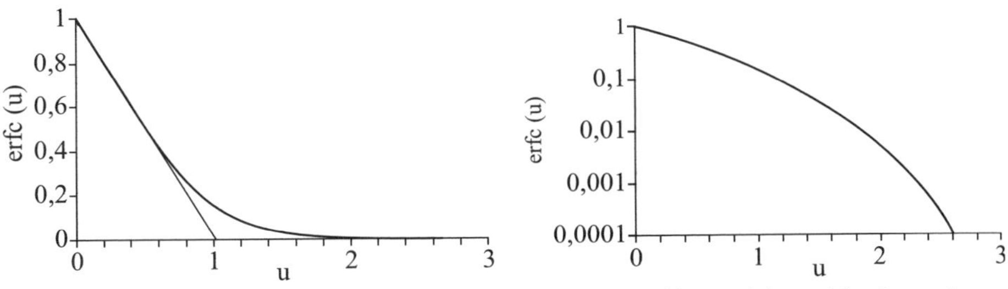

a) Linear scale b) Semi-logarithmic scale

Figure 3.2. *Complementary error function*

Existing numerical tables give the values of the complementary error function. Its variation according to $u = x / 2\sqrt{Dt}$ is given in Figure 3.2, in linear and semi-logarithmic scale. As we can see in Figure 3.2a, the tangent at the origin cuts the abscissa-axis at the point *u=1*, i.e. $x = 2\sqrt{Dt}$. The impurities dose ranging between this point and the sample surface approximately represents 90% of the total dose penetrated in the sample ($Q = (2/\sqrt{\pi})C_0\sqrt{Dt}$). We can thus consider, with a good approximation that the depth affected by the diffusion is about $2\sqrt{Dt}$, which explains why this quantity is often called the diffusion depth.

– *Redistribution of a buried Gaussian*: this is the simplest approximation of the redistribution of a dopant profile introduced by ion implantation. As a first approximation, the corresponding initial profile is Gaussian:

$$C_{(x,0)} = C_{t=0}^{\max} \exp\left(-\frac{(x - Rp)^2}{2\Delta Rp^2}\right) \qquad [3.15]$$

Rp is the mean projected range and ΔRp is the standard deviation. If *Rp* is sufficiently large so that the impurity profile does not interact to a significant degree with the surface during the diffusion, and if the diffusivity can be considered constant, the profile obtained at the end of a time *t* is always a Gaussian centered in *x=Rp*, therefore given by a relation similar to equation [3.15]. Its maximal value and standard deviation vary with time, according to the following laws:

$$C_t^{\max} = C_{t=0}^{\max} \frac{\Delta Rp_{t=0}}{\Delta Rp_t} \qquad [3.16]$$

and:

$$\Delta Rp_t^2 = \Delta Rp_{t=0}^2 + 2Dt \qquad [3.17]$$

3.2.1.4. *Boltzmann-Matano analysis*

When *D* depends on the concentration or more generally varies with the coordinates of space or time, equations [3.9] or [3.10] are no longer valid, and the general continuity equation [3.8] cannot be integrated any more simply.

However, an estimate of the variation of the diffusivity with *C* can be obtained from the experimental diffusion profile by the Boltzmann-Matano method: by the change of variable $\lambda = x / \sqrt{t}$, the partial differential equation [3.8] is easily transformed into a simple differential equation:

$$\lambda \frac{dC}{d\lambda} = -2 \frac{d}{d\lambda}\left(D \frac{dC}{d\lambda}\right) \qquad [3.18]$$

If the boundary conditions are simply expressed according to the only variable λ (which is the case for example for a diffusion with constant surface concentration, but not for the redistribution of a thin film or a Gaussian), a double integration leads to the following relation:

$$D_{(C^*)} = -\frac{1}{2t}\frac{\int_0^{C^*} x dC}{(dC/dx)_{C^*}} \qquad [3.19]$$

As seen in Figure 3.3, the numerator of this expression is the area ranging between the profile $C(x, t)$, the axis x and the line $C=C^*$, whereas the denominator is the tangent to the profile at the point $C=C^*$.

This geometrical interpretation of relation [3.19] makes the determination of D possible by graphic means from the experimental profile.

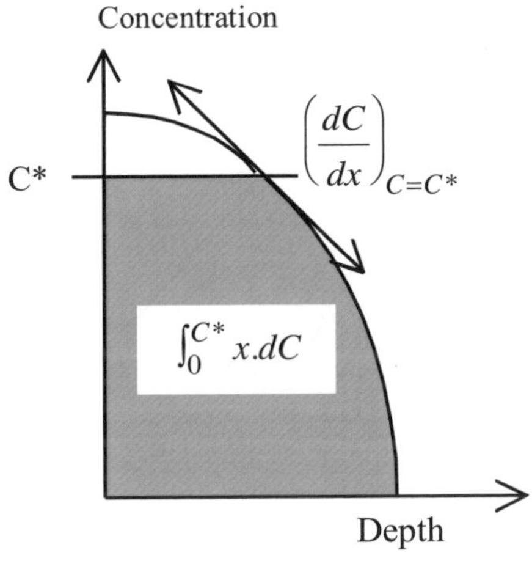

Figure 3.3. *Boltzmann-Matano analysis*

3.2.2. *Elementary mechanisms*

3.2.2.1. *Atomic mechanisms*

Figure 3.4 sums up the variation with the temperature of the diffusivity of various impurities experimentally measured in intrinsic silicon. We note that, depending on the type of impurity, the experimental values have very different orders of magnitude, ranging, for a temperature of about 900°C, from $10^{-5}/10^{-4}$ cm^2s^{-1} for impurities such as copper or cobalt, up to values of about $10^{-16}/10^{-15}$ cm^2s^{-1} for the common dopants (P, B, As). Correlatively, the activation energies (see section 3.2.2.2) involved vary from 0.2/0.5 eV to 4 or 5 eV. This variety of orders of magnitude of the diffusion coefficient simply reflects the different microscopic mechanisms by which an impurity atom can move in a crystalline solid. These mechanisms are closely related to the atomic structure (network) of the concerned solid and to the equilibrium position of the impurity in the crystal lattice of the matrix.

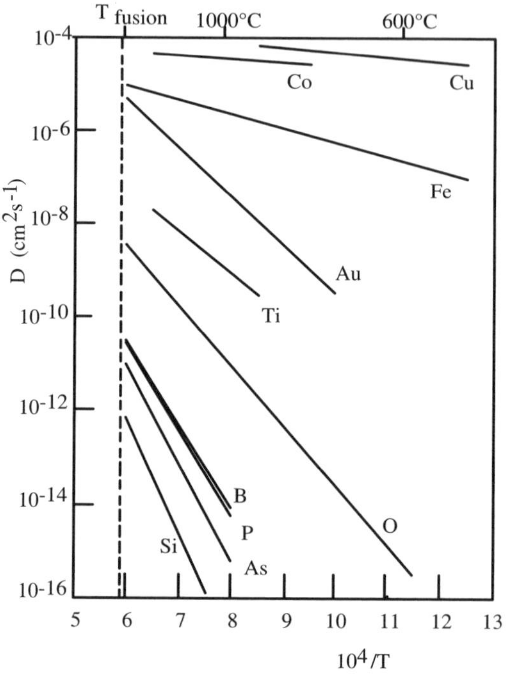

Figure 3.4. *Diffusivities in intrinsic Si*

According to this equilibrium position (stable site), we distinguish substitutional impurities, whose stable site is a crystal lattice site of the matrix, and interstitial impurities, whose stable site is between the network sites (in a void between the matrix atoms).

Interstitial impurities: this is the simplest mechanism, illustrated in Figure 3.5, during which the impurity gradually "hops" from interstitial site to interstitial site and can thus easily cover significant distances. This mechanism thus leads to high diffusion coefficient. In silicon, light elements (H, He, etc.) and transition metals of the 4th line of the periodic table (Ti, V, Cr, etc.) behave according to this pattern. Although it does not directly concern this chapter, essentially devoted to the dopant behavior, we will nonetheless point out that direct interstitial diffusion is often slowed down by the interactions of the diffusing species with the other present impurities (in particular the dopants), leading to the formation of more or less stable complexes and to an important reduction of the effective diffusion coefficient. The interested reader will be able to refer, for example, to the references [MAT 89] and [HEI 99], respectively discussing the case of hydrogen and copper. In the case of transition metals, these interactions must necessarily be taken into account for a good understanding of the technological steps, aiming at trapping these impurities (gettering)

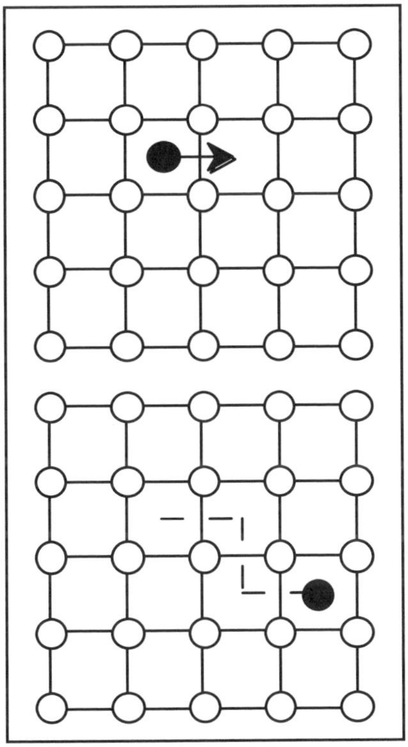

Figure 3.5. *Simple interstitial diffusion*

Substitutional impurities: the diffusion of these impurities is only possible by involving the structural point defects of silicon. These defects are either vacancies (vacant sites in the network), or self-interstitial (surplus silicon atoms in the network), which gives place to two types of elementary diffusion mechanisms:

– The vacancy mechanism is illustrated in Figure 3.6: to diffuse, the impurity must wait until a vacancy reaches its neighborhood. The diffusion is then carried out by sites exchange between the impurity and the vacancy. In the most general case, this vacancy can then move away from the impurity atom, by exchanges with the neighboring silicon atoms, and the impurity atom must expect the arrival of another vacancy to perform an additional hop. On the other hand, because of a strong binding energy between the vacancy and the impurity, the considered vacancy can happen to remain in close proximity with the impurity and the corresponding complex can diffuse as such in the network. Later on we will reconsider this diffusion concept with "pairs".

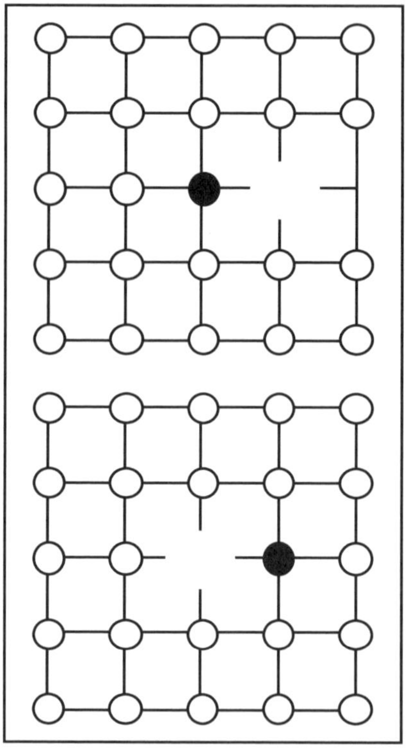

Figure 3.6. *Vacancy diffusion*

– Figure 3.7 shows the diffusion mechanism assisted by self-interstitials. In the silicon network, self-interstitials have the structure schematized in (1): the interstitial remains linked to the network and is in fact a "dissociate self-interstitial", i.e. two silicon atoms share a single network site. When this self-interstitial is in the proximity of a substitutional impurity, one of the silicon atoms can interact with the impurity, which leads to the formation of an interstitial impurity (schematized in (2)). This is in fact a "mixed dissociate interstitial", then able to diffuse according to the hopping sequence described in Figure 3.7. In the example illustrated here, the mixed interstitial is dissociated (with reforming of the self-interstitial) after a hop of the impurity. In fact, in the effective 3D silicon network, the mixed interstitial (the interstitial impurity) can carry out several hops and thus diffuse on long distances, before dissociating. All occurs then as if it was the diffusion of an impurity/self-interstitial "pair".

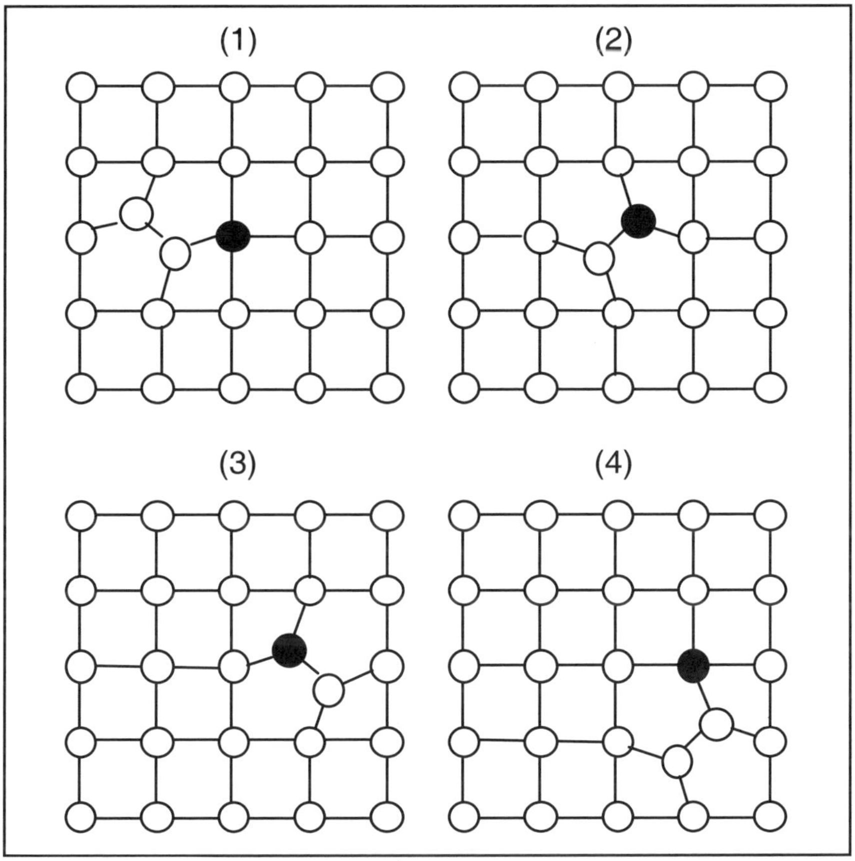

Figure 3.7. *Diffusion assisted by self-interstitials*

Dissociative mechanisms: these mechanisms occur for impurities, the stable site of which is a substitutional site (where they are relatively immobile). But these impurities have a noticeable probability of existence in an interstitial site (with a very high diffusion coefficient). The resulting effective diffusivity, with values intermediate between those of the substitutional and pure interstitial impurities, is mainly governed by the mechanism allowing the considered impurity to go from one type of site to another. According to the type of point defect implied in these exchanges (see Figure 3.8), we distinguish two dissociative mechanisms:

– The "Frank-Turnbull" mechanism: schematized in Figure 3.8a, this mechanism was initially proposed to explain the copper diffusion in germanium [FRA 56]. It involves the substitutional impurity/interstitial impurity exchange via network vacancies, according to the reaction:

$$A_i + V \leftrightarrow A_s \qquad [3.20]$$

– The "*kick-out*" mechanism: the previous mechanism was regarded for a long time as suitable to describe the diffusion of the "heavy" transition metals (Pt, Au, etc.) in silicon. However, it was then shown that it was not convenient for a proper description of all the experimental observations. A new mechanism, involving self-interstitials, known as "*kick-out*", was proposed by Gösele [GOS 80]. In this mechanism, illustrated in Figure 3.8b, a self-interstitial can eject a substitutional impurity of its site, leading to the formation of an interstitial impurity. The opposite reaction would lead to the return of the impurity into the substitutional site:

$$I + A_s \leftrightarrow Si_s + A_i \qquad [3.21]$$

In the previous equations, A indicates the impurity (the index informing about its site), V designates a vacancy and I, a self-interstitial. Without going into the details, let us say that only a combination of the two previous mechanisms is suitable for a comprehensive description of the diffusion profiles of gold in silicon [MAT 92].

It is moreover interesting to notice that the diffusion mechanism of the substitutional impurities assisted by self-interstitials is identical to the kick-out mechanism, if the hypothesis is made that the interstitial impurity remains linked to the network. In fact, these two mechanisms, introduced and described in different contexts, lead to the same continuity equations and it is in general not easy (nor often even relevant) to distinguish them.

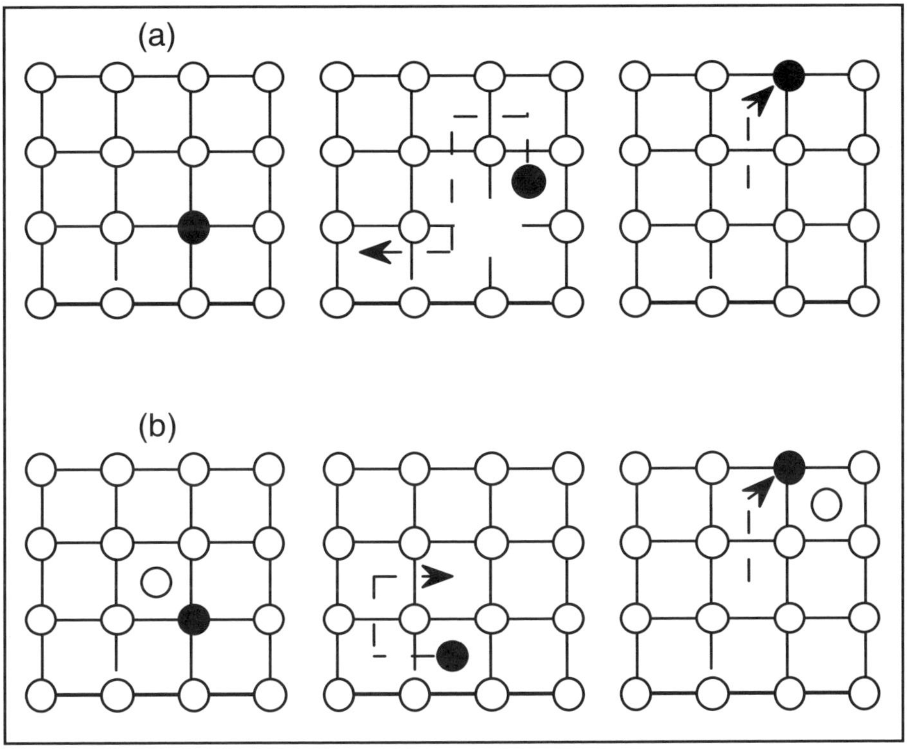

Figure 3.8. *Dissociative mechanisms: (a) "Frank-Turnbull"; (b) "kick-out"*

3.2.2.2. *Atomic theory of the diffusion coefficient*

As seen in Figure 3.4, the experimental variations of the diffusion coefficients with the temperature can in general be easily described by an Arrhenius law:

$$D = D_0 \exp(-\frac{Q}{kT}) \qquad [3.22]$$

T is the absolute temperature (in Kelvin) and k is the Boltzmann constant. The parameter Q is called the diffusion activation energy. If we consider that during its displacement, the diffusing species jump from one stable site to another with a frequency ν, this energy Q is also the activation energy of ν:

$$\nu = \nu_0 \exp(-\frac{Q}{kT}) \qquad [3.23]$$

This activation energy concept is easily illustrated using Figure 3.9, which schematically describes the fluctuation of Gibbs free enthalpy (G) of the crystal during the displacement of the diffusing species. When the diffusing species is on its stable site (S), G is minimal (situation of thermodynamic equilibrium). Because of the thermal agitation, the atom vibrates with a frequency ν_0 close to the Debye frequency of the crystal (about 10^{13} s^{-1}). This displacement must be accompanied by a lattice deformation, and thus by an increase of G. In a certain point, called the saddle point (C), halfway between two stable sites, G passes by a maximum. If, thanks to thermal agitation, the diffusing species reaches this saddle point, it can then "fall" into the other stable site, without additional energy cost. The activation energy is thus the difference in Gibbs free enthalpy of the crystal between its value at the saddle point and the equilibrium value.

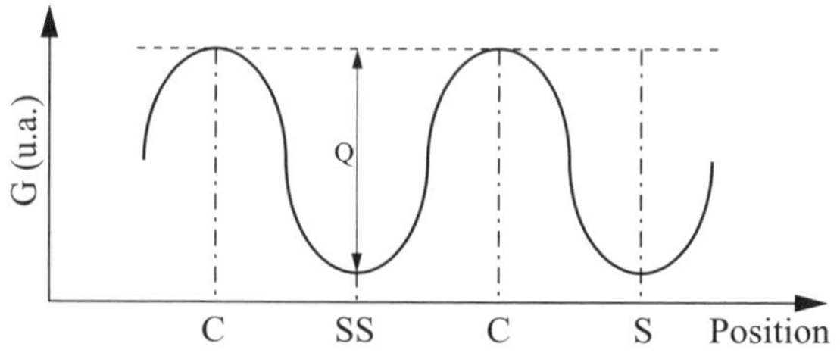

Figure 3.9. *Activation energy concept*

For a given atomic mechanism, the activation energy, as well as the pre-exponential term D_0, can be connected to the physical parameters describing the elementary hop. First of all, it is necessary, independently of the mechanism involved, to establish the relation between the diffusivity and the jump frequency. It is obvious that, whatever the elementary mechanism, an atom can move in a crystal only by a series of elementary hops. If these hops take place in a completely random way (each hop being independent of the previous hop), we speak about a "random walk". We can then show (for example [PHI 91], [TUC 74]) that, in a cubic crystal, the diffusion coefficient is connected to the jump frequency by the relation:

$$D = \frac{1}{6} a^2 \nu \qquad [3.24]$$

where a is the length of the elementary hop, i.e. the distance between the first neighboring atoms (a= 2.35 Å for silicon). In reality, atomic hops do not occur purely randomly. Indeed, a hop could be correlated to the previous hop. This results in the introduction of a corrective multiplicative factor f into equation [3.24]. This correlation factor is lower than unity, and its value depends on the crystal lattice structure and on the atomic diffusion mechanism. We can now understand the microscopic origin of the orders of magnitude of the activation energies for various diffusion mechanisms.

Substitutional impurity: for the vacancy mechanism or for the mechanism assisted by self-interstitials, two conditions must be filled to allow an atomic hop of the impurity to happen. First, it is necessary that the defect is in the immediate proximity (first neighbor) of the impurity, which is characterized by a probability P_X. It is then necessary that the defect exchange its site with the impurity, which occurs with a frequency ν_X. The hopping frequency of the impurity is thus the product of these two terms:

$$\nu = P_X \nu_X \qquad [3.25]$$

with, according to the already explained approach (see Figure 3.9):

$$\nu_X = \nu_0 \exp(-\frac{\Delta G_X^m}{kT}) = \nu_0 \exp(\frac{\Delta S_X^m}{k}) \exp(-\frac{\Delta H_X^m}{kT}) \qquad [3.26]$$

ΔG_X^m is the migration free enthalpy of the defect X, which is naturally broken down (*G*=*H*-*TS*) into a migration entropy (ΔS_X^m) and a migration enthalpy (ΔH_X^m). If we assume the defects are distributed in a purely statistical way in the crystal, the probability of having a defect X in the proximity of the impurity is simply the proportion of crystal sites occupied by a defect X. However, for most impurities, there is a binding energy (E_X^b) between the impurity and the defects, so that the probability P_X is given by:

$$P_X = \frac{C_X}{N_S} \exp(\frac{E_X^b}{kT}) \qquad [3.27]$$

where C_X is the volume atomic concentration of defect X at temperature *T,* and N_S the volume density of network sites (5×10^{22} cm^{-3} for silicon). Simple thermodynamic considerations enable us to show that, at a temperature *T* different from the absolute zero *(0 K)*, any crystal contains a non-null equilibrium concentration of defects, given by:

$$C_X^* = N_S \exp(-\frac{\Delta G_X^F}{kT}) = N_S \exp(\frac{\Delta S_X^F}{k}) \exp(-\frac{\Delta H_X^F}{kT}) \qquad [3.28]$$

where ΔG_X^F is the free enthalpy of formation of the defect X, ΔS_X^F and ΔH_X^F being the corresponding entropy and enthalpy.

By combining the previous various relations, we then easily show that the diffusion coeficient is well described by relation [3.22] with:

$$D_0 = f\,\frac{1}{6}a^2\nu_0 \exp(\frac{\Delta S_X^F + \Delta S_X^m}{kT}) \quad [3.29]$$

and:

$$Q = \Delta H_X^F + \Delta H_X^m - E_X^b \quad [3.30]$$

By introducing the diffusion coefficient of the defect ($D_X = (1/6)a^2\nu_X$), we easily obtain the diffusivity of the substitutional impurity according to the macroscopic parameters of the defect:

$$D = f\,\frac{D_X C_X}{N_S}\exp(\frac{E_X^b}{kT}) \quad [3.31]$$

This proves that the diffusion coefficient of a substitutional impurity is directly proportional to the concentration of the point defect, which it uses as a "vehicle". This basic concept is at the base of a rigorous interpretation of the diffusion "anomalies", which we will see later on.

Interstitial impurity: in this case, the analysis is much simpler. Indeed, the point defects are not involved here and, as long as its concentration remains low (which is generally the case for impurities in silicon), the impurity has a probability equal to 1 to find a vacant interstitial site within its proximity. We then simply obtain:

$$D_0 = \frac{1}{6}a^2\nu_0 \exp(\frac{\Delta S_i^m}{kT}) \quad [3.32]$$

and:

$$Q = \Delta H_i^m \quad [3.33]$$

ΔS_i^m and ΔH_i^m being the migration entropy and enthalpy of the interstitial impurity.

3.2.3. *Semiconductor specificities*

The previous sections describe the diffusion of an impurity in a crystalline solid, independently of the type of solid considered. In this section, we discuss the specificities to be taken into account in the case of diffusion in semiconductors. These additional effects are essentially related to the modulation of the position of the Fermi level in the band gap, under the doping influence.

3.2.3.1. *Doping influence*

It was very quickly observed in experiments that the value of the diffusion coefficient in semiconductors is affected by the doping level of the material. A typical and perfectly interpreted example is the influence of a high dopant concentration on the self-diffusion coefficient of germanium [VAL 57]. In this particular case, an N doping increases the diffusivity, whereas a P doping decreases it. This phenomenon is easily explained by considering the specificity of the semiconductor state, regarding the structure of point defects. The semiconductor state is characterized by the existence of a forbidden band (the "gap"), i.e. an energy domain where the electronic states density is null. But this perfect forbidden band structure is closely related to the crystal lattice periodicity. Thus, any local break of this periodicity (induced for example by the presence of an impurity or a point defect) can induce a possible electronic state inside the gap, spatially restricted to the immediate proximity of the defect. We then say that the defect introduces a "deep level". The occupation of this level by an electron is obviously governed by the Fermi-Dirac statistics and thus depends on the position of the Fermi level in the gap (itself controlled by doping). For example, we know that the vacancy in germanium introduces an acceptor level into the gap, so that the vacancy can be neutral (V^0) or negatively charged (V^-), depending on whether the level is empty or occupied. From Fermi-Dirac statistics, we easily obtain the relative concentrations of these states of charge, according to doping:

$$[V^-] = [V^0]\exp(-\frac{E_A - E_F}{kT}) = \delta^-(\frac{n}{n_i})[V^0] \quad [3.34]$$

with:

$$\delta^- = \exp\left(-\frac{E_A - E_{Fi}}{kT}\right) \quad [3.35]$$

In the previous equations, E_A is the position of the deep level introduced by the vacancy, E_F is the Fermi level, and n is the density of free electrons in the conduction band. The index i indicates the intrinsic values. The total vacancy concentration is thus given by:

$$[V]^{tot} = [V^0] + [V^-] = [V^0]\left\{1 + \delta^-\left(\frac{n}{n_i}\right)\right\} \quad [3.36]$$

As the thermal equilibrium imposes the concentration value of *neutral* vacancies, it results from the previous relation, that the total vacancies concentration strongly depends on the doping level. The defects are used as diffusion "vehicle", whatever their state of charge. Therefore, the diffusion coefficient depends on the doping level.

We can now easily interpret the results concerning the germanium self-diffusion: a strong doping with donors increases the value of n, and thus the total vacancy concentration, while a doping with acceptors decreases n, inducing a decrease of the total vacancies concentration. This is in perfect agreement with the experimentally observed fluctuations [VAL 57].

In the case of silicon, the situation is slightly different. Indeed, the silicon point defects (vacancies or self-interstitials) have an amphoteric behavior: they introduce deep acceptor and donor levels, so that the total defect concentration is increased by the two doping types, donor or acceptor. Later on we will see that these effects must be accounted for to explain the diffusion profiles of the silicon dopants.

3.2.3.2. *Influence of the induced electric field*

Another important effect must imperatively be taken into account in the semiconductors, in the presence of a dopant gradient. Let us consider for example a non-polarized silicon sample containing phosphorus (donor element) in variable concentration. This concentration gradient of donors necessarily implies the existence of a concentration gradient of free electrons. In order to maintain the equilibrium (absence of electric current), the electron flux, induced by this concentration gradient is balanced by the establishment of a space charge built-in electric field, given by the conventional formula:

$$E = -\frac{kT}{q}\frac{1}{n}\frac{\partial n}{\partial x} \qquad [3.37]$$

Let us now consider the dopant flux. At high temperature (necessary condition so that the dopant is mobile), the donor atoms are entirely ionized, so that phosphorus is a positively charged ion (P^+). In the presence of an electric field, it is thus necessary to use the generalized form of the flux, given by equation [3.6]. By using equation [3.37], we then obtain:

$$J_P = -D\left\{\frac{\partial [P^+]}{\partial x} + \frac{[P^+]}{n}\frac{\partial n}{\partial x}\right\} \qquad [3.38]$$

where the Einstein relation ($D/\mu = kT/q$) is used to express the mobility as a function of D. In addition, in the absence of external polarization, the free electrons concentration is locally connected to the phosphorus concentration by the equation of electrical neutrality:

$$n = \frac{n_i^2}{n} + [P^+] \qquad [3.39]$$

Thus, the gradient of n can be expressed according to the gradient of $[P^+]$, and it is then possible to put the flux equation in the form of Fick's first law, with an effective diffusivity depending on local doping:

$$J_P = -D^{eff} \frac{\partial [P^+]}{\partial x} = -hD \frac{\partial [P^+]}{\partial x} \qquad [3.40]$$

with:

$$h = \frac{2n^2}{n^2 + n_i^2} \qquad [3.41]$$

The corrective term h is the acceleration factor related to the electric field induced by the dopant gradient, and its value varies from 1 at low doping ($n = n_i$) to 2 at the most at very high doping ($n >> n_i$).

In the case of an acceptor gradient (for example boron in Si), we obtain a completely similar relation, by replacing the free electrons concentration with the holes concentration (p):

$$h = \frac{2p^2}{p^2 + n_i^2} \qquad [3.42]$$

Lastly, the two effects described above (influence of doping and built-in electric field) are sensitive only for dopant concentrations higher than the density of intrinsic carriers at the diffusion temperature (about several 10^{18} to 10^{19} cm^{-3} towards 800 – 1,100°C in Si). For lower concentrations, the concentration of the free carriers remains equal to the density of intrinsic carriers and the corrective terms are equal to 1.

3.3. Dopant diffusion in single-crystal silicon

3.3.1. *Predeposition methods*

The predeposition term designates a technological step during which a controlled quantity of dopant is locally introduced into the substrate. This generally is a dopant of the type opposite to the substrate dopant. This leads to the realization of various P/N junctions, necessary to the device architecture. These steps, crucial for the optimization of the device electrical characteristics, require a reproducibility and a perfect control of the introduced dopant doses, and of their spatial distribution. That is essentially why, over several decades, ion implantation has become the standard technique to carry out predeposition of dopants in microelectronics. This specific technique is the subject of a complete chapter in this book and will not be discussed here.

Nevertheless, some specific devices, such as photovoltaic cells for example, do not support an ion implantation step, which would deteriorate the life time of minority carriers too much and which is also more expensive. For this type of devices, we continue to use the same techniques as those initially used in microelectronics, namely diffused predepositions. We distinguish two types of predepositions, according to whether the dopant transportation towards the silicon wafer occurs in vapor phase or simply in solid phase.

3.3.1.1. *Predepositionin vapor phase*

The principle of the predeposition technique in vapor phase, schematized in Figure 3.10, is relatively simple to implement. It simultaneously treats a batch of several wafers, which is an unquestionable economical interest. The wafers to be doped are put inside a diffusion furnace, made up of a quartz tube surrounded by heating elements, in which we maintain a certain partial pressure of the doping gas. Let us note that a neutral carrying gas (nitrogen) permanently circulates in the tube, in order to avoid any pollution by elements coming from the surrounding air. This nitrogen must be very pure, in order not to pollute the furnace. To decrease even more contamination risks, diffusion tubes with double inner walls are often used. An ultrapure gas flow (in general nitrogen) circulates between those inner walls.

The simplest dopant sources to implement are gases such as arsine (AsH_3) or phosphine (PH_3) for N doping (with donors), or diborane (B_2H_6) for P doping (with acceptors). However, these gases are dangerous or even lethal at very low concentration. In general, we thus prefer using liquid sources, such as $POCl_3$ or BBr_3. They are liquid at room temperature, but easily vaporized to be introduced into the diffusion furnace.

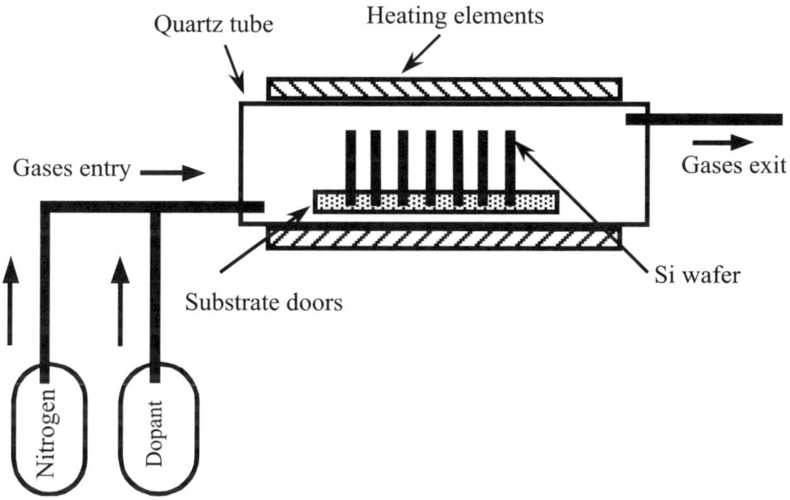

Figure 3.10. *Schematic representation of a diffusion furnace for predepositionin the gaseous phase*

We can also use solid sources, such as certain glasses, containing dopants, like boron nitride or a glass doped with phosphorus. These sources are in the form of slices and are in general introduced into the furnace, alternately with the wafers to be doped. During the high temperature thermal process, the dopant contained in these source materials evaporates and thus maintains a partial pressure of dopants in the proximity of the silicon wafers.

Whatever the type of source used, during the high temperature predeposition step, part of the doping atoms, transported by the gaseous phase, enters from the surface in the silicon volume. After the sources withdrawal, once the desired total dose is reached, the doping elements are further diffused towards the bulk of the substrate, by an additional thermal step (at high temperature) to obtain the desired profile.

3.3.1.2. *Deposition in the solid phase*

In this technique, the dopant source, consisting of a thin film of a material containing the adequate impurity, is directly deposited on the surface of the silicon wafer.

The most convenient is to use "doping pastes" in the form of sol-gel solutions of strongly doped glasses (borosilicates, phosphosilicates), easily deposited by spin-coating, with the same principle as the deposition of photosensitive resists for lithography.

After stoving to evaporate the solvent, the wafers covered with doped glass are introduced into a diffusion furnace to undergo the high temperature thermal treatment, inducing the dopant diffusion from the film deposited towards the bulk of silicon. The use of this type of source also allows us to use rapid thermal annealing techniques, similar to those used to activate the dopants after the ion implantation step.

3.3.2. *Main experimental observations*

3.3.2.1. *Diffused predepositions in neutral atmosphere*

One of the striking facts concerning dopant diffusion in silicon is the tremendous influence of the dopant concentration on the shape of the diffusion profiles. This influence is easily highlighted on the experimental profiles corresponding to diffused predepositions. Let us recall that this technique was used for a long time to introduce dopants into silicon. Although now completely given up in favor of ion implantation (at least for ULSI microelectronic technologies), the corresponding results established the majority of our knowledge on fundamental mechanisms governing dopant diffusion.

This is why we recall these results here. Since the diffused predeposition step corresponds to diffusion with a constant surface concentration, Fick's laws indicate that the concentration profile should simply be a complementary error function, if the diffusivity is a constant.

In reality, the experimental results are much more complex.

Phosphorus: for example Figure 3.11a shows the various profiles obtained after a one hour predeposition at 1,000°C for various surface concentrations [WIL 77]. For surface concentrations lower than the density of intrinsic carriers, the profile is in accordance with the one expected for a simple diffusion mechanism (complementary error function). On the other hand, when the surface concentration increases, strong nonlinearities are highlighted. Firstly, (surface concentrations of several 10^{19} cm^{-3}) there is a local increase in the diffusion coefficient at high concentrations, leading to "rectangular" profiles in semi-logarithmic scale. Then, gradually, a slight inflexion appears on the profiles, with a zone with a strong gradient on the surface, followed by a curve "tail" corresponding to a large increase in the effective diffusion coefficient. When the surface concentration exceeds 2 to 3 x 10^{20} cm^{-3}, an additional phenomenon appears: a concentration plateau, corresponding to a very fast diffusion, visible in the surface zone, is followed by the high gradient zone and the curve tail, as previously. The variation of the effective diffusion coefficient, according to the local phosphorus concentration, obtained by the Boltzmann-Matano method, is shown in Figure 3.11b, for these highest surface concentrations. This figure highlights the strong enhancement of the diffusion in the bulk compared to the diffusivity extracted from conditions at low surface concentrations (intrinsic diffusion), as well as the strong acceleration in the surface region. These two extremes are separated by a zone, where we observe a local decrease in the diffusivity, corresponding to the inflexion area on the experimental profile.

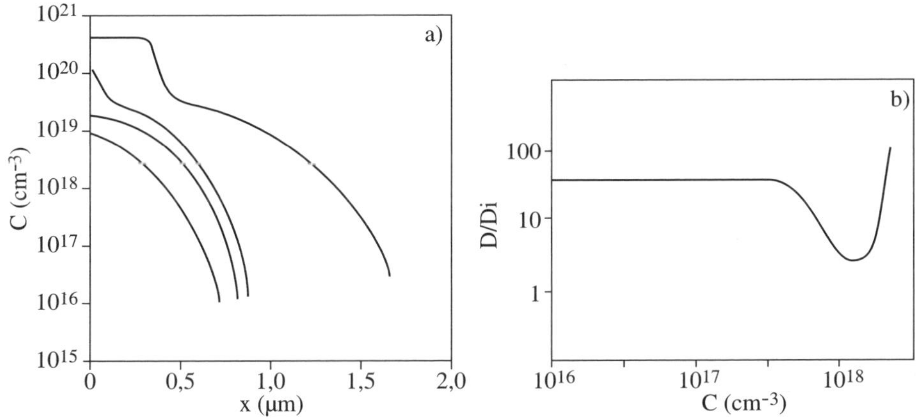

Figure 3.11. *a) Profiles of phosphorus predeposition at 1000°C. b) Boltzmann-Matano analysis for very high surface concentrations*

Arsenic: as in the case of phosphorus, arsenic predeposition profiles at low surface concentrations have a form correctly described by a simple complementary error function, allowing us to extract the intrinsic diffusivity of the impurity. However, when the surface concentration exceeds the intrinsic carriers density, we observe deviations compared to the ideal curve, different from those observed with phosphorus. Indeed, diffusions at temperatures of about 1,000°C lead to profiles of nearly rectangular shape (in semi-logarithmic scale), such as Figure 3.12a (according to [FAI 73]).

Moreover, as we can notice in the figure, there is a significant difference between the total arsenic concentration and the electrically active concentration, measured by spreading resistance. The result of the Boltzmann-Matano analysis of this experimental profile is given in Figure 3.12b, and shows a clear decrease of the effective diffusion coefficient at high arsenic concentrations, preceded by a domain where the diffusivity is sensibly proportional to the local arsenic concentration, the effective diffusion coefficient tending to its intrinsic value at lower arsenic concentrations.

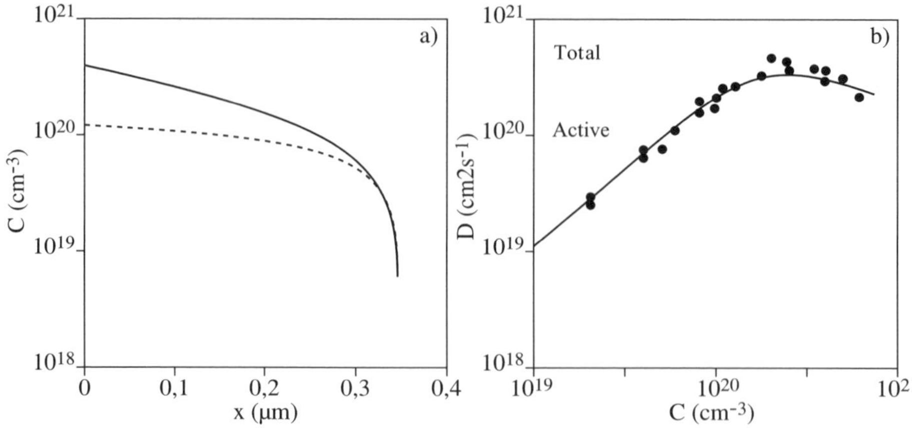

Figure 3.12. *a) Typical profile of arsenic predeposition at high concentration (1,000°C, 1 hr) and b) corresponding Boltzmann-Matano analysis*

Boron: as for the other dopants, boron predeposition profiles are in conformity with the complementary error function for the intrinsic conditions, whereas nonlinearities exist at higher concentrations. In the case of boron, these nonlinearities appear with sensibly rectangular profiles, such as those of Figure 3.13a (according to [QUO 72]). The corresponding Boltzmann-Matano analysis (see Figure 3.13b) reveals a slight increase of the diffusion coefficient in the bulk, compared to the intrinsic values, and a variation sensibly proportional to the boron concentration at higher concentrations.

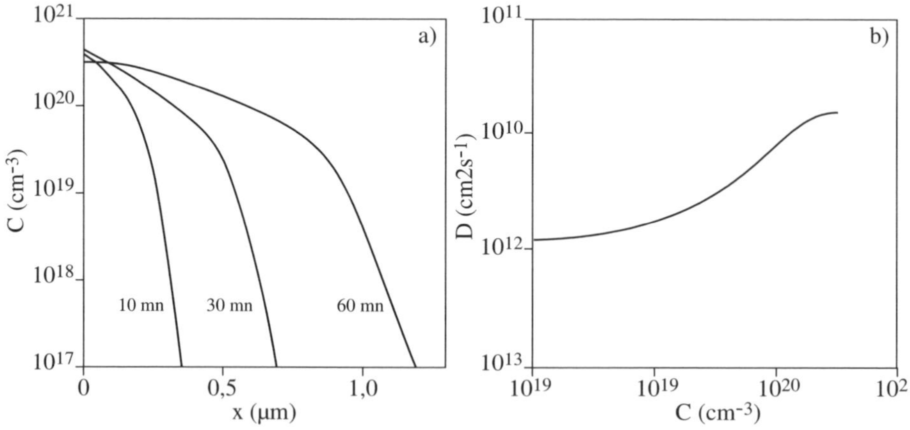

Figure 3.13. *(a) Typical profile of boron predeposition at high surface concentration at 1000°C and (b) corresponding Boltzmann-Matano analysis for the 60 min profile*

3.3.2.2. *Couplings between impurities*

Interesting pieces of information concerning the mechanisms involved during the dopant diffusion, are obtained when we observe the coupling effects between impurities. As an example, we will consider the processes sequence necessary to manufacture a bipolar “planar” N-P-N transistor. An initial localized diffusion of the acceptor (boron for example) is carried out in an N-type substrate, to form the basis of the transistor. The emitter is then formed by localized diffusion, through a smaller window, of a donor (phosphorus or arsenic) in high concentration. During the second step, there is of course a simultaneous diffusion of the donor and the acceptor. The interactions taking place during these sequential diffusions can separate into two distinct effects, however occurring at the same time.

The first of these effects is caused by the electric field, induced by the donor diffusion, in accordance with what is explained in section 3.2.3.2, and results in the appearance of a dip in the boron profile close to the junction, as illustrated in Figure 3.14 (according to [BLA 72]). To understand the formation of this dip, it is sufficient to realize that the electric field induced by the arsenic gradient is positively directed, and it thus induces a negative contribution to the total flux of the boron atoms, which are negatively charged (see equation [3.6]). The effect can be sufficiently strong to efficiently drift the boron atoms towards the surface. On the other hand, beyond the junction, the concentrations are relatively low, so that the carrier concentration has its intrinsic value and there is no more induced field. Boron atoms thus diffuse exclusively under the effect of their gradient, therefore towards the bulk. These opposite effects in surface and bulk necessarily induce a depletion in boron concentration close to the junction.

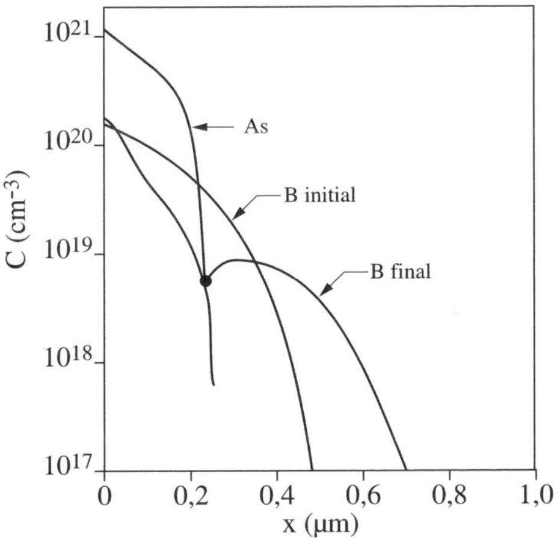

Figure 3.14. *Coupled boron-arsenic diffusion*

The second effect, illustrated in Figure 3.15a, is known as the "emitter push effect" and is essentially observable for N-P-N structures with phosphorus emitters: the high concentration phosphorus diffusion used for the fabrication of the emitter is accompanied by a more important penetration of the base-collector junction in the zones undergoing phosphorus diffusion at the surface. It is now established (see for example [WIL 77]) that this pushing effect has the same physical origin as the appearance of the tails in the phosphorus profiles. These effects are associated with a local increase of the impurity diffusivities, due to a strong supersaturation of point defects, related to P diffusion. This interpretation in terms of point defects is comforted by the studies of redistribution enhancement of a buried boron layer, caused by a high concentration P diffusion at the surface (see Figure 3.15b).

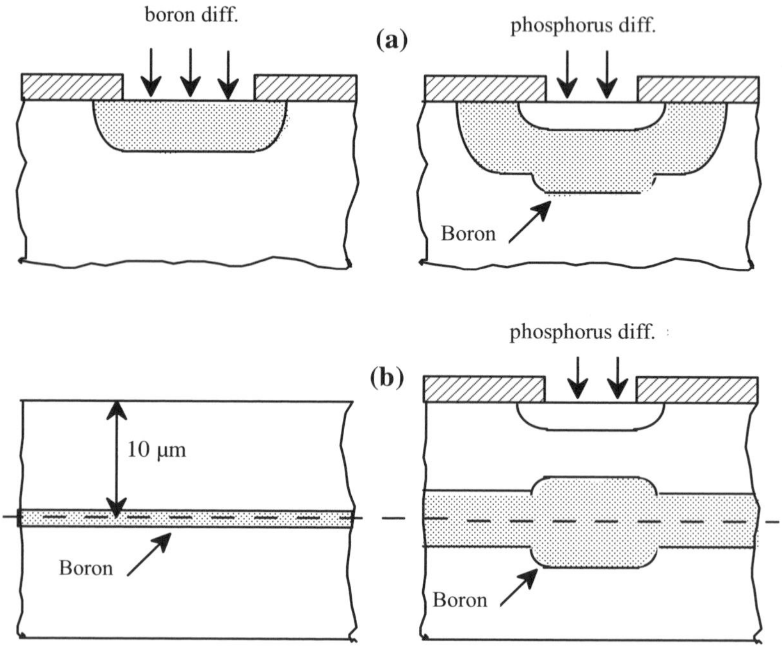

Figure 3.15. *Influence of a localized phosphorus diffusion on boron redistribution: (a) emitter push effect; (b) buried layer extension*

This type of experiment makes it evident that acceleration occurs on long distances (several tens of microns), without any direct interaction between the boron layer and the surface phosphorus, and that the order of magnitude of this acceleration is the same as the one observed in the tails of the P profile. It is thus now perfectly accepted that phosphorus diffusion in Si, at high concentration, generates non-equilibrium point defects. Since effects of a similar nature, although of lower intensity, are observed with arsenic emitters, or in boron for P-N-P transistors, it is now considered that the mechanism inducing these defect supersaturations is general for all the common dopants.

3.3.2.3. *The oxidation influence*

Dopant diffusion in silicon is strongly affected by the atmosphere in which annealing is carried out. In particular, an oxidizing annealing generally involves a diffusion acceleration (OED, oxidation enhanced diffusion) of the common dopants (B, P, As) and a slowing down (ORD, oxidation retarded diffusion) of the antimony. A retarded diffusion is also observed for phosphorus, arsenic and boron under some experimental conditions: Si (111) oxidation at very high temperature during very long times [HIL 81]. The main results are as follows:

– for each dopant, the effect depends on the substrate direction, OED being stronger for the faces (111) than (100);

– OED is more important for oxidation in a wet atmosphere than for a dry oxygen atmosphere, directly related to the oxidation rate increase;

– the addition of chlorinated compounds diminishes OED;

– the relative importance of OED increases when the temperature decreases;

– finally, OED (or ORD) is a long range effect, associated with a supersaturation (or undersaturation) of point defects.

In fact, Hu [HU 74] highlighted that all these pieces of information are in perfect correlation with those concerning the stacking faults behavior (growth or decrease) induced by oxidation. We can thus deduce that the two phenomena are related to the same disturbances on point defects. As we know, thanks to electron microscopy, stacking faults induced by oxidation are of "extrinsic" nature (due to a self-interstitials excess). Therefore, oxidation injects self-interstitials in silicon and the diffusion of boron, phosphorus and arsenic is assisted, at least partially, by self-interstitials (mechanism schematized in Figure 3.7). Moreover, due to the recombination between the defects, a self-interstitial supersaturation necessarily induces a vacancy undersaturation. This explains the antimony diffusion retardation. Indeed, its diffusion, because of its size, is almost exclusively vacancy-assisted. However, as the diffusion of the other dopants can also be retarded by oxidation, it is now accepted that dopant diffusion is assisted by the two types of point defects, vacancies and self-interstitials. The appearance of OED or ORD simply depends on the relative importance of each mechanism for the considered dopant and of the self-interstitial supersaturation level (or the vacancy undersaturation level) peculiar to each experimental condition.

3.3.3. *Modeling*

3.3.3.1. *"Normal" diffusion*

As long as the experimental characteristics observed on the diffusion profiles do not involved non-equilibrium defects, it is possible to simply model the dopant diffusion, using the mechanisms described above. In particular, profiles with concentrations lower than the intrinsic carriers density are perfectly described thanks to Fick's second law, with a constant diffusion coefficient. This coefficient is called the intrinsic diffusivity and is usually noted D_i. The variation of these intrinsic coefficients with the temperature is shown in Figure 3.4, and the corresponding values of activation energies and of the pre-exponential terms are given in Table 3.1 for common dopants. These values represent the average of various experimental sources ([MAT 84] and included references).

Dopant	D_0 (cm^{2}s^{-1})	Q (eV)
P	3.19	3.66
As	31.65	4.14
B	2.64	3.60

Table 3.1. *Intrinsic diffusion in Si*

The effects related to the modulation of the Fermi level position (variation of *D* with the doping, built-in electric field; see section 3.2.3) are sufficient to report what occurs with higher concentrations, as long as there is no injection of non-equilibrium defects. This simplified modeling is thus rather well adapted for arsenic and boron, in an extensive range of experimental conditions. We then reach expressions of the effective diffusivity of the form:

$$D_{As} = h\left\{D^0 + D^-\left(\frac{n}{n_i}\right)\right\} \qquad [3.43]$$

$$D_B = h\left\{D^0 + D^+\left(\frac{p}{n_i}\right)\right\} \qquad [3.44]$$

where h is the acceleration factor, due to the induced electric field (equations [3.41] or [3.42]) and D^0, D^-, D^+ are the individual contributions of the various defects charge states. In intrinsic conditions ($n=p=n_i$, and $h=1$), these equations reduce to:

for As: $D_i = D^0 + D^-$ [3.45]

for B: $D_i = D^0 + D^+$ [3.46]

So that the effective diffusivity can finally be put in the form usually used:

$$D_{As} = h\frac{1+\beta^-(n/n_i)}{1+\beta^-}D_i \tag{3.47}$$

and:

$$D_B = h\frac{1+\beta^+(p/n_i)}{1+\beta^+}D_i \tag{3.48}$$

where the parameters β represent the relative weight of the contribution to the diffusion of the charged defects, compared to neutral ones.

Lastly, to be more exhaustive, let us indicate that to simulate the decrease of the effective diffusion coefficient at higher arsenic concentrations, associated with the existence of a considerable proportion of electrically inactive arsenic, it is necessary to admit the existence of complex, neutral and immobile clusters, including several arsenic atoms. This results in an additional term in the expression of the effective diffusion coefficient. For example, if we consider the As_2V complex, often considered, associating two arsenic atoms to a vacancy [FAI 73], we obtain:

$$D_{As} = h\frac{1+\beta^-(n/n_i)}{1+\beta^-}\frac{D_i}{1+KC_V[As^+]} \tag{3.49}$$

where K is the mass action law constant of the formation reaction of the immobile complex.

To be able to describe the oxidation influence, it is moreover necessary to introduce disturbances on the populations of point defects. By taking into account the fact that the diffusion coefficients are proportional to the concentration of the point defects used as vehicle for the diffusion, for impurities such as dopants, the diffusion of which is assisted by the two types of defects, we can express the effective diffusion coefficient in oxidizing condition (or more generally in the presence of a *homogeneous* disturbance of populations of point defects), and for low dopant concentrations (intrinsic conditions), in the form:

$$D_i^{ox} = f_I D_i \frac{C_I}{C_I^*} + (1-f_I)D_i \frac{C_V}{C_V^*} \tag{3.50}$$

In the equation above C_I and C_V are the self-interstitials and vacancy concentrations. The asterisks indicates the values of thermal equilibrium, and f_I represents the relative weight of the mechanism assisted by self-interstitials for the intrinsic diffusion ($0 < f_I < 1$). The values of f_I for each dopant, and thus the nature of the defect preferentially assisting their diffusion, was, for a long time, the object of controversy. However, the results obtained by the group of the university of Stanford (United States) [URA 99] now unambiguously conclude that, at about 1,000°C, the boron and phosphorus diffusion is almost exclusively assisted by self-interstitials ($f_I > 0.97$) and that the antimony diffusion is almost purely vacancy-assisted ($f_I < 0.02$). The arsenic has an intermediate situation, with a significant contribution of the two types of defects ($0.35 < f_I < 0.55$).

We could believe that, to model the oxidation influence in extrinsic conditions (high dopant concentrations), it is enough to replace in equations [3.47] or [3.48], the intrinsic diffusivity, D_i with the value given by equation [3.50], including all the charge states in the point defects concentrations. Although this approach has sometimes been used, it makes the implicit assumption that the relative weight of the vacancy mechanisms or the mechanisms assisted by self-interstitials (i.e. the value of f_I) is not affected by doping. This is not *a priori* justified, since the deep levels introduced by each defect are not the same ones. Experimental results also seem to indicate that the relative amplitude of OED depends on the doping level [JOH 93, MIY 85].

We see there an example of limitation of the description of dopant diffusion in terms of effective diffusivity. Despite the increasing complexity introduced into the expression of the effective diffusivity, this type of approach does not entirely convey all the couplings existing between the point defects and the dopants. It thus does not give a rigorous interpretation of the various diffusion "anomalies". A proper modeling goes through the explicit description of the microscopic mechanisms involved. This is the subject of the following section.

3.3.3.2. *"Anomalous" diffusion: diffusion by pairs*

We saw in section 3.3.1 that an essential feature of dopant diffusion, particularly visible in the phosphorus case, is to cause the creation of non-equilibrium point defects. The most appropriate mechanism to explain it was initially proposed by Yoshida [YOS 74] and Fair and Tsai [FAI 77], in the context of the vacancy mechanism. We now know that the dominant mechanism is assisted by self-interstitials [URA 99], but the basic concept remains the same. We will thus present it below, independently of the nature of the defect. In fact, a full description can be made by accounting for the two types of defects [MAT 84].

The fundamental idea of this type of model is that, because of a very high binding energy between the dopant atom and the point defects, it is always the same defect X (vacancy or self-interstitial), which remains in the proximity of a given dopant atom (A), to ensure its successive hops. We can then consider that we have a "pair" dopant-defect (AX), diffusing as such in the matrix and that the dopant diffusion is only due to the migration of these AX pairs. Let us note, for example, that such phosphorus-vacancy (PV) pairs are experimentally observed in irradiated silicon and are known under the name of "E centers" [WAT 64].

If we assume that these pairs are in local equilibrium with the isolated substitutional dopants and with the free defects, their concentration is proportional to that of the dopant. A dopant's gradient thus necessarily involves a gradient of dopant-defect pairs (even for a constant initial concentration of free defects) and thus a diffusion of these pairs under the effect of their gradient. The pairs flux towards the bulk of the sample causes then a local increase in the pairs concentration, in zones with low initial free dopant concentration, and thus an unbalance is established between the various species. In order to restore the equilibrium, part of the pairs must dissociate themselves to release the substitutional dopant atoms. At the same time, this dissociation involves a release of the point defect constituent of the pair, and thus a generation of free non-equilibrium defects. When dopant concentration is high, the concentration of dissociating AX pairs is much more important than the thermal equilibrium concentration of free defects. There is then creation of a strong local free defects supersaturation.

As silicon single crystals are of very good crystalline quality, there are almost no annihilation sites for these non-equilibrium defects, which can thus diffuse far in the crystal volume. This explains the wide range of the diffusion acceleration, associated with the presence of these defects in supersaturation. However, a large proportion of the released defects will diffuse towards the sample surface, where they will be annihilated. At first order, the surface can be regarded as a perfect "sink", maintaining at this place the thermal equilibrium of the free defects concentrations. The supersaturation in the bulk is thus accompanied by an important flux of point defects toward the surface. Because of the strong coupling between the defects and the dopant atoms, via AX pairs, this defects flux drags with it a considerable proportion of dopant atoms, thus creating a negative contribution to the total impurities flux. This explains the appearance of the inflexion in the phosphorus profiles with the strong gradient zone on the surface (see Figure 3.11a) [MAT 82a].

To better understand what occurs, it is for once simpler to use the mathematical transcription of the model. To make it simpler, we neglect here any influence of the charge states and other built-in electric field effects, to focus on the specificity of the diffusion by pair. We thus consider an impurity A forming stable and mobile pairs, by reaction with a point defect X, following the reaction:

$$A + X \leftrightarrow AX \tag{3.51}$$

The pairs concentration, in the hypothesis of local equilibrium, is then related to those of the substitutional dopants and of the free defects by:

$$[AX] = K[A][X] \tag{3.52}$$

If we make the assumption that only the pairs are mobile, the total dopant flux is equal to the pairs flux, that is to say:

$$J_A = J_{AX} = -D_{AX}\frac{\partial[AX]}{\partial x} \tag{3.53}$$

where D_{AX} is the diffusivity of the pair AX. By derivation of the relation [3.52], we thus obtain:

$$J_A = -D_{AX}K[X]\frac{\partial[A]}{\partial x} - D_{AX}K[A]\frac{\partial[X]}{\partial x} \tag{3.54}$$

This equation highlights a drift term that is very similar to the one created by the electric field, when we consider the influence of the charge states (i.e. the formation of dopant-electrons "pairs"), the concentration gradient of defects playing here a role equivalent to the carriers gradient (electric field). A factoring of the previous relation put it into the form of Fick's first law, connecting the A flux to its concentration gradient, with an effective diffusivity given by:

$$D_A^{eff} = D_{AX}K[X]\left\{1+\frac{A}{X}\frac{\partial[X]/\partial x}{\partial[A]/\partial x}\right\} \tag{3.55}$$

The term before the bracket is the effective diffusion coefficient accessible by conventional models. We find yet again the proportionality to the concentration of free defects, which correctly accounts for the diffusion acceleration (or of the slowing down) in the presence of a defect supersaturation (or undersaturation). On the other hand, relation [3.55] clearly highlights an *additional coupling via gradients of point defects*, which only gives a proper description of the experimental observations. In particular, for phosphorus predepositions, it is clear that because of the recombination of the defects at the surface of the sample, the defect and dopant gradients have opposite signs in the surface region. The corresponding term in the brackets of relation [3.55] thus induces a *negative* contribution to the effective diffusion coefficient, thus explaining the inflexion appearance on the profiles. We will see further that accounting for this coupling by the defects gradients (point defects "wind" effect) is the only way to explain diffusion "anomalies" having significant consequences on the device electric behavior.

A proper modeling of most of the experimental observations concerning the dopant predepositions is feasible from the concept of pair diffusion, provided that we take into account the two types of pairs, dopant-vacancy and dopant-self-interstitial, and the charge states of the various species concerned, as well the built-in electric field. Corresponding simulations are then done by the simultaneous resolution of the continuity equations, for the dopant and for the two types of point defects, coupled between them by terms of generation/recombination, corresponding to the formation/dissociation of the pairs and to the recombination between vacancies and self-interstitials. The comprehensive presentation of these simulations will not be discussed in this chapter, and interested readers should to refer to [MAT 84].

However, the model presented above is not convenient at all for modeling the plateau appearance at very high phosphorus concentrations, corresponding to a very strong diffusion acceleration. On the contrary, within the framework of the model of pair diffusion alone, if the dopant concentration is increased, a higher coupling through the defect gradient is expected, and thus the more the inflexion should be marked [MAT 82a]. Another diffusion mechanism must thus occur for concentrations higher than 2-3 x 10^{20} cm^{-3}. This is due to the fact that, in this concentration range, the phosphorus-vacancy (PV) pairs can no longer be regarded as isolated entities during their entire diffusion cycle [MAT 82b].

To understand the diffusion mechanism of these PV pairs, let us consider Figure 3.16a, representing such a pair in the silicon network: a vacancy is located at (1) in the immediate proximity of a substitutional phosphorus atom. As long as the vacancy and the phosphorus atom alternatively exchange their sites, there is no macroscopic phosphorus displacement. So that such a displacement occurs, it is necessary for the vacancy to reach site (1'), while successively passing by sites (2), (3) and (2'), and to finally exchange with the phosphorus atom. This sequence can then be repeated on one of the other hexagons surrounding the phosphorus, thus ensuring the macroscopic PV pair diffusion.

We thus see that, to be able to diffuse, the PV pair must partially dissociate: the saddle point corresponds to the vacancy in the third neighbor position from the substitutional phosphorus. If the phosphorus concentration is sufficiently low, the residual binding energy ensures the consistency of the PV pair and the vacancy remains linked to the considered phosphorus atom.

Let us now assume that the concentration of substitutional atoms is rather high, so that we have the situation schematized in Figure 3.16b (for simplicity reasons, we drew a planar 2D hexagonal network, but the reasoning remains true for the actual silicon network): a PV pair is formed by the association of a vacancy V and of an impurity atom P_1, but a second atom P_2 (or P_2') is located as the fifth neighbor of P_1. In this case, the site (3), the third neighbor of P_1, is also in the position of second neighbor (2') of P_2.

The system thus has the same energy whether the vacancy is in this site or in site (2). This vacancy can thus directly diffuse from P_1 to P_2 with a potential barrier lower than the one existing for the diffusion of an isolated PV pair.

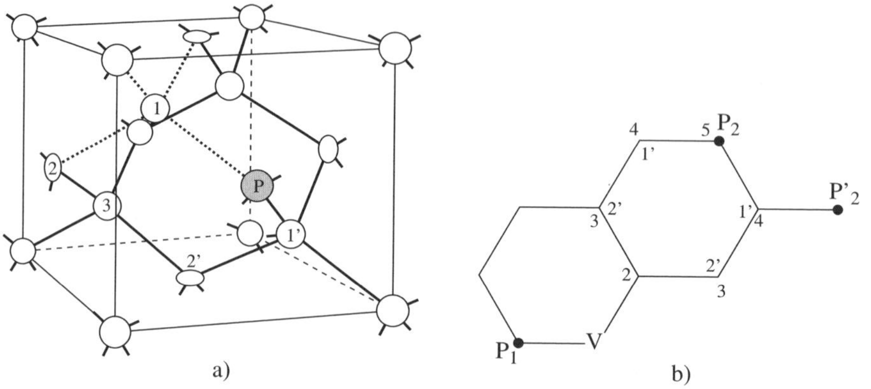

Figure 3.16. *PV pair diffusion: (a) low phosphorus concentrations; (b) very high phosphorus concentrations*

Let us consider the "network" of the phosphorus atoms in the silicon matrix. Two phosphorus atoms are considered, connected by an easy vacancy diffusion path, if they are in the position of fifth neighbors or closer in the silicon network. By connecting the substitutional atoms that way, we produce clusters in which the vacancies can move with a high mobility. Although this network of fifth neighbors is not steady, it is obvious that, for sufficiently high impurity concentrations, large clusters can be obtained. We expect percolation phenomena [GEN 76]: for concentrations exceeding a C^* threshold, there is an "infinite" cluster of phosphorus atoms in position of fifth neighbors from each other through the silicon network. Inside this cluster, vacancies can freely move from one impurity atom to another and this accelerated vacancies diffusion induces a diffusion acceleration of the dopant, by a classical vacancy diffusion mechanism (see Figure 3.6).

Simple geometrical estimates show that the percolation threshold on the "network" of the fifth neighbors is about several 10^{20} cm^{-3}. We thus see that the plateau appearing on phosphorus predeposition profiles is easily explained by this percolation phenomenon. Easily observable in the case of phosphorus, because of the characteristic profiles shape, this phenomenon is completely general and occurs for other dopants, if the concentration exceeds the concentration threshold. In the arsenic case, it is observable with more difficulty because of the formation of the immobile complexes, so that experimental profiles are the result of a compromise between the two mechanisms [MAT 83]. For boron, the percolation phenomenon is not usually observable for standard diffusion conditions, because the solubility limit of this impurity is lower than the percolation threshold. On the other hand, these phenomena can probably play a role for non-equilibrium conditions, corresponding to the introduction of very strong doses by ion implantation, as is increasingly the case in modern fabrication technologies of integrated circuits. Theoretical calculations made with the Monte Carlo method confirm the validity of the percolation model, by highlighting a co-operative behavior, creating a very important increase in the vacancy component for dopant concentrations exceeding several 10^{20} cm^{-3} [DUN 97, FED 94]. In addition, we should realize that the diffusion acceleration is also applied to any other substitutional species present in the material, even in low concentration, as soon as the percolation is induced by a high phosphorus concentration, or another substitutional impurity having an intense interaction with vacancies, as it was experimentally shown [LAR 93].

3.4. Examples of associated engineering problems

3.4.1. *Redistribution of the implanted dopants: transient enhanced diffusion*

One of the major problems technologists are confronted with, is the realization of ultra-thin junctions, of depths of several tens of nanometers for deep-submicron devices (of gate length lower than 50 nm). The problem is particularly important for P/N junctions, carried out by boron ion implantation. Indeed, the necessary annealing for the dopant activation is accompanied by an abnormally high redistribution of the dopant profile. This phenomenon is illustrated in Figure 3.17 (according to [SED 88]), showing the evolution during annealing at 950°C of boron profiles implanted in Si. This figure clearly highlights an important profile redistribution after a 10 second annealing. Such redistribution corresponds to an effective diffusivity exceeding the equilibrium value by several orders of magnitude. On the other hand, we see that the profile remains unchanged for subsequent annealing (the profile at 30 s is almost identical to the profile at 10 s), until the moment when, for definitely more important durations, we find again the normal thermal diffusion process (profile at 20 minutes). These results, confirmed by many other studies, show that this diffusion acceleration is a transient phenomenon. The transient time scale depends on the annealing temperature, with an activation energy close to the self-diffusion temperature (approximately 4.5 - 5 eV) [MIC 86].

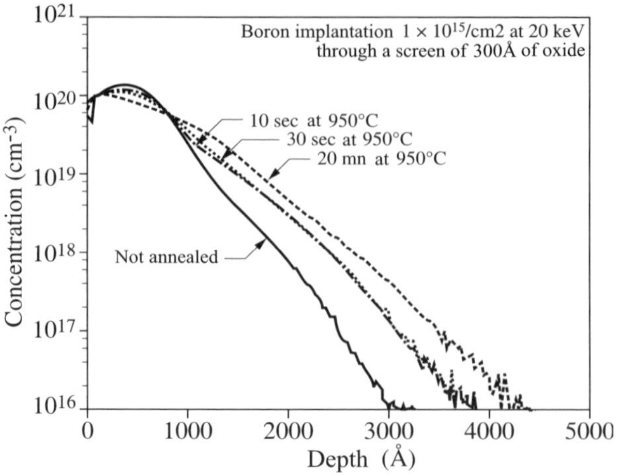

Figure 3.17. *Transient enhanced diffusion of implanted boron (from [SED 88])*

Because of its tremendous technological impact, this phenomenon, known as TED (transient enhanced diffusion), has been studied by many authors. It is now established that this accelerated diffusion is associated with the self-interstitials, created by ion implantation, which survived the direct recombination with the vacancies. In particular, many authors noted a correlation between the transient diffusion and a certain type of interstitial clusters, the "{113}" defects, whose dissolution during annealing seemed to be the self-interstitials source inducing transient diffusion [STO 97]. However, transient diffusion phenomena have also been observed in the absence of {113} defects [ZHA 95], so that the phenomenon is obviously more complex. Meticulous experimental observations of the evolution of the implantation defects, by high-resolution electron microscopy, established that transient diffusion is the result of the self-interstitials supersaturation, created by implantation, but that there is then a competition between the defects out-diffusion towards the surface where they are annihilated, their couplings with boron, the nucleation and then the interstitial clusters evolution. More precisely, various cluster types (simple aggregates, {113}, perfect or faulted stacking faults) are in competition with one another [CLA 03]. Based on these observations, models describing the evolution of these various interstitial clusters are now available. With them, we can follow the temporal evolution of the residual supersaturation of self-interstitials, responsible for the TED phenomenon [LAM 05, ORT 01, ORT 04].

In addition to the enhanced transient diffusion, it was very quickly noted that the redistribution of the implanted boron layers gave way to another phenomenon, a strongly penalizing phenomenon: after the activation annealing, boron is not completely active, even for concentrations lower than its solubility limit. This phenomenon is generally explained by the formation of mixed complexes (clusters) of boron and self-interstitials (B_nI_m), known as BIC (boron interstitials clusters) [STO 97]. *Ab initio* theoretical calculations seem to justify the existence of such complexes and give indications on their possible charge states and formation energies according to their composition (n, m) [LEN 00]. The behavior (formation/dissolution) of these BIC also plays a role in the transient diffusion by influencing the kinetics of self-interstitials evolution [BOU 03, BOU 05, COL 01, PIC 06].

From a diffusion modeling point of view, it is possible to take into account the formation/dissolution of the various interstitial clusters and of the corresponding extended defects ({113}, stacking faults), as well as the BIC, by additional terms of generation/recombination in the continuity equations of the various involved species (dopants and point defects) [ORT 01, ORT 04]. It is then possible to obtain completely satisfactory simulations of the redistribution profiles of implanted boron and of activation degree [BOU 03, BOU 05, PIC 06], but with complex numerical calculations demanding high computer resources.

3.4.2. *Engineering of ultra-thin junctions*

TED phenomena and dopant deactivation described in the previous section are very penalizing for the realization of ultra-thin junctions necessary for the fabrication of the ultimate MOS transistors, for which junction depths of several tens of nanometers only, associated with a very intense dopant activation, are essential. Because of these challenges, several technological approaches are studied to go around these problems. We present below a brief description of the main ideas considered. All these approaches have as common objective to minimize the interactions between boron and self-interstitials.

3.4.2.1. *Engineering of point defects*

An interesting approach consists of using a high-energy (from several hundred keV to several MeV) silicon implantation before boron implantation. This high-energy implantation creates in the surface area, where the boron is implanted, a zone supersaturated in vacancies, contributing to destroying self-interstitials responsible for TED [NEJ 03]. The suppression of self-interstitials also minimizes the formation of BIC, which leads to better boron activation [SMI 06].

3.4.2.2. *Fluorine co-implantation*

Several works indicate a clear decrease of the TED when fluorine is co-implanted with the dopant. This was observed for boron [DOW 98, SHA 01], but also for arsenic [LIN 93] and phosphorus [MOG 97]. Nevertheless, the related precise mechanism is not yet clearly established. Although technologically easy to implement in the case of boron (gaseous BF_3 is the standard source for implantation), this approach suffers from several negative effects. First of all, it is well-known that the presence of fluorine sensibly accelerates boron diffusion in silica, which leads to integration problems in a complete process flow, because of the increasing risks of boron penetration in the gate oxide or in the spacers. Moreover, some authors indicate that fluorine tends to deactivate boron [DRO 03, SHA 01].

3.4.2.3. *Carbon co-implantation*

Carbon being an inevitable residual impurity of Cz silicon, its behavior in silicon has been studied for a long time (see [KIM 89] for the main results). Among the many possible reactions, substitutional carbon (C_s) can be ejected from its stable site by reaction with a self-interstitial to form an interstitial carbon (C_I). This carbon interstitial, itself mobile, reacts in its turn with another substitutional carbon, to form an immobile pair C_IC_S. This behavior makes carbon an effective trap for self-interstitials and a potential candidate to reduce TED. This was experimentally demonstrated and studied in detail in the case of boron [CAG 08, PAW 06a] and phosphorus [CAG 07, CAG 08, KU 06, PAW 06b], implanted in pre-amorphized Si by germanium implantation (PAI). It results from these studies that, thanks to an optimization of the placement (implantation energy) and of the carbon dose used, it is possible to completely remove the boron and phosphorus transient diffusion, as shown in Figure 3.18 [CAG 08].

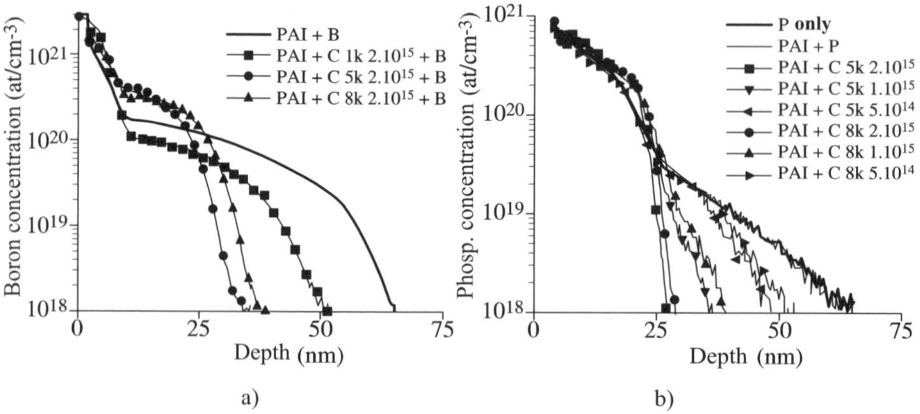

Figure 3.18. *Influence of co-implanted carbon on the form of boron (a) and phosphorus (b) profiles – according to [CAG 08]*

Figure 3.18b, relative to phosphorus, highlights the fact that by an optimized selection of the experimental conditions, selection based on the specific mechanisms of the P diffusion described previously, it is possible to completely eliminate the accelerated phosphorus diffusion tail. It is thus possible to obtain, with this impurity, profiles as abruptly as with arsenic [CAG 07, CAG 08]. This has an unquestionable advantage insofar as phosphorus being a lighter impurity, its implantation introduces less defects than arsenic, which can be crucial for ultimate technologies, using increasingly lower thermal budgets, for which the residual defect annealing is a true challenge.

3.4.3. *Reverse short channel effect*

This example is given to illustrate the technological importance of the diffusional coupling phenomena, via defect gradients for the optimization of strongly submicron devices. Let us recall that the threshold voltage of a MOS transistor controlled among other things by the concentration of the dopant present in the channel zone located right under the gate oxide, varies with the device length (lateral distance between the source and drain zones). Purely electrostatic considerations imply that this threshold voltage, all other things being equal, must decrease along with a diminution of the MOS length. This is due to the fact that the width of depletion zones induced by the drain and source junctions are no longer negligible in front of the MOS length, so that the quantity of substrate carriers to be "evacuated" by gate polarization is reduced for small size devices. On the other hand, experimental results show in most cases an increase in the threshold voltage, with a decreased gate length, before the expected natural drop. This reverse short channel effect (RSCE) can only be explained by considering disturbances of the channel doping profile induced by the fabrication step of the sources/drains [RAF 93].

Indeed, these junctions are manufactured by ion implantation of a high-dose dopant (in general As for a N-MOS), followed by an activation annealing. During this activation annealing, excess self-interstitials introduced by implantation evolve in a complex way, as described previously. This being said, the diffusion length of the free defects is relatively important, so that the interstitials supersaturation spreads over distances largely exceeding the junctions areas. There is in particular a very strong defects injection up to the channel zone, where they will induce considerable effects on the present dopant atoms (boron for an N-MOS).

Let us consider a vertical section of the device at an abscissa point z along the channel (see Figure 3.19a). In the bulk there is a self-interstitial supersaturation coming from the source and drain zones, whereas the surface maintains the thermal equilibrium concentration . There is thus a strong defects gradient between the bulk and the surface. This gradient ("defect wind") necessarily induces a boron atoms diffusion towards the surface (see section 3.3.3.2, equation [3.54]), so that even if the boron profile is initially flat, the final profile will show a boron accumulation at the surface (see Figure 3.19a). However, the level of self-interstitials supersaturation coming from source and drain zones vary with z along the channel, and the phenomenon intensity thus varies along the channel. We thus reach an inhomogeneous distribution of the boron concentration profile, as schematized on Figure 3.19b, showing a longitudinal section along the channel, at the close proximity of the interface. Using this figure, we easily understand the appearance of the reverse short channel effect. If the channel is long enough, the distance (δl) affected by the lateral disturbance of the boron concentration is negligible. The average dopant concentration in the channel is not changed and the threshold voltage is not affected. On the other hand, with the size decrease of the device, all other things being equal, this transition zone is no longer negligible, and the average channel doping increases, involving an increase in the threshold voltage, as experimentally observed.

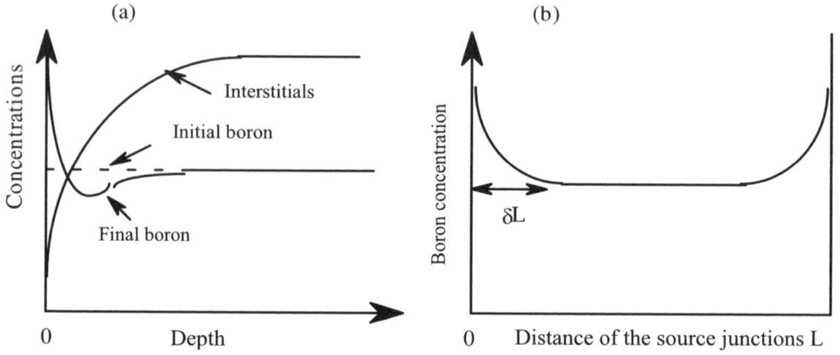

Figure 3.19. *Schematic section of the dopants and defects profiles in the channel zone during the activation annealing of the implanted junctions: (a) vertical section in the proximity of the source junction; (b) longitudinal section in the proximity of the interface with gate oxide*

3.5. Dopant diffusion in germanium

Given up for long years due to the benefit of silicon for traditional MOS technologies (the Si/SiO_2 interface being the most powerful), the use of germanium is again considered for the ultimate devices. Indeed, the best mobility of charge carriers in this material is an interesting idea for faster devices, which in any event will have to use a gate dielectric different from SiO_2 (dielectric with strong permittivity) [ITR 09]. The optimization of these devices will thus go through a good control of the technology of this "new" material. In particular, it will be necessary to have efficient models able to precisely describe dopant diffusion in germanium, to feed the simulation software for technological computer aided design (TCAD tools).

Unfortunately, the basic studies on germanium were given up at the beginning of the 1960s, after the advent of silicon, and have only very recently begun again. Consequently, the theoretical understanding degree on dopant diffusion in Ge and the availability of efficient simulation models, is still far from what is at the disposal of technologists for silicon. The aim of this section, which does not claim to be exhaustive, is to give a short review of the state of knowledge, still partial, on this topic.

3.5.1. *Thermal diffusion process*

The first experimental pieces of data concerning the thermal diffusion process of common dopants in germanium is from the 1950s. A complete study, using the available techniques of the time, was carried out by W.C. Dunlap [DUN 54]. He measured the donors (Sb, As, P) and acceptors (Ga, Al, In and B) diffusion in the range of 650 - 900°C, using radioactive tracers and the p-n junction method. These first results indicate a definitely faster diffusion of the donors than the acceptors, with values of the diffusion coefficient at 700°C, of about 10^{-14} cm^2s^{-1} for boron and gallium, 10^{-12} cm^2s^{-1} for phosphorus and $6.5x10^{-12}$ cm^2s^{-1} for arsenic. Another outstanding fact is that all the dopants have about the same activation energies of diffusion (about 2.5 eV), except for boron which has a definitely higher activation energy. This is not figured in the article of W.C. Dunlap, which allocated this difference to an experimental problem. This 2.5 eV energy is significantly lower than the activation energy of germanium self-diffusion, about 3 eV [LET 56, VOG 83, WER 85]. This highlights a considerable attractive binding energy between these dopants and the defect responsible for diffusion. More recent data, taking advantage of more sophisticated investigation means, such as secondary ions mass spectrometry (SIMS), can give precise determinations of the dopant profiles and lead to a more precise image of the experimental reality. The main results are summarized below.

3.5.1.1. *Self-diffusion*

All the experimental pieces of data available lead us to conclude almost certainly that germanium self-diffusion takes place by a vacancy mechanism. The most convincing study is from Mr. Werner *et al.*, analyzing the influence of doping and of the hydrostatic pressure on the self-diffusivity of germanium [WER 85]. In this study, these authors confirmed that the self-diffusivity of Ge varies with the doping concentration, consistently with the existence of an acceptor level for the vacancy. They also clearly highlighted that the germanium self-diffusivity decreases with pressure. This last result, corresponding to a positive activation volume, is the signature of a vacancy mechanism.

This conclusion is in addition completely consolidated by the *ab initio* theoretical calculations concerning the formation and migration energies of the elementary defects (vacancy and self-interstitial) in Ge. The most recent outcomes [VAN 07] give, for the neutral defects, a formation energy of 2.35 (± 0.1) eV for the vacancy and of 3.25 (± 0.25) eV for the self-interstitial. The calculated migration energies are respectively about 0.6 (± 0.1) eV for the vacancy and approximately twice as much for the self-interstitial. We then reach (see equation [3.30] of section 3.2.2.2) an estimate of 2.95 (±0.2) eV for the activation energy of self-diffusion by the vacancy mechanism (in complete harmony with the experimental value of 3.09 eV [WER 85]), and larger than 4 eV for an interstitial mechanism. It would thus seem that, contrary to the silicon case, the self-interstitial plays a negligible role in germanium self-diffusion.

3.5.1.2. *Donors diffusion*

An exhaustive study of donor diffusion (P, As and Sb) was recently undertaken by S. Brotzman and H. Bracht in the temperature range of 600 - 920°C, and for various dopants surface concentration [BRO 08]. The main results are as follows:

– the effective diffusion coefficient of the donors varies with the doping level. This behavior of extrinsic diffusion is correctly described by a model implying only singly negatively charged dopant-vacancy pairs (XV^-), i.e. doubly negatively charged vacancies ($V^=$). These results seem consistent with the *ab initio* theoretical calculations concerning the states of charge of the vacancy and the behavior of the dopant-vacancy pairs in Ge [CHR 08, SPI 08];

– the intrinsic diffusion coefficient of these three impurities are significantly higher than the self-diffusivity. At a given temperature, the intrinsic diffusion coefficient increases with the impurity size. For example, at 700°C, the values obtained are 1.6×10^{-14} cm^2s^{-1} for P, 3×10^{-13} cm^2s^{-1} for As and 1×10^{-12} cm^2s^{-1} for Sb;

– activation energies of the intrinsic diffusion coefficient are lower than the activation energy of the self-diffusion, with slightly different values depending on the dopant, of 2.85 (± 0.04) eV for P, 2.71 (± 0.06) eV for As and of 2.55 (± 0.03) eV for Sb. This decrease with the atom size is naturally interpreted as an increase in the binding energy between the substitutional donor and the vacancy.

It should however be mentioned that the interpretation in terms of a mechanism implying exclusively doubly negative vacancies, has not yet reached a consensus. Some authors indicate the need to take into account interactions with neutral vacancies and doubly negatively charged vacancies to correctly describe the arsenic profiles redistribution [VAI 00]. Moreover, recent simulation attempts of the redistribution of phosphorus profiles on the basis of an interaction with doubly negatively charged vacancies, with or without accounting for neutral vacancies, were unsuccessful [CAN 08]. However, very good adjustments were obtained by taking into account a contribution of the triply charged vacancies, introducing a cubic variation of the effective diffusivity with the free electrons concentration [CAN 10]. Other authors, on the basis of the influence of the pressure on the implanted profiles redistribution, even challenge the vacancy mechanism for arsenic diffusion [MIT 96]. It is thus clear that complementary studies are to be done to definitively clarify the atomic diffusion mechanisms of the donors in germanium.

3.5.1.3. *Boron diffusion*

The most recent studies concerning boron diffusion in germanium lead to extremely low diffusivity values, about 10^{-17} to 10^{-15} cm^2s^{-1}, for temperatures between 800 and 900°C [UPP 04]. Correlatively, the activation energy is about 4.6 eV, much higher than the self-diffusion energy. Such a difference in activation energies results in thinking that the elementary diffusion mechanisms (implied point defects) are not similar, which seems to indicate that boron diffusion, contrary to the donors case, would primarily be assisted by self-interstitials [UPP 04]. Nevertheless, at this stage, we cannot completely exclude the hypothesis of a vacancy mechanism with a very strong repellent interaction (negative binding energy) between the vacancy and the boron atom. Such a repellent interaction, but at a less important degree, is for example invoked to explain why silicon, isoelectronic impurity, diffuse in germanium with an activation energy higher than the self-diffusion energy, but via a vacancy mechanism [SIL 06].

3.5.2. *Implanted dopants and junctions engineering*

In addition to the fundamental studies, briefly summarized above, the possible interest of germanium as a basic material for upcoming technological CMOS generations has also motivated studies aiming to find a way to the fabrication of good quality junctions, using common microelectronics techniques, i.e. by ion implantation followed by an activation annealing. The interested reader will find an excellent summary of the main issues and results in an article by E. Simoen *et al.* [SIM 06] briefly summarized below.

3.5.2.1. *P^+/N junctions*

Despite a rather low solubility limit of chemical boron in germanium, lower than 10^{19} cm^{-3} [JON 87], it has been proved rather simple to elaborate P^+ layers by ion boron implantation, having a sufficient conductivity to manufacture efficient devices. On the one hand, this is due to the very low diffusivity of B, as previously indicated, preventing the redistribution of the implanted profile during activation annealing. On the other hand, this is also due to a dynamic annealing phenomenon leading to dopant self-activation during implantation [JON 87]. It is even possible to reach total boron activation by ion implantation at room temperature, without subsequent annealing [UPP 01]. Thermal annealing can then reach active concentrations beyond the limit of boron solubility. Active concentrations of a few 10^{20} cm^{-3} are easily reachable by RTA (rapid thermal annealing).

In addition, it was shown that it was possible to further increase the active boron concentration, up to a level close to 5 x 10^{20} cm^{-3}, by a simple annealing at 400°C of a profile implanted in germanium pre-amorphized by high-dose Ge ion implantation [CHA 05]: during the solid phase epitaxy, leading to germanium recrystallization, boron atoms are trapped in substitutional sites. Although these solid solutions are supersaturated in boron, they prove rather stable and resist to 60 seconds annealing at 600°C [SIM 06], which is sufficient considering the thermal budget planned for ultimate MOS technologies. This should probably be correlated with the very low boron diffusivity, leading to very slow kinetics for reactions in solid phase (like germination or precipitates growth) involving this impurity. Nevertheless, the defects created by implantation and pre-amorphization are eliminated with difficulty by these annealings with low thermal budget [SIM 05]. A compromise thus still remains to be found to obtain acceptable leakage currents.

3.5.2.2. *N^+/P junctions*

Due to their atomic mass being higher than that of boron, donor (P, As, Sb) ion implantation, at the doses required by the technology, induces very significant damage of the germanium substrate, generally leading to the formation of an amorphized P type surface layer [BEN 92]. The restoration of the crystalline quality and the activation essential to the making of the N^+ layer then require annealing with high thermal budgets. In the case of antimony, the damage is too important to be efficiently annealed: there is in fact a formation of cavities resulting from the vacancies clustering created by implantation [JAN 06]. For arsenic, the problem is a little less important, but generally the annealed layers have a very important surface roughness [JAN 06].

Because of that, phosphorus proves to be the best applicant for the creation of N^+/P junctions. Nevertheless, the realization of efficient junctions by phosphorus implantation and annealing still faces a certain number of difficulties, summarized below:

– an important outdiffusion (dopant evaporation from the surface) occurs during the activation annealing, leading to a very strong dose loss of the dopant [CAN 08, POO 85]. An adequate encapsulation (SiO_2) minimizes this problem, but does not completely avoid it;

– a very fast phosphorus diffusion in extrinsic condition leads to a significant broadening of the implanted profile, which raises the problem of the ultra-thin junctions necessary for the upcoming technological generations;

– the electrically active phosphorus concentration saturates at rather low values of several 10^{19} cm^{-3}, even for strong implantation doses. The origin of this difference between the active concentration and the chemical solubility is not yet perfectly solved.

In order to get around these difficulties, various technological approaches are studied, like laser annealing for better dopants activation (incorporation beyond the equilibrium limit) [TSO 06] or defect engineering following a logic similar to the one applied for silicon [SIM 06].

Therefore, we can note that the development of well-established technology for the creation of operational junctions in germanium still requires more thorough research. Moreover, the essential studies remain to be completed to lead to models of dopant diffusion in Ge suitable for reliable technological simulations. This will be made possible, like in the case of silicon, by the understanding and modeling of the couplings between dopants and (point or complex) defects of the material.

3.6. Conclusion

Throughout this chapter, emphasis was put on the need, for a proper model of dopant diffusion in silicon and for the optimization of junction engineering, to take into account increasingly complex couplings phenomena between dopant atoms and material point defects. In particular, the few examples developed in section 3.4 show the extreme importance of these couplings phenomena in the determination of the real dopant profiles inside the structure of the devices, and thus *in fine* on the device electric characteristics.

Moreover, it is obvious that with the systematic reduction of the device size, a simple 1D description is no longer sufficient, the 2D, or even 3D simulations, becoming essential for a description of experimental realities (the short reverse channel effect is a typical example of this fact). We thus understand why the optimization of the elementary devices is increasingly demanding regarding the modeling of dopant diffusion, in order to have reliable predictive tools in the domain of technological simulation.

3.7. Bibliography

[BEN 92] BENOURHAZI K., PONPON J.P., "Implantation of phosphorus and arsenic ions in germanium", *Nuclear Instruments and Methods in Physics Research Section B*, no. 71, p. 406-411, 1992.

[BLA 72] BLANCHARD B. *et al.*, "Compte-rendu du colloque Caractérisation des matériaux et des technologies des semi-conducteurs", Grenoble, 1972.

[BOU 03] BOUCARD F., Modélisation de la diffusion des dopants dans le silicium pour la réalisation de jonctions fines, PhD Thesis, Louis Pasteur University, Strasbourg, 2003.

[BOU 05] BOUCARD F., ROGER F., CHAKAROV I., ZHUK V., TEMKIN M., MONTAGNER X., GUICHARD E., MATHIOT D., "A comprehensive solution for simulating ultra-shallow junctions: from high dose/low energy implant to diffusion annealing", *Materials Science Engineering B*, no. 124, p. 409-414, 2005.

[BRO 08] BROTZMANN S., BRACHT H, "Intrinsic and extrinsic diffusion of phosphorus, arsenic, and antimony in germanium", *Journal of Applied Physics*, no. 103, p. 033508, 2008.

[CAG 07] CAGNAT N., MATHIOT D., LAVIRON C. J., "Ultrathin n+/p junction in preamorphized silicon by phosphorus and carbon coimplantation engineering: influence of C location", *Journal of Applied Physics*, no. 102, p. 106102, 2007.

[CAG 08] CAGNAT N., Implantation ionique et jonctions fines - Caractérisation, ingénierie des défauts et application aux technologies 65 et 45 nm, PhD Thesis, Louis Pasteur University, Strasbourg 1, 2008.

[CAN 08] CANNEAUX T., MATHIOT D., ROQUES S., SCHMITT S., DUBOIS C. "Diffusion of phosphorus implanted in germanium", *Materials Science Engineering B*, no. 154, p. 68-71, 2008.

[CAN 10] CANNEAUX T., MATHIOT D., PONPON J.P., LEROY Y., "Modeling of phosphorus diffusion in Ge accounting for a cubic dependence of the diffusivity with the electron concentration", *Thin Solid Films*, no. 518, p. 2394-2397, 2010.

[CAR 59] CARSLAW H.S., JAEGER J.C., *Conduction of Heat in Solids*, Clarendon Press, Oxford, 1959.

[CHA 05] CHAO Y.L., PRUSSIN S., WOO J.C.S., SCHOLZ R., "Preamorphization implantation-assisted boron activation in bulk germanium and germanium-on-insulator", *Applied Physics Letters*, no. 87, p. 142102, 2005.

[CHR 08] CHRONEOS A., BRACHT H., GRIMES RW., UBERUAGA B.P., "Vacancy-mediated dopant diffusion activation enthalpies for germanium", *Applied Physics Letters*, no. 92, p. 172103, 2008.

[CHU 05] CHUI C.O., KULIG L., MORAN J., TSAI W., SARASWAT K.C., "Germanium n-type shallow junction activation dependences", *Applied Physics Letters*, no. 87, p. 091909, 2005.

[CLA 03] CLAVERIE A., COLOMBEAU B., DE MAUDUIT B., BONAFOS C., HEBRAS X., BEN ASSAYAG G., CRISTIANO F., "Extended defects in shallow implants", *Applied Physics Letters A*, no. 76, p. 1025-1033, 2003.

[COL 01] COLOMBEAU B., SMITH A. J., COWERN N. E. B., PAWLAK B. J., CRISTIANO F., DUFFY R., CLAVERIE A., ORTIZ C. J., PICHLER P., LAMPIN E., ZECHNER C., "Current Understanding and Modeling of B Diffusion and Activation Anomalies in Preamorphized Ultra-Shallow Junctions", in P. PICHLER, A. CLAVERIE, R. LINDSAY, M. ORLOWSKI, W. WINDL (eds), *Silicon Front-End Junction Formation – Physics and Technology, Materials Research Society Symposium Proceedings*, no. 810, p. C3.6.1, 2001.

[CRA 75] CRANCK J., *The Mathematics of Diffusion*, 2nd edition, Clarendon Press, Oxford, 1975.

[DOW 98] DOWNEY D.F., CHOW J.W., ISHIDA E., JONES K.S., "Effect of fluorine on the diffusion of boron in ion implanted Si", *Applied Physics Letters*, no. 73, p. 1263-1265, 1998.

[DRO 03] DROBNY V.FF, CHANG K., ROBINSON D., "Activation behavior of BF_2^+ implants in RTP annealed silicon", in B. BROWN, T.L. ALFORD, M. NASTASI, M.C. VELLA (eds), *14th Conference on Ion Implantation Technology Proceedings*, IEEE, New York, p. 686, 2003.

[DUN 97] DUNHAM M., "Lattice Monte Carlo Simulations of Vacancy-Mediated Diffusion and Aggregation Using Ab Initio Parameters", in T. DIAZ DE LA RUBIA, S. COFFA, P.A. STOLK, C.S. RAFFERTY (eds), *Defects and Diffusion in Silicon Processing, Materials Research Society Symposium Proceedings*, no. 469, p. 353-358, 1997.

[DUN 54] DUNLAP JR W.C., "Diffusion of impurities in germanium", *Physics Review*, no. 94, p. 1531-1540, 1954.

[FAI 73] FAIR R.B., WEBER G.R., "Effect of complex formation on diffusion of arsenic in silicon", *Journal of Applied Physics*, no. 44, p. 273-279, 1973.

[FAI 77] FAIR R.B., TSAI J.C.C., "Quantitative model for diffusion of phosphorus in silicon and emitter dip effect", *Journal of the Electrochemical Society*, no. 124, p. 1107-1118, 1977.

[FED 94] FEDOTOV S.A., "Diffusion of electrically active Sb atoms in heavily-doped silicon - Monte-Carlo simulation with regard to collective phenomena", *Physica Status Solidi (b)*, no. 86, p. 375-382, 1994.

[FRA 56] FRANCK F.C., TURNBULL D., "Mechanism of Diffusion of Copper in Germanium", *Physical Review*, no. 104, p. 617-618, 1956.

[GEN 76] DE GENNES P.G., "La percolation: un concept unificateur", *La Recherche*, no. 72, p. 919-927, 1976.

[GOS 80] GOSELE U., FRANK W., SEEGER A., "Mechanism an kinetics of the diffusion of gold in silicon", *Appled Physics*, no. 23, p. 361-368, 1980.

[HEI 99] HEISER T., ISTRATOV A.A., FLINK C., WEBER E.R., "Electrical characterization of copper related defect reactions in silicon", *Materials Science and Engineering B*, no. 58, p. 149-154, 1999.

[HIL 81] HILL C., "Measurement of local diffusion coefficients in planar device structures", in H.R. HUFF, R.J. KREIGLER, Y. TAKEISHI (eds), *Semiconductor Silicon*, The Electrochemical Society, Princeton, p. 988-996, 1981.

[HU 74] HU S.M., "Formation of stacking faults and enhanced diffusion in the oxidation of silicon", *Journal of Applied Physics*, no. 45, p. 1567-1573, 1974.

[ITR 09] "International Technology Roadmap for Semiconductors", 2009, available at http://www.itrs.net/.

[JAN 06] JANSSENS T., HUYGHEBAERT C., VANHAEREN D., WINDERICKX G., SATTA A., MEURIS M., VANDERVORST W., "Heavy ion implantation in Ge: Dramatic radiation induced morphology in Ge", *Journal of Vaccum Science and Technology B*, no. 24, p. 510-514, 2006.

[JOH 93] JOHN J.P., LAW M.E., "Oxidation enhanced diffusion of phosphorus in silicon in heavily doped background concentrations", *Journal of the Electrochemical Society*, no. 140, p. 1489-1491, 1993.

[JON 87] JONES K.S., HALLER E.E., "Ion-implantation of boron in germanium", *Journal of Applied Physics*, no. 61, p. 2469-2477, 1987.

[KIM 89] KIMMERLING L.C., ASOM M.T., BENTON, J.L., DREVINSKI P.J., CEFER C.E. "Interstitial defect reaction in Si", *Materials Science Forum*, no. 38, p. 141-150, 1989.

[KU 06] KU K.C. *et al.*, "Effects of germanium and carbon coimplants on phosphorus diffusion in silicon", *Applied Physics Letters*, no. 89, p. 112104, 2006.

[LAM 05] LAMPIN E., ORTIZ C.J., COWERN N.E.B., COLOMBEAU B., CRISTIANO F., "Combined master and Fokker-Planck equations for the modelling of the kinetics of extended defects in Si", *Solid-State Electronics*, no. 49, p. 1168-1171, 2005.

[LEN 00] LENOSKY T.J., SADIGH B., THEISS S.K., CATURLA M.J., DE LA RUBIA T.D., "Ab initio energetics of boron-interstitial clusters in crystalline Si", *Applied Physics Letters*, no. 77, p. 1834-1836, 2000.

[LET 56] LETAW H. JR., PORTNOY W.M., AND SLIFKIN L., "Self-diffusion in germanium", *Physics Review*, no. 102, p. 636-639, 1956.

[LIN 93] LIN D.G., ROST T.A., "The impact of fluorine on CMOS channel length and shallow junction formation", *International Electron Device Meeting*, Washington DC, 1993.

[MAT 82a] MATHIOT D., PFISTER J.C., "Influence of the non-equilibrium vacancies the diffusion of phosphorus into silicon", *Journal of Applied Physics*, no. 53, p. 3053-3058, 1982.

[MAT 82b] MATHIOT D., PFISTER J.C., "High-concentration diffusion of P in Si – A percolation problem", *Journal de Physique Lettres*, no. 43, p. L453-L459, 1982.

[MAT 83] MATHIOT D., PFISTER J.C., "Diffusion of arsenic in silicon – Validity of the percolation model", *Applied Physics Letters*, no. 42, p. 1043-1044, 1983.

[MAT 84] MATHIOT D., PFISTER J.C., "Dopant diffusion in silicon – A consistent view involving non-equilibrium defects", *Journal of Applied Physics*, no. 55, p. 3518-3530, 1984.

[MAT 89] MATHIOT D., "Modelling of hydrogen diffusion in n-type and p-type silicon", *Physics Review B*, no. 40, p. 5867-5870, 1989.

[MAT 92] MATHIOT D., "Gold, Self-diffusion and dopant diffusion in silicon", *Physics Review B*, no. 45, p. 13345-13355, 1992.

[MIC 86] MICHEL A.E., "Diffusion modelling of the redistribution of ion implanted impurities", in T.O. SEDGWICK, T.E. SEIDEL, B.Y. TSAUR (eds), *Rapid Thermal Processing, Materials Research Society Symposium Proceedings*, no. 52, p. 3-13, 1986.

[MIT 96] MITHA S., AZIZ M.J., SCHIFERL D., POKER D.B., "Activation volume for arsenic diffusion in germanium", *Applied Physics Letters,* no. 69, p. 922-924, 1996.

[MIY 85] MIYAKE M., "Oxidation-enhanced diffusion of ion-implanted boron in heavily phosphorus-doped silicon", *Journal of Applied Physics*, no. 58, p. 711-715, 1985.

[MOG 97] MOGUL H.C., ROST T.A., LIN D.G., "Advantages of LDD-only implanted fluorine with submicron CMOS technologies", *IEEE Transactions on Electron Devices*, no. 44, p. 388-394, 1997.

[NEJ 03] NEJIM A., SEALY B., "Retarding transient enhanced diffusion of boron in silicon with high energy silicon co-implants", *Semiconductor Science and Technology*, no. 18, p. 839-844, 2003.

[LAR 93] LARSEN A.N, LARSEN K.K., ANDERSEN P.E., SVENSSON B.G., "Heavy doping effects in the diffusion of group IV and group-V impurities in silicon", *Journal of Applied Physics*, no. 73, p. 691-698, 1993.

[ORT 01] ORTIZ C, MATHIOT D., "A new kinetic model for the nucleation and growth of interstitial clusters in silicon", in M.A. FOAD, J. MATSUO, P. STOLK, M.D. GILES, K.S. JONES (eds), *Si Front-End Processing – Physics and Technology of Dopant-Defect Interactions III, Materials Research Society Symposium Proceedings*, no. 669, p. J5.6.1, 2001.

[ORT 04] ORTIZ C.J., PICHLER P., FUHNER T., CRISTIANO F., COLOMBEAU B., COWERN N.E.B., CLAVERIE A., "A physically based model for the spatial and temporal evolution of self-interstitial agglomerates in ion-implanted silicon", *Journal of Applied Physics*, no. 96, p. 4866-4877, 2004.

[PAW 06a] PAWLAK B.J., JANSSENS T., BRIJS B., VANDERVORST W., COLLART E.J.H., FELCH S.B., COWERN N.E.B., "Effect of amorphization and carbon co-doping on activation and diffusion of boron in silicon", *Applied Physics Letters*, no. 89, p. 062110, 2006.

[PAW 06b] PAWLAK B.J., DUFFY R., JANSSENS T. VANDERVORST W., FELCH S.B., COLLART E.J.H., COWERN N.E.B., "Suppression of phosphorus diffusion by carbon co-implantation", *Applied Physics Letters*, no. 89, p. 062102, 2006.

[PHI 91] PHILIBERT J., *Atom Movements. Diffusion and Mass Transport in Solids*, Editions de Physique, Les Ulis, 1991.

[PIC 06] PICHLER P. *et al.*, "Diffusion and activation of dopants in silicon and advanced silicon-based materials", *Physica Scripta*, no. T126, p. 89-96, 2006.

[POO 05] POON C.H., TAN L.S., CHO B.J., DU A.Y., "Dopant loss mechanism in n^+/p germanium junctions during rapid thermal annealing", *Journal of the Electrochemical Society*, no. 152, p. G895-G899, 2005.

[QUO 72] QUOIRIN J.B., Détermination de profil de diffusion de bore dans le silicium à l'aide de l'analyseur ionique Castaing-Slodzian, PhD Thesis, Grenoble University, 1972.

[RAF 93] RAFFERTY C.S., VUONG H.H., ESHRAGHI S.A., GILES M.D., PINTO M.R., HILLENIUS S.J., "Explanation of reverse short channel effect by defect gradients", *International Electron Device Meeting*, Washington DC, 1993.

[SED 88] SEDGWICK T.O., MICHEL A.E., DELINE V.R., COHEN S.A., LASKY J.B., "Transient boron-diffusion in ion-implanted crystalline and amorphous silicon", *Journal of Applied Physics*, no. 63, p. 1452-1463, 1988.

[SHA 01] SHANO T., KIM R., HIROSE T., FURUTA Y., TSUJI H., FURUHASHI M., TANIGUCHI K., "Realization of ultra-shallow junction: suppressed boron diffusion and activation by optimized fluorine co-implantation", *International Electron Device Meeting*, Washington DC, 2001.

[SIL 06] SILVESTRI H.H., BRACHT H., HANSEN J.L., LARSEN A.N., HALLER E.E., "Diffusion of silicon in crystalline germanium", *Semiconductor Sciences and Technology*, no. 21, p. 758-762, 2006.

[SIM 05] SIMOEN E., SATTA A., MEURIS M., JANSSENS T., CLARYSSE T., BENEDETTI A., DEMEURISSE C., BRIJS B., HOFLIJK I., VANDERVORST W., CLAEYS C., "Defect removal, dopant diffusion and activation issues in ion-implanted shallow junctions fabricated in crystalline germanium substrates", *Solid State Phenomena*, no. 108-109, p. 691-696, 2005.

[SIM 06] SIMOEN E. *et al.*, "Ion-implantation issues in the formation of shallow junctions in germanium", *Material Science in Semiconductor Processing*, no. 9, p. 634-639, 2006.

[SMI 06] SMITH A.J., COWERN N.E.B., GWILLIAM R, SEALY B.J., COLOMBEAU B., COLLART E.J.H., GENNARO S., GIUBERTONI D., BERSANI M., BAROZZI M., "Vacancy-engineering implants for high boron activation in silicon on insulator", *Applied Physics Letters*, no. 88, p. 082112, 2006.

[SPI 08] SPIEWAK P., VANHELLEMONT J., SUEOKA K., KURZYDLOWSKI K.J., ROMANDIC I., "First principles calculations of the formation energy and deep levels associated with the neutral and charged vacancy in germanium", *Journal of Applied Physics*, no. 103, p. 086103, 2008.

[STO 97] STOLK P.A., GOSSMANN H.J., EAGLESHAM D.J., JACOBSON D.C., RAFFERTY C.S., GILMER G.H., JARAIZ M., POATE J.M., LUFTMAN H.S., HAYNES T.E., "Physical mechanisms of transient enhanced dopant diffusion in ion-implanted silicon", *Journal of Applied Physics*, no. 81, p. 6031-6050, 1997.

[TSO 06] TSOUROUTAS P., TSOUKALAS D., FLORAKIS A., ZERGIOTI I., SERAFETINIDES A.A., CHERKASHIN N., MARTY B., CLAVERIE A., "Laser annealing for n^+/p junction formation in germanium", *Material Science in Semiconductor Processing*, no. 9, p. 644-649, 2006.

[TUC 74] TUCK B., *Introduction to Diffusion in Semiconductors*, IEE monograph series 16, Peter Peregrinus Ltd, Stevenage, 1974.

[UPP 01] UPPAL S., WILLOUGHBY A.F.W., BONAR J.M., EVANS A.G.R., COWERN N.E.B., MORRIS R., DOWSETT M.G., "Diffusion of ion-implanted boron in germanium", *Journal of Applied Physics*, no. 90, p. 4293-4295, 2001.

[UPP 04] UPPAL S., WILLOUGHBY A.F.W., BONAR J.M., COWERN N.E.B., GRASBY T., MORRIS R.J.H., DOWSETT M.G., "Diffusion of boron in germanium at 800-900 degrees C", *Journal of Applied Physics*, no. 96, p. 1376-1380, 2004.

[URA 99] URAL A., GRIFFIN P., PLUMMER J.D., "Fractional contributions of microscopic diffusion mechanisms for common dopants and self-diffusion in silicon", *Journal of Applied Physics*, no. 85, p. 6440-6446, 1999.

[VAI 00] VAINONEN-AHLGREN E., AHLGREN T., LIKONEN J., LEHTO S., KEINONEN J., LI W., HAAPAMAA J., "Identification of vacancy charge states in diffusion of arsenic in germanium", *Applied Physics Letters*, no. 77, p. 690-692, 2000.

[VAL 57] VALENTA M.W., RAMASASTRY C., "Effect of heavy doping on the self-diffusion of germanium", *Physics Review*, no. 106, p. 73-75, 1957.

[VAN 07] VANHELLEMONT J., SPIEWAK P., SUEOKA K., "On the solubility and diffusivity of the intrinsic point defects in germanium", *Journal of Applied Physics*, no. 101, p. 036103, 2007.

[VOG 83] VOGEL G., HETTICH G., MEHRER H., "Self-diffusion in intrinsic germanium and effects of doping on self-diffusion in germanium", *Journal of Physics C: Solid State Physics*, no. 16, p. 6197-6204, 1983.

[WAT 64] WATKINS G.D., CORBETT J.W., "Defects in irradiated silicon: electron paramagnetic resonance and electron-nuclear double resonance of the Si-E center", *Physics Review*, no. 134, p. A1359-A1377, 1964.

[WER 85] WERNER M., MEHRER H., "Effect of hydrostatic pressure, temperature, and doping on self-diffusion in germanium", *Physics Review. B*, no. 32, p. 3930-3937, 1985.

[WIL 77] WILLOUGHBY A.F.W., "Interactions between sequential dopant diffusion in silicon - Review", *Journal of Physics. D*, no. 10, p. 455-480, 1977.

[YOS 74] YOSHIDA M., ARAI M., NAKAMURA H., TERUNUMA Y., "Excess vacancy generation mechanism at phosphorus diffusion into silicon", *Journal of Applied Physics*, no. 45, p. 1498-1506, 1974.

[ZHA 95] ZHANG L. H., JONES K. S., CHI P. H., SIMONS D. S., "Transient enhanced diffusion without {311} defects in low energy B+ - implanted silicon", *Applied Physics Letters*, no. 67, p. 2025-2027, 1995.

Chapter 4

Epitaxy of Strained Si/Si_{1-x} Ge_x Heterostructures

4.1. Introduction

4.1.1. *General introduction*

This chapter discusses the reduced pressure-chemical vapor deposition (RP-CVD) of Si/SiGe (C) heterostructures for nanoelectronics and opto-electronics. The chapter is structured as follows. Section 4.1 will present: (i) the development since the beginning of the industrial CVD of Si and SiGe; (ii) the epitaxy tool used in CEA-LETI to carry out studies described in this chapter; and (iii) some general concepts of epitaxy which will be useful later on. The epitaxy of strained Si/$Si_{1-x}Ge_x$ and Si/$Si_{1-y}C_y$ heterostructures as conducting channels of pMOS and nMOS transistors, will be detailed in sections 4.2 and 4.3. The selective epitaxial growth on SOI substrates of Si or SiGe(:B) recessed and raised sources and drains will be described in sections 4.4 and 4.5.

In section 4.6, we will present the structural properties of (i) SiGe virtual substrates and the tensily-strained Si layers grown on top and (ii) the sSOI substrates resulting from them. We will also focus on the electric gains (thanks notably to dual channels) in nMOS and pMOS transistors built on top. The structural, electric and optical devices properties of thick Ge layers and of Ge islands on Si(001) will be described in section 4.7. We will highlight in section 4.8, the exciting possibilities offered from an architectural point of view by the lateral selective etching of SiGe

---

Chapter written by Jean-Michel HARTMANN.

sacrificial layers. Finally, the most outstanding results will be summarized in section 4.9 and some prospects will also be presented.

4.1.2. *Chemical vapor deposition from the beginning*

This section is freely inspired by the recent article by W.B. De Boer [BOE 08], one of the founding fathers of modern industrial epitaxy, who worked at ASM in the years 1980-1990 and who took part in the development of the Epsilon tool.

4.1.2.1. *Introduction*

Chemical vapor deposition (CVD) of Si is a relatively old technique of the semiconductor industry. It was originally used in bipolar transistors in order to deposit at high temperatures (typically 1,000-1,200°C) a high resistivity (i.e. intrinsic) Si thick layer on strongly doped (by ion implantation) buried layers. We then have a high avalanche-breakdown current and a low collector resistance. Thanks to very low carbon and oxygen levels, the crystalline quality of the deposited Si layer is equal to or higher than that of the substrate. Potential difficulties like auto doping (exodiffusion of the dopants implanted beforehand in undesired zones), the appearance of slip lines (temperature non-uniformity on the surface of the substrate), etc., are known and controlled to a certain extent.

Despite its cost, epitaxy has also been introduced into the fabrication of metal-oxide-semi-conductor (MOS) transistors. The use of an epitaxial layer as the channel beneath the gate of small-size MOS transistors can indeed prove to be interesting for threshold voltage adjustment and carrier mobility boosting. It can in particular lead to the suppression of parasitic short-circuits between adjacent individual transistors, to the reduction of errors in memory cells, to higher quality gate dielectric, etc.

The most recent low-temperature epitaxy methods are very different from the conventional methods mentioned above. Deposition temperatures (typically 600-700°C) are indeed such that dopant exodiffusion is no longer a problem. We can thus obtain sophisticated doping profiles with a control within the nanometer of planar SiGe layers with a thickness of ten nanometers or less, etc. From the industrial point of view, optimizing low-temperature epitaxy methods lead to heterojunction bipolar transistors (HBT). The SiGe:C base layer with the suitable boron doping profile is then inserted between the emitter and the collector layers (Si *in situ* doped with arsenic or phosphorus), rather than ion implanted. We can also remember the recent adoption in mass production of recessed, *in situ* B doped SiGe sources and drains in order to boost (thanks to uniaxial compressive strain) the hole mobility in the Si channel of short gate length p-type MOS devices.

4.1.2.2. *CVD in general*

Atmospheric pressure-chemical vapor deposition (AP-CVD) was used in the 1960s in order to deposit SiO_2, Si_3N_4 and Poly-Si layers. The need for minimizing deposition temperatures and for improving the methods led to the development of the low pressure CVD (LP-CVD) in the middle of the 1970s. The transport of active chemical species towards the surface of the Si substrate is thus improved (with the increase in their mean free path), leading to a much better deposition uniformity on a wafer and wafer to wafer and to a decrease by two orders of magnitude of gases consumption. The layout of the reactors has evolved, going from cold walls (wafers parallel to the gases flow) to hot walls (wafers vertically laid out side by side and perpendicular to the gases flow), as illustrated in Figure 4.1. The time deposition, longer in LP-CVD, is compensated by the more significant number of processed wafers, leading to a drastic process/wafer cost cut.

A new technology, the so-called Plasma-Enhanced CVD (PE-CVD), was developed a short time after LP-CVD. It enabled to reduce the deposition temperature even more by dissociating the decomposition of the gaseous precursors and their actual deposition on the surface. SiO_2, Si_3N_4 and Poly-Si layers can thus be deposited at temperatures lower than or equal to 400°C, i.e. several hundred degrees below the temperatures commonly used in LP-CVD.

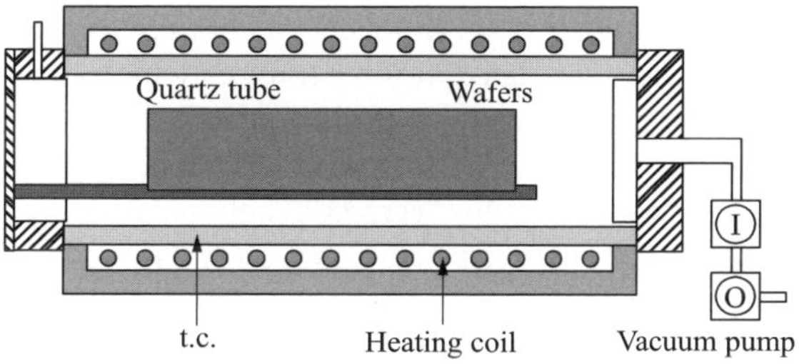

Figure 4.1. *Schematic description of a horizontal LP-CVD reactor*

4.1.2.3. *High temperature conventional Si epitaxy*

Conventional Si epitaxy methods (with atmospheric pressure and with chlorinated precursors such as $SiCl_4$) were (and are still) characterized by very high deposition temperatures (1,000-1,200°C), inducing a marked inter-diffusion of the dopants between the epitaxial layer and the substrate. The interfacial zone, where the doping level changes will vary from a few hundred nm up to several microns, depending on the process conditions and the doping species used. The situation is complicated by the doping atoms, released by the surface or by the rear face of the wafer, which are re-deposited in the epitaxial layer ("auto-doping"), enlarging the depletion layer. The adoption of reduced pressures (50-100 Torr) at the end of the 1970s drastically reduced the n-type dopants (i.e. As or P) auto-doping and became standard for bipolar transistor epitaxy (on As doped buried layers). The pre-epitaxy surface preparation was also improved. High temperature HCl etching of a few thousand Å of Si, used in the 1960s and 1970s as surface preparation for bipolar transistors, was replaced by a H_2 bake, which can remove the native oxide.

4.1.2.4. *Low-temperature Si and SiGe epitaxy*

The high deposition temperatures used in "conventional" epitaxy have always been troublesome. A large number of researchers tried at the beginning of 1980s to reduce them. The ideas explored were as follows: (i) plasma-assisted epitaxy (decomposition of the gaseous precursors using a remote plasma and not on the surface of the wafer); (ii) adoption of gaseous precursors (like SiH_4) decomposing at lower temperature; (iii) the use of carrier gases other than hydrogen, such as argon or helium, etc. The problem with low temperature deposition was (and to a certain extent remains) the crystalline quality of the layer.

Meyerson *et al.* (IBM), thinking that the residual oxide on the surface was the limiting factor for the low-temperature growth of good quality epitaxial layers, proposed at the beginning of 1980s to call upon Ultra-high vacuum CVD (UHV-CVD), in order to avoid surface re-oxidation during epitaxy. The native oxide was removed *ex situ*, the wafers were loaded in the epitaxy tool and good crystalline quality Si layers were deposited at temperatures which can be lower than 500°C. It was very innovative at the time. Similar quality Si layers were at the same time obtained by Sedgewick *et al.* (IBM) with atmospheric pressure at temperatures of about 700°C. Process conditions were conventional and included an *in situ* H_2 bake before epitaxy. What changed the situation was the adoption of advanced Pd purifiers (maximal water and oxygen concentrations: ten parts per million) for carrier gases, as well as a greater purity of precursor gases.

The difference between the extremely low partial pressures of oxygen and water necessary in UHV-CVD and the much less important constraints in more conventional CVD industrial equipment is due to the presence of large amounts of ultra-pure hydrogen, the carrier gas. The surface coverage by hydrogen atoms is then high at low temperature, passivating the surface with respect to impurity adsorption and incorporation, enabling the epitaxy of Si or SiGe layers of good structural and electronic quality.

Some of the problems of the high temperature conventional epitaxy disappear at low temperatures. The appearance of slip lines is no longer a problem. Boron auto-doping disappears and very abrupt boron doping profiles can be obtained. Si/SiGe heterostructures with morphologically as well as chemically abrupt interfaces can be deposited. Phosphorus doping is however more difficult at low-temperature and arsenic doping profiles are uncontrollable. Carbon is otherwise troublesome in high temperature epitaxy. An important surface concentration in C atoms generates the presence of SiC crystallites at the interface and thus of epitaxy defects. At low-temperature, C atoms can however be introduced in a substitutional way into the Si crystal lattice, inhibiting the diffusion of B dopants. This property is industrially used in HBTs in order to maintain the square B doping profile of the base layer (which is then a SiGeC:B alloy) during subsequent high temperature process steps. Low-temperature CVD, which is characterized by comparatively low growth rates, is complementary to the high temperature CVD and can be currently carried out in the same equipment.

4.1.2.5. *Development of CVD equipment*

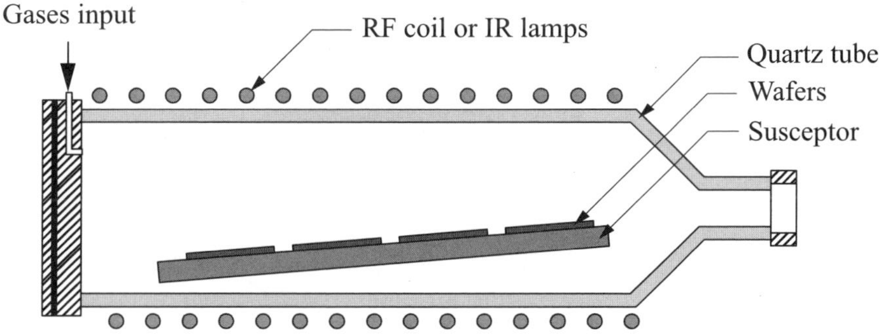

Figure 4.2. *Horizontal typical CVD reactor*

The horizontal atmospheric pressure reactor schematically described in Figure 4.2, was the choice equipment at the beginning of the semiconductor industry. The wafers were placed on a graphite susceptor and the gases flowed in a quartz chamber. The heating was ensured by RF filaments wrapped around the chamber, or by high intensity lamps focused on the susceptor, which radiated through quartz. For some applications, a resistive heating was directly installed in the susceptor. The walls were cooled, in order to minimize deposition on them. This was usually done with air, but also with water (tube with double inner surfaces). The most sophisticated reactors were characterized by a susceptor tilted upward, in gas output, in order to increase its speed and to compensate for the depletion in the gaseous phase. The reactors evolved during the transition from the AP-CVD to the CVD at reduced pressure (RP-CVD). We can notably mention the barrel reactors appearance, with a susceptor in the shape of a pyramid in a vertical tube (see Figure 4.3). Contrary to a horizontal reactor (which would immediately collapse), barrel reactors could be used at reduced pressure (50 - 100 Torr), thanks to vacuum pumps. After it was demonstrated that reduced pressures decreased the n-type auto-doping, this type of reactor became standard for the epitaxy of bipolar transitors and was used for many years.

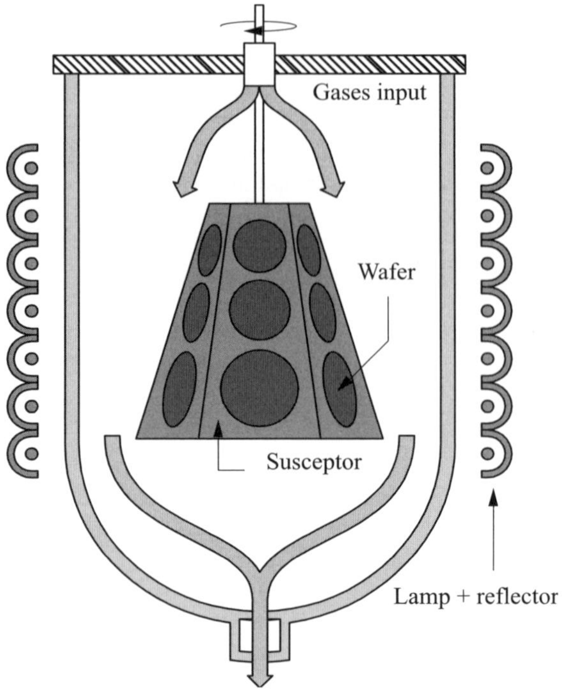

Figure 4.3. *RP-CVD barrel reactor*

UHV-CVD reactors were introduced in the 1980s for low-temperature depositions, with a certain number of adjustments for the management of the ultra-high vacuum and to obtain very low partial pressures of contaminants. At the same time, manufacturers of industrial epitaxy reactors started to develop the next generations of equipment, i.e. large size high capacity reactors (increase of the wafer diameter), hardly different from the one described schematically in Figure 4.3. They were then focused on obtaining several μm thick p-type Si layers on p+ Si layers.

Another type of reactor was developed during these years. This reactor closely looked like the horizontal lamp reactor schematically shown in Figure 4.2, with the following specificities: a single wafer chamber with rotation of the wafers during the process, in order to compensate for gas phase depletion and thus generate a good deposition uniformity. A heating by lamp was combined with a low thermal inertia susceptor to achieve an acceptable output. Loading/unloading (also called "loadlock") chambers, as well as a transfer chamber were added, to improve the cleanliness of the deposition chamber, essential for good quality of the epitaxial layers.

Indeed, the walls and valves of the deposition chamber are no longer in contact with the ambient air during loading/unloading steps, and thus not in contact with oxygen or moisture, for weeks or even months. More powerful than batch (i.e. multi-wafers) reactors in terms of uniformity and crystalline quality, these single-wafer reactors became competitive or even superior in terms of cost per wafer for diameters higher than or equal to 150 mm. The ability to operate at low pressure was added later on.

We will find a schematic description in Figure 4.4.

Low-temperature Si and SiGe epitaxies were tried in this type of reactor, with more convincing results, even with atmospheric pressure. Two types of deposition reactors were thus in competition for low-temperature depositions at the beginning of the 1990s: dedicated UHV-CVD batch reactors and single wafer RP-CVD reactors as in Figure 4.4, also able to operate at high temperature. It was not really difficult to choose between these two technologies, given the flexibility and the ease of use of single-wafer reactors.

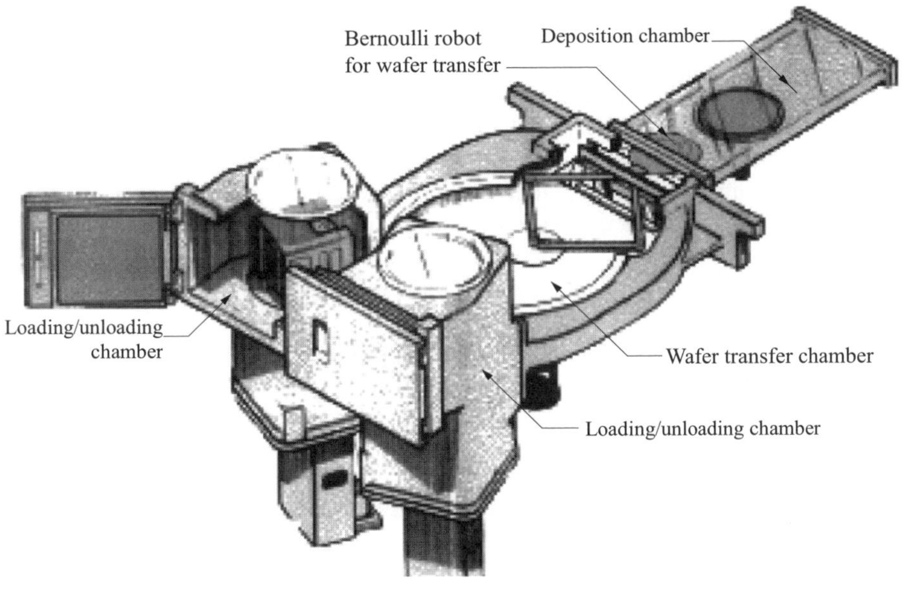

Figure 4.4. *Example of a modern epitaxy reactor: the ASM Epsilon*

4.1.3. *The Epi Centura epitaxy tool*

In 1999, CEA-LETI acquired an Epi Centura 200 mm epitaxy mainframe from Applied Materials. This is what we call a cluster tool, i.e. a Centura platform on which two RP-CVD chambers called A and B are connected (see Figure 4.5). We find on the platform two loading/unloading chambers ("Loadlocks A and B"), which can hold up to twenty five 200 mm plates. These loadlocks, exposed to the air of the clean room during the wafer loading/unloading, are pumped down then re-filled with N_2 several times prior to 20 pressure stabilization, i.e. the pressure of the transfer chamber. Given nitrogen, as well as hydrogen, the carrier gases used during epitaxy are of high purity (about 99.999999% after going through gases purification cartridges), the wafers stored in loadlocks are in an inert environment, which has its importance after an ex-situ "HF-last" wet cleaning. The "frog" robot located in the transfer chamber (in which 15 standard liters per minute of N_2 flow continuously) enables the *horizontal* movement of wafers between the various chambers. Wafer removal from / deposition on the robot quartz blade or the peripheral chambers (i.e. movements *according to z*) is individually ensured in each one of those. The wafer cooling and centering chamber makes it possible: (i) to take into account upcoming movements between chambers, the minor alignment defects of the wafers on the quartz blade of the transfer robot and (ii) the postdeposition cooling of the wafers (thanks to a thick chuck with water circulation) before putting them back in the plastic baskets of the load locks. Before describing more precisely the epitaxy chambers, we will note the presence of a certain number of peripheral pieces of equipment, vital for the functioning of such a facility, such as: (i) the electronic control cabinet which also delivers the power to the lamps, (ii) a pressure-lowering cabinet for the various gases used, (iii) pumping lines and four pumps (in the basement) to control the pressure in the different chambers, (iv) a burning/scrubbing system for used gases (in the basement downstream of the pumps), (v) cabinets containing the various gas bottles, etc. It is indeed an expensive installation (approximately 4 million dollars for the whole) and it is complex to implement.

We will find in Figure 4.6 a schematic description of the two epitaxy chambers connected to our system. Approximately 40 cm in diameter, they are physically delimited by two transparent quartz domes (in light gray on the figure), solidarized to a circular metal support using clamp rings (both in dark gray on the figure) crushing Viton or Teflon gaskets (in order to withstand high temperatures). During epitaxy, the substrate lies horizontally on a slightly hollowed graphite susceptor covered with SiC (horizontal thick line on the figure). A rather complex system of arms enables, thanks to the robot fastened at the bottom of the lower dome: (i) the independent or joint wafer and susceptor upward and downward movements (in order to load or unload the wafer) and (ii) the rotation of the {susceptor + wafer} assembly (30 turns per minute), in order to achieve a satisfactory spatial homogeneity of the deposited layers. The carrier and process gases are injected in a laminar way just above the surface of the wafer (input and output through quartz inserts), with the possibility to adjust the center to edge uniformity with micrometer screws. Their thermal decomposition on the surface is ensured with the help of two banks of 20 lamps able to deliver up to 2 kW each (a bank above and one below the chamber). In order to ensure a better spatial uniformity of the temperature, the heating zones are divided into two parts. Eight lamps out of the 20 of each bank point towards the center of the wafer, the 12 remaining towards its periphery (vertical and oblique beams on the figure). The control of the temperature is ensured with the help of two infra-red pyrometers, one looking directly at the surface of the wafer (#1), the other looking at the backside of the susceptor on which the wafer lies (#2).

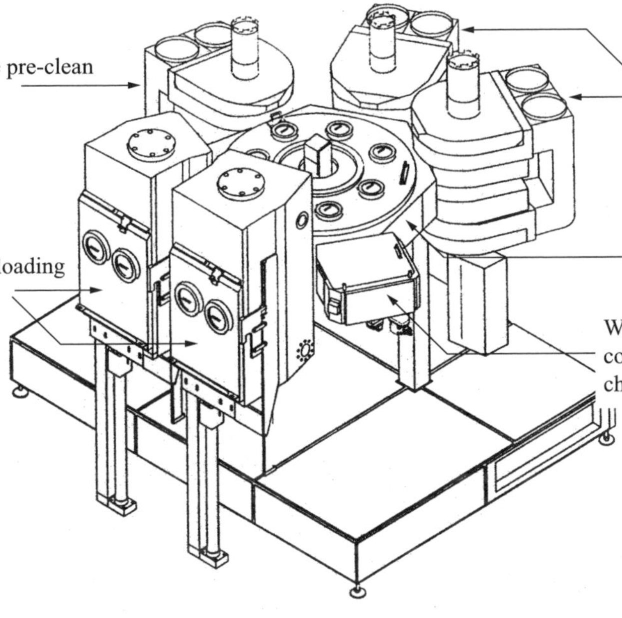

Figure 4.5. *Schematic description of the Epi Centura epitaxy cluster tool. The prototype low-temperature H_2 surface preparation chamber is no longer in service*

In a simplified way, these pyrometers operate according to the principle of the blackbody: the power radiated P is collected and converted in temperature T, using a law of power $P = \sigma T^4$, with $\sigma \sim 0.65$ for Si and ~ 0.8 for graphite covered by SiC. Concretely, the lower pyrometer is usually used to control the growths. The large surface emissivity fluctuations related to the use of blanket or patterned substrates, the upper dome wall clogging during deposition, the presence of a buried oxide, etc., indeed make the higher pyrometer temperature readings unreliable. The cooling of the whole chamber is ensured by: (i) water circulation loops inserted in metallic elements and (ii) a very important volume of air circulating (using a ventilator) in a loop around the chamber quartz walls. The growth pressures, accessible with this configuration of chamber (and pumps) are in the range of 10 - 600 Torr (atmospheric pressure = 740 Torr). The pressure is controlled thanks to a 1,000 Torr Baratron gauge and a butterfly valve at the output of thechamber. The growth temperature range is between 350°C (minimal temperature for a reliable reading of the temperature by the pyrometers) and 1,100°C (maximal temperature not generating long-term deterioration of the air tightness gaskets).

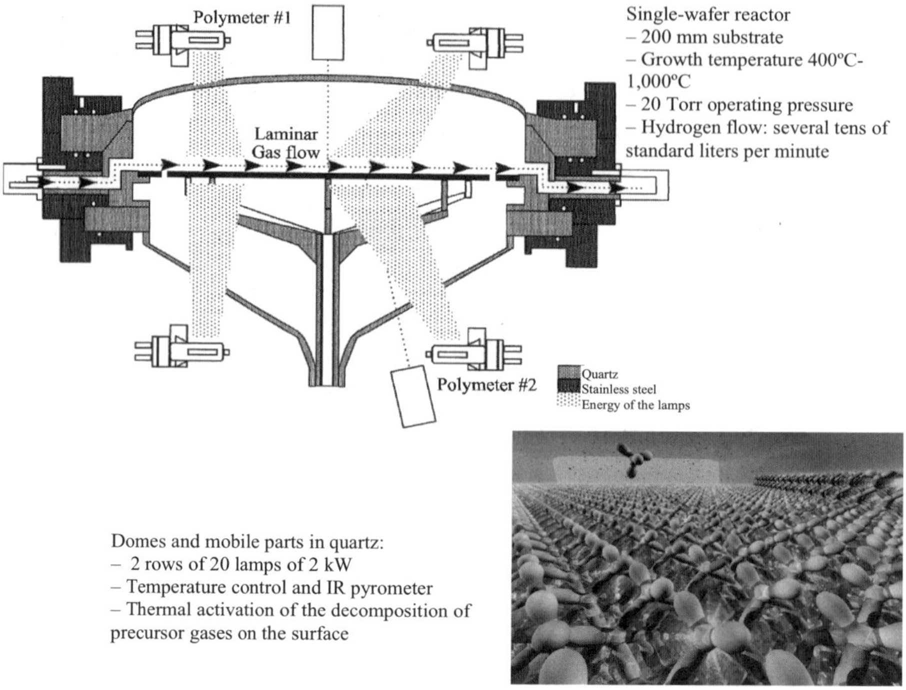

Figure 4.6. *Schematic description of the epitaxy chamber and artistic sight of a Si surface above which SiH_4 molecules are arriving*

Epitaxy chambers are cold under N_2 or hot under H_2. The latter is indeed used as a carrier gas during growth. The important H_2 mass-flow used (a few tens of standard liters per minute (slm), typically) lower the partial pressure of the gaseous impurities (typical rise in pressure of a chamber isolated from its pumping line: ~ 40 μTorr/min.) with values compatible with an epitaxy (i.e. some 10^{-10} Torr). Precursor Si gases are either pure dichlorosilane (SiH_2Cl_2), or pure silane (SiH_4). We will use SiH_2Cl_2, if we wish to have a growth which is selective versus SiO_2 or Si_3N_4

on patterned substrates. We will then deposit in the Si windows, not on these masking materials. SiH_4 will be implemented when the growth does not have to be selective. We will then deposit single-crystal Si in the Si windows and polycrystalline Si on SiO_2 or Si_3N_4. The Ge and C precursory gases are respectively germane (GeH_4) diluted at 2% in H_2 and monomethylsilane ($SiCH_6$) diluted at 5% in H_2. In order to n-type dope the crystalline matrix, i.e. to insert donor atoms which provide it with electrons, we will use phosphine (PH_3), a precursor gas of P. Diborane (B_2H_6), the precursor gas of B, will be used for the p-type doping (i.e. insertion of atoms that capture electrons from the matrix, providing it with holes). Concentrations in PH_3 and in B_2H_6, 2,000 parts per million (ppm) in the bottles (the dilution gas being H_2), can be, should the need arise, modulated between 5.54 ppm and 367 ppm, thanks to mixers. There is thus a factor ~ 4,000 between the minimum and maximum mass-flows of pure diborane and phosphine that can be delivered in the chambers. The last gas to be available is gaseous hydrochloric acid (HCl). It is either used in order to increase selectivity on patterned wafers, or implemented at high temperature (1,200°C), high pressure (300 Torr) and with mass-flows of more than 10 slm, in order to clean post-epitaxy the domes' quartz walls. We will note in the following of this chapter that the absolute values of the mass-flows are never given (in order to protect our know-how, considering that such a tool is commercially available). We can nevertheless say here that the *maximal* values of all the mass-flow controllers used to inject pure (SiH_2Cl_2, SiH_4 or HCl) or diluted (GeH_4, $SiCH_6$, PH_3 or B_2H_6) active gases in the chambers, range between 100 and 1,000 standard cubic centimetres per minute (sccm).

4.1.4. *Some general concepts of epitaxy*

Before talking about the results, we will recall here some definitions and concepts specific to epitaxy. Carrying out an *epitaxy* consists of depositing on a *single-crystal* substrate, a *single-crystal* layer adopting exactly the same crystalline structure as the structure of the substrate (case of III-V or IV-IV semiconductors crystallizing in the sphalerite or in the diamond structure, such as GaAs, Si, Ge, etc.), or a crystalline structure dictated by the substrate underneath, sometimes by means of rotations of the crystalline directions (metals on insulating substrates, etc.). We speak about *homoepitaxy* when the substrate and the deposited layer are of the same nature (for example: Si on Si). We speak about *heteroepitaxy*, when the substrate and the deposited layer are not of the same nature (for example: SiGe on Si).

Si/SiGeC stacks *in situ* doped or not (we will speak about *heterostructures*) studied during these works, have most of the time been deposited on a Si substrate with a (100) surface (bulk-type or not (presence of a buried oxide layer), blanket or patterned, etc.). It thus appears useful at this stage to point out the essential features of the elementary components of a SiGeC alloy. Si, Ge and C all three crystallize in the diamond phase, i.e. two overlapping sphalerite structures shifted one from the other by a quarter of the large diagonal (see Figure 4.7).

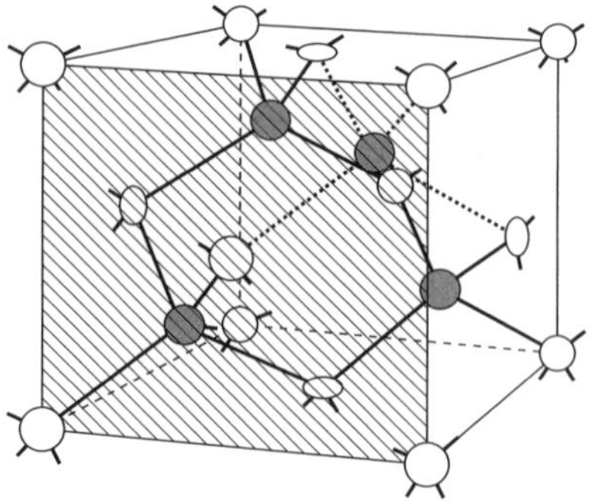

Figure 4.7. *Diamond structure and the usual (100) growth plane*

The lattice parameter of C (i.e. the length of one of the cube edges in Figure 4.7) is much lower than that of Si, itself lower than the one of Ge (a_c = 3.567 Å ⇔ a_{Si} = 5.43105 Å ⇔ a_{Ge} = 5.65785 Å). This increase in the lattice parameter is accompanied by a significant decrease in the energy band gap: Eg (C) = 5.48 eV ⇔ Eg (Si) = 1.11 eV ⇔ Eg (Ge) = 0.66 eV). We thus conceive intuitively that mixing Si, Ge and C in a SiGeC alloy will induce important energy band gap modifications between these limiting values, as well as the built-in stresses (see sections 4.2, 4.3, 4.6 and 4.7). We will however keep in mind that it is extremely difficult to incorporate C in substitution into a Si or Ge array. The miscibility of C in Si is indeed only 10^{-4}% at thermodynamic equilibrium (i.e. at 1,400°C), with an unfortunate tendency to form SiC precipitates for higher concentration values! On the other hand, Si and Ge are miscible in all proportions. During a heteroepitaxy, there will be in all probability a discrepancy in the lattice parameter between the deposited layer and the substrate. There will thus be an accumulation of elastic energy in it. For thin deposited thicknesses, atomic columns of the substrate will be extended in the epitaxial layer, then called *pseudomorphic*. A layer will be in *tension* (in *compression*), when its lattice parameter a_L in its bulk state is lower (higher) than the one of the substrate. The in-plane lattice parameter being, for a pseudomorphic layer, equal to the one of the substrate (i.e. $a_L^{//} = a_S$), we have for a tensily-strained layer $a_L^{\perp} < a_L < a_L^{//} = a_S$, and for a compressively-strained layer, $a_L^{\perp} > a_L > a_L^{//} = a_S$. Indeed, the perpendicular lattice parameter is given by $a_L^{\perp} = a_L + 2c_{12}/c_{11}(a_L - a_S)$, with $2c_{12} / c_{11}(Si) = 0.77$, $2c_{12} / c_{11}(Ge) = 0.74$ and $2c_{12} / c_{11}(C) = 0.61$. However, when the thickness of the deposited layer is more important, it will at some point become energetically favorable to inject misfit dislocations in order to accommodate the lattice parameter mismatch and minimize the total energy stored. The layer will be called *relaxed.* The thickness, to which the transition from pseudomorphic to plastically relaxed layer occurs, will be called *critical thickness for plastic relaxation*. The various configurations of the {SiGe on Si} system is schematically illustrated in Figure 4.8.

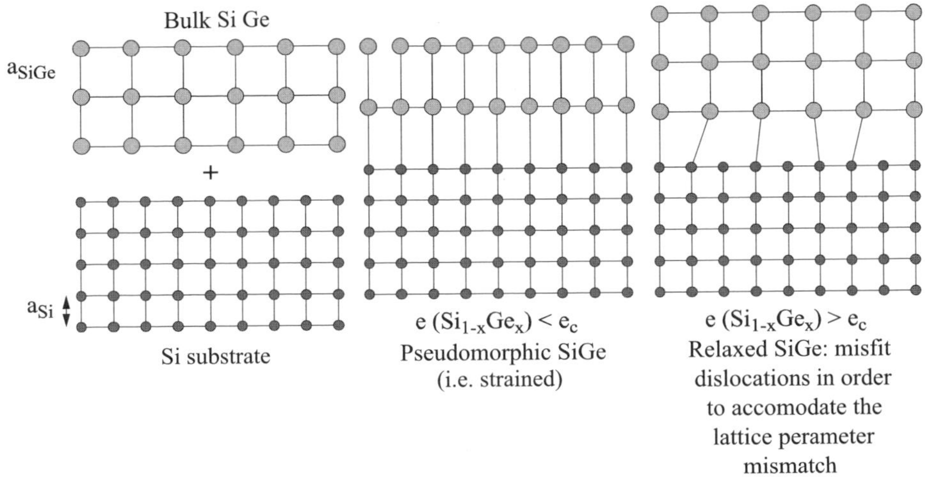

Figure 4.8. *Schematic illustration of atomic arrangement during an epitaxy of a SiGe layer on a Si substrate*

We represented in Figure 4.9 the critical thickness for plastic relaxation of SiGe on Si(100) as a function of the Ge concentration. The lowest curve corresponds to the critical thickness for plastic relaxation at thermodynamic equilibrium. It gives us the limiting values of the thickness not to be exceeded for SiGe layers deposited at (or submitted later on to) high temperatures. The highest curves, as well as the symbols, give us the critical plastic thicknesses associated with SiGe layers deposited at lower temperatures (and with significant growth rates). In fact, there are three thickness domains: (i) a domain of effective stability of the compressive stress below the critical thickness curve at the thermodynamic equilibrium; (ii) a *metastability* domain (hatched zone between the curves of Figure 4.9), where it will be necessary to pay attention to thermal budget minimization during growth and subsequent process steps, so that the layers remain in compression; and (iii) a high thickness domain for which SiGe layers will be partially, or even fully relaxed. It will be advisable to add for the high Ge concentrations and/or the high growth temperatures, an additional concept, i.e. the concept of *critical thickness of elastic relaxation* (located below the critical plastic thickness), above which the SiGe surface will become rough.

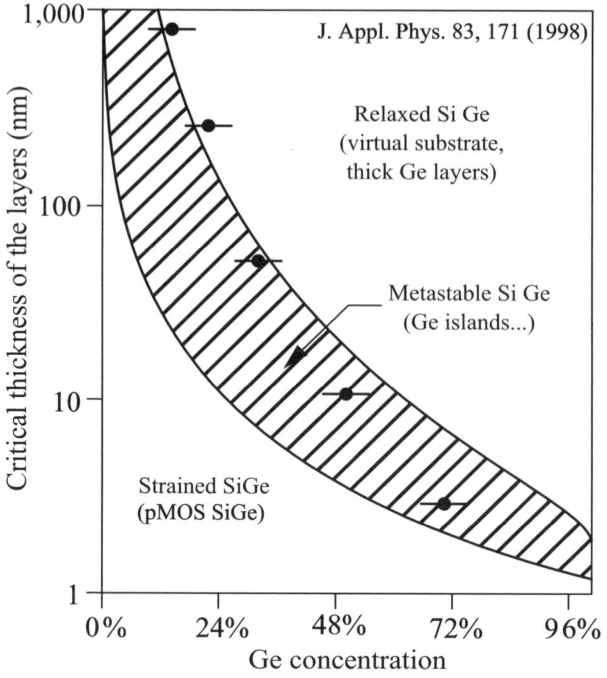

Figure 4.9. *Critical thickness for plastic relaxation (in nm) as a function of Ge concentration for SiGe layers deposited on Si(100) substrates. It should be noted that the z scale is logarithmic!*

SiGe has indeed this specificity, when it is compressively-strained on Si, to try to minimize the elastic energy stored by ripples of the growth front, as we will see in the following sections.

We will not describe furthermore the structural, as well as electric or optic properties of the SiGeC alloys. For further reading refer to the excellent works [CLA 07, CRE 06, DIM 07, KAS 00].

4.2. Engineering of the pMOSFET transistor channel using pseudomorphic SiGe layers

4.2.1. *Introduction*

It can prove extremely interesting to insert under the gate of a pMOS transistor $Si/Si_{1-x}Ge_x/Si$ stacks (see Figure 4.10), in order to benefit from holes mobility gains (factor of up to 2-3), associated with this type of heterostructure [SCH 97, WHA

98]. Because of the concentration in Ge and the compressive strain in the $Si_{1-x}Ge_x$ channel induced by its pseudomorphic deposition on Si(001), a type II band alignment occurs between Si and $Si_{1-x}Ge_x$, with a small conduction band edge shift (about ten meV) and an important shift ΔEv of the valence band edge (the $Si_{1-x}Ge_x$ layer behaving like a quantum well for the holes). Several studies have been devoted to the determination of $\Delta Ev = Ev(Si_{1-x}Ge_x) - Ev(Si)$ according to the Ge concentration *x*. Van de Walle and Martin predicted in 1986 the following theoretical linear dependence: $\Delta Ev = 0.84x$ (eV) [WAL 86]. From low-temperature photoluminescence measurements on $Si_{1-x}Ge_x$ layers in compression, encapsulated by Si, Dutartre *et al.* [DUT 91] and Robbins *et al.* [ROB 92] obtained at the beginning of the 1990s, the following parabolic relations: $\Delta Ev = 1.01x - 0.835x^2$ (eV) and $\Delta Ev = 0.874x - 0.376x^2$ (eV).

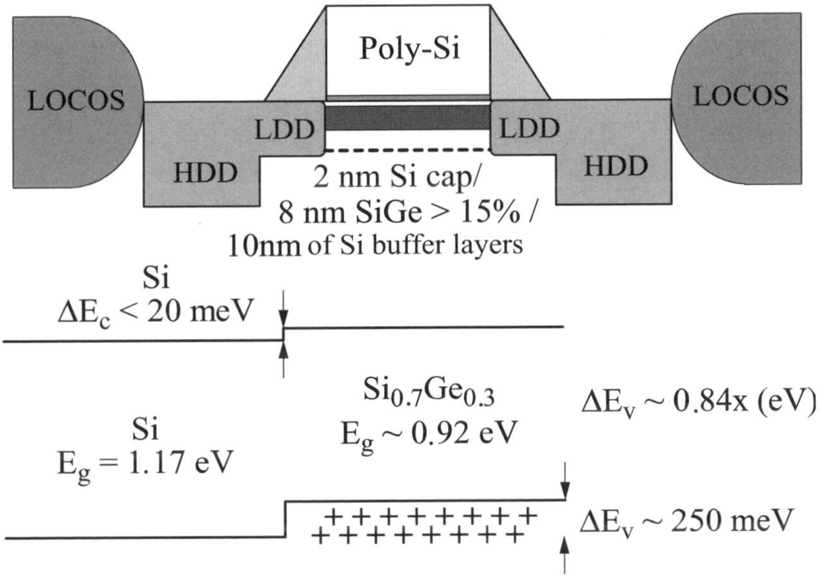

Figure 4.10. *Schematic representation of a pMOSFET transistor of small length gate (sub. 0.1 μm) incorporating in the channel zone a Si/SiGe/Si stacking and an alignment of the valence and conduction bands between Si and* $Si_{0.7}Ge_{0.3}$

From the epitaxy point of view, a certain number of pre-requisites must be answered, so that integration proceeds satisfactorily. First of all, it is essential that the growth takes place selectively compared to the insulating oxide, whatever the chosen insulation scheme (i.e. local oxidation of silicon – LOCOS, shallow trench isolation – STI, etc.). For that, we will have to use a chlorinated growth chemistry, i.e. SiH_2Cl_2 + GeH_4 (with possibly the addition of HCl), combined with a low growth pressure (i.e. typically several tens of Torr). Once functioning points are acquired on blanket wafers (i.e. growth rate and Ge concentration for given mass-flows of the precursor gases, temperatures and pressures), the significant SiGe growth rate increase and the slight Ge content increase occurring when switching from blanket to patterned substrates with different size Si windows will have to be quantified (the so-called “global” and “local” loading effects). Finally, it will be advisable to reduce for high Ge contents the growth temperature, in order to avoid any elastic relaxation of the stresses accumulated in the $Si_{1-x}Ge_x$ layer through the formation of a rough growth front. We will cover these various points in the following sections. We will then present the hole mobility gains that can be obtained in pMOSFETS transistors thanks to such stacks.

4.2.2. *Growth kinetics of Si and SiGe in chlorinated chemistry*

During the last ten years we have thoroughly explored Si and SiGe growth kinetics in chlorinated chemistry on (001) surfaces [HAR 02a, HAR 02b, HAR 03a, HAR 03b, HAR 04a, HAR 05a, HAR 07a]. As an example, we will look at the impact of the HCl mass-flow on the Si growth rate, at various temperatures and with dichlorosilane as a gaseous precursor. In the same way, we will quantify the influence of the HCl mass-flow on the SiGe growth rate and Ge concentration at 700°C, with dichlorosilane and germane as precursor gases. Lastly, we will briefly deal with the impact of growth temperature on the SiGe growth kinetics.

4.2.2.1. *Growth kinetics of Si in chlorinated chemistry*

Figure 4.11 presents the growth rate of Si as a function of the reverse absolute growth temperature for dichlorosilane only or for dichlorosilane + gaseous hydrochloric acid [HAR 02b]. These growth rates have been obtained via differential weighing of thick Si layers deposited on 200 mm Si(001) substrates. First of all, we notice the existence of two regimes, known as “high temperature”, controlled by the amount of precursor gases sent to the sample surface, and another one known as “low-temperature”, controlled by the desorption of the H or Cl atoms present on the surface. In the “high temperature” regime, the growth rate is dictated by the precursor gas mass-flow, as illustrated by the factor two on the growth rate between DCS/H_2 = 0.0025 and DCS/H_2 = 0.00125 mass-flow ratios. To add HCl almost does not decrease the growth rate in this regime, as we can see in Figure 4.2. An activation energy equal to 4.1 - 4.2 $kcal.mol.^{-1}$ is associated with the low increase of the growth rate in this regime. The growth rate in the “low-temperature” regime, which strongly drops when the growth temperature decreases, is controlled by the number of dangling bonds not passivated by H or Cl atoms which are available for the adsorption and decomposition of the SiH_2Cl_2 molecules.

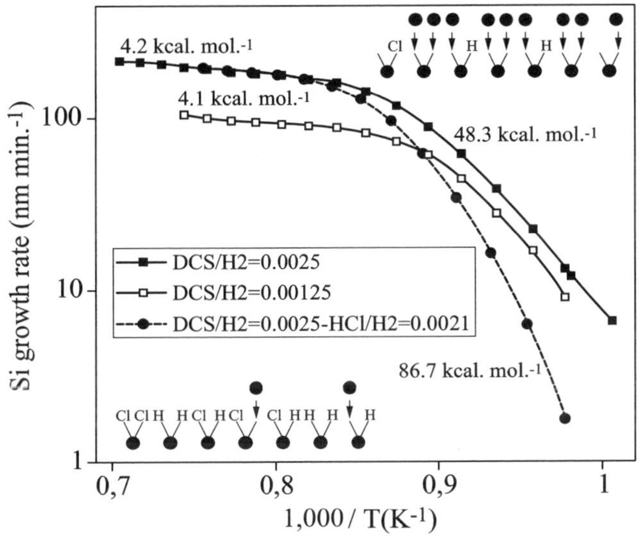

Figure 4.11. *Arrhenius plot of the Si growth rate as a function of the reverse absolute growth temperature, for various SiH_2Cl_2 / H_2 and HCl / H_2 mass-flow ratios*

For this reason, we can note that the activation energy associated with the strong increase with temperature of the Si growth rate is for SiH_2Cl_2 only very close to the Si-H bond energy, i.e. 47 kcal.mol.$^{-1}$ [SIN 89]. Meanwhile, the activation energy associated with the growth of Si thanks to a SiH_2Cl_2 + HCl chemistry, is close to the energy associated with the Si-Cl bond, i.e. 90 kcal.mol.$^{-1}$ [GUP 90]. Adding HCl induces, for a fixed SiH_2Cl_2 mass-flow and at a given temperature, a significant drop in the growth rate. We modelled, based on the relationship suggested by Kongetira *et al.* [KON 97], the growth rate dependence on temperature and SiH_2Cl_2 and HCl mass-flows in the "low-temperature" regime (i.e. T < 1,100K). We found the following relationship [HAR 03a] (PDCS, PH_2 and PHCl being the partial pressures in Torr of the gases in question):

$$GR(nm/min.) = 6.4x10^{10} e^{-\frac{49.3 kcal.mol.^{-1}}{k_B T}} (P_{DCS})^{0.36} (P_{H_2}) - 1.4x10^{9} e^{-\frac{34.3 kcal.mol.^{-1}}{k_B T}} (P_{HCl})^{0.52}$$

4.2.2.2. *Growth kinetics of SiGe in chlorinated chemistry*

We have also studied the SiGe growth kinetics at 750°C [HAR 02a, HAR 02b], 700°C [HAR 03b], 650°C [HAR 04a], 600°C [HAR 07a] and finally 550°C [HAR 05a] as a function of the SiH_2Cl_2, GeH_4 and HCl mass-flows. For these studies, we used high-resolution X-Ray diffraction (see Figure 4.12), to extract, for pseudomorphic SiGe layers deposited on Si(001), their Ge concentration (from the angular deviation between the peak associated with the Si substrate and the peak associated with the SiGe layer) and their thickness (from the angular width of the peak associated with the SiGe layer). In all cases, we could confirm the good crystalline quality of the layers (i.e. their pseudomorphic nature), thanks to the presence of thickness fringes on both sides of the SiGe layer peak. The nominal thickness of the SiGe layer was of course reduced when its Ge concentration increased, so that it always remained below the critical thickness for plastic relaxation [HUA 98].

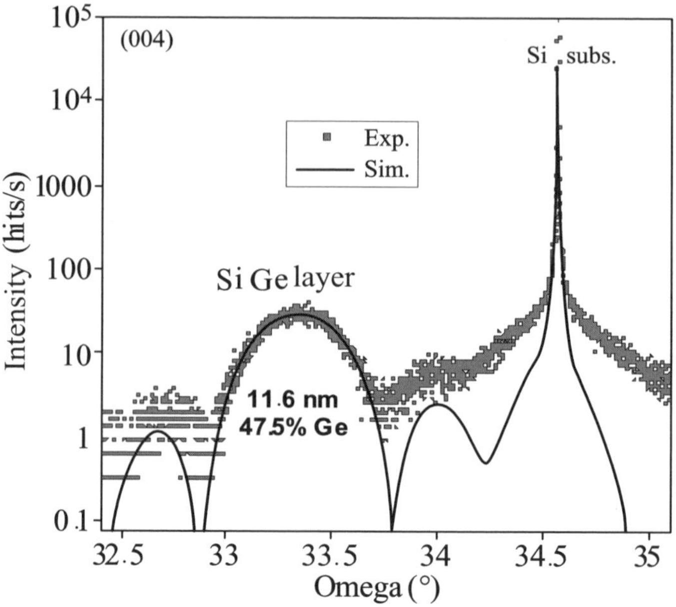

Figure 4.12. *Standard ω-2θ scan around the (004) X-Ray diffraction order for a 11.6 nm thick $Si_{0.525}Ge_{0.475}$ layer deposited at 550°C on Si (001)*

We will find in Figure 4.13 the Ge concentrations and the SiGe growth rates obtained at 700°C, 20 Torr for fixed SiH_2Cl_2 mass-flows ($F(SiH_2Cl_2)/F(H_2) = 0.0025$ or 0.01) and variable GeH_4 and HCl mass-flows [HAR 03b]. The dependence of the Ge concentration on the $F(GeH_4)/(F(SiH_2Cl_2) + F(GeH_4))$ mass-flow ratio is sub-linear. The addition of HCl leads, for given SiH_2Cl_2 and GeH_4 mass-flows, to a clear Ge concentration increase. Lastly, there is a strong discontinuity between the $F(SiH_2Cl_2)/F(H2) = 0.0025$ and $F(SiH_2Cl_2)/F(H_2) = 0.01$ Ge concentration curves, which seems related to a reaction of order 2 regarding the SiH_2Cl_2 adsorption, compared to a reaction of order 1 for the GeH_4 adsorption [ITO 95].

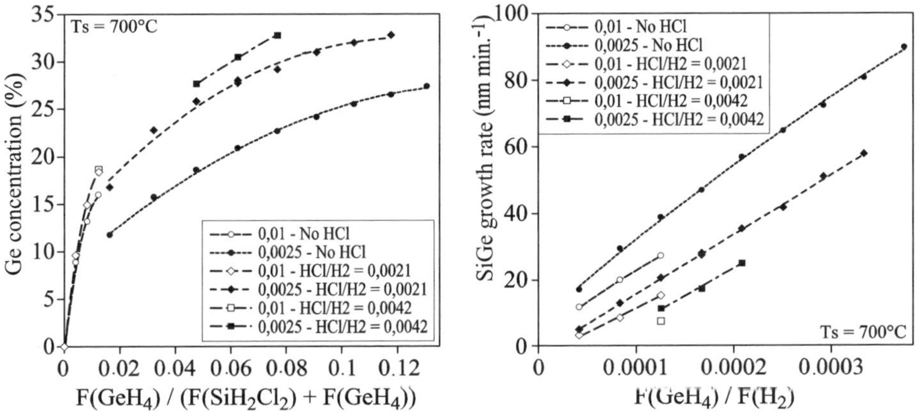

Figure 4.13. *Ge concentration (on the left) and SiGe growth rate (on the right) at 700°C, 20 Torr as functions of either the $F(GeH_4)/(F(SiH_2Cl_2) + F(GeH_4))$ or the $F(GeH_4)/F(H_2)$ mass-flow ratios, this for two SiH_2Cl_2 mass-flows ($F(SiH_2Cl_2)/F(H_2) = 0.0025$ or 0.01) and various HCl mass-flows (see figure inserts)*

We note a very marked, almost linear-increase of the SiGe growth rate, when the $F(GeH_4)$ /$F(H_2)$ mass-flow ratio increases. This increase is related to a major decrease in the H and Cl atoms surface coverage when Ge atoms are chemically adsorbed on the surface. The latter indeed act as preferential desorption centers of the H or Cl atoms on the surface, freeing sites for the adsorption of SiH_2Cl_2 and GeH_4 molecules. Indeed, energies associated with the Ge-H and Ge-Cl bonds, 37 kcal. mol.$^{-1}$ [SUR 84] and 51 kcal. mol.$^{-1}$ [KAM 98], are much lower than the binding energies of Si-H and Si-Cl, 47 kcal. mol.$^{-1}$ [SIN 89] and 90 kcal. mol.$^{-1}$ [GUP 90]. Adding HCl induces the formation of chlorides, chlorosilanes and chlorogermanes in the gaseous phase, decreases the quantity of Si or Ge atoms available for incorporation, and thus generates an important drop in the growth rate [CHA 87, KNU 94].

We illustrated in Figure 4.14 the influence of the growth temperature on the growth kinetics of SiGe with a SiH_2Cl_2 + GeH_4 chemistry (without HCl) [HAR 03b]. An increase in the growth temperature generates, for given SiH_2Cl_2 and GeH_4 mass-flows, a very clear increase in the growth rate, correlated to a significant decrease of the Ge concentration. From Arrhenius plots of the SiGe growth rate functions of the reverse absolute temperature, it is possible to go back up to the associated activation energy in the temperature range 550°C-650°C. This energy decreases in an almost linear way from 57 to 41 kcal. mol.$^{-1}$, when the concentration (at 600°C) goes from 18% to 44% [HAR 07a].

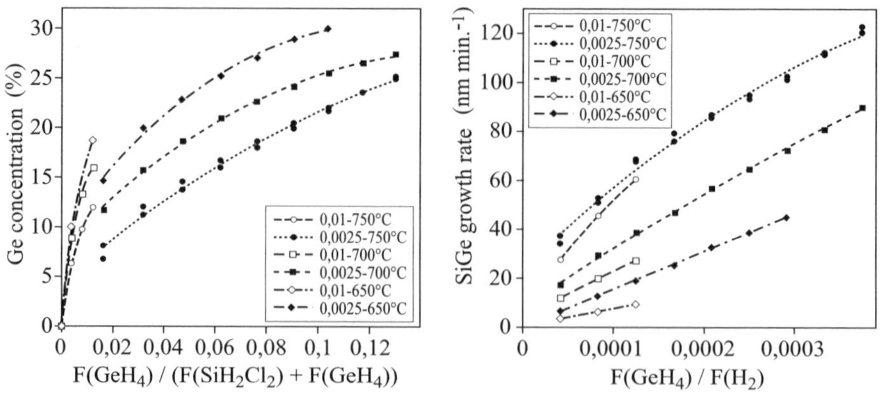

Figure 4.14. *Ge concentration (on the left) and SiGe growth rate (on the right) at 750°C, 700°C or 650°C as functions of either the $F(GeH_4)/(F(SiH_2Cl_2) + F(GeH_4))$ or the $F(GeH_4)/F(H_2)$ mass-flow ratios, this for two SiH_2Cl_2 mass-flows ($F(SiH_2Cl_2)/F(H_2)$ = 0.0025 or 0.01). Growth pressure was 20 Torr*

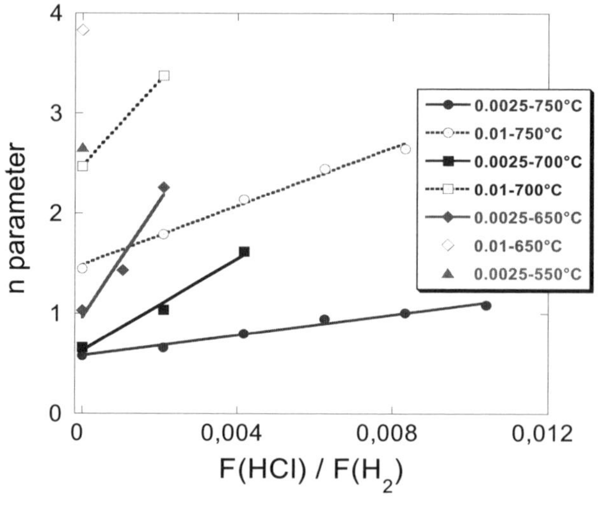

Figure 4.15. *Parameter n as a function of the $F(HCl)/F(H_2)$ mass-flow ratio, at various growth temperatures and for two $F(SiH_2Cl_2)/F(H_2)$ mass-flow ratios: 0.0025 and 0.01*

Regarding the x concentration in Ge, its dependence according to the $F(GeH_4)/(F(SiH_2Cl_2)$ mass-flow ratio can satisfactorily be modelled by the following relation: $x^2/(1-x) = n * F(GeH_4)/(F(SiH_2Cl_2)$ [SUE 00]. We reported in Figure 4.15 the n values thus obtained. The higher n is, the higher the concentration in Ge will be for given precursor gas mass-flows. It is confirmed that the higher the HCl flow is and the lower the growth temperature is (in the range 550°C-750°C), the higher *n* will be.

As an additional element, we will note an almost linear increase of *n* with the $F(HCl)/F(H_2)$ mass-flow ratio, with a steeper slope when the growth temperature decreases. In addition, we will note that it is possible to model the dependence in temperature of *n* with a given HCl flow, with laws of the $n = n_0 * exp(-E_a/k_BT)$ type, with negative "activation energies". As an example, we find $n = 7.6x10^{-4} exp(13.3\ kcal.\ mol^{-1}/k_BT)$ for $F(SiH_2Cl_2)/F(H_2) = 0.0025$ and $F(HCl)/F(H_2) = 0$ in the range 550°C-750°C. This value of -13.3 $kcal.mol.^{-1}$ is very close to the value (-14 kcal. $mol.^{-1}$) for growth pressures lower than 20 Torr found by Sue and Lee [SUE 00].

4.2.3. *Transposition on patterned substrates*

As mentioned in the introduction, during the transition from blanket to patterned substrates with variable windows sizes, we are confronted with several phenomena proving to be extremely annoying to obtain planar SiGe layers of controlled thickness and concentration.

First of all, there are loading effects of two natures: global (strong increase of the growth rate and slight increase of the Ge concentration, while going from blanket to patterned substrates) and local (fluctuation of the growth rate and the concentration, when the Si windows size fluctuates) [BOD 97, GIA 03, HAR 03c, HAR 04a, HAR 05a, ITO 95, KOL 09, MEN 01]. As an example, we have reported in Figure 4.16, the standardized growth rates (i.e. the growth rate in a window divided by the growth rate on blanket substrate) obtained at 700°C, at given SiH_2Cl_2 and GeH_4 flows and for variable HCl flows [HAR 03c]. First, we note that, for null HCl flow, we have an increase by up to a factor 4 of the growth rate, due to a very important surface coverage (with this mask) by SiO_2: 92%. Adding HCl however makes it possible to appreciably reduce the increase of the growth rate. Lastly, we will note that the standardized growth rate fluctuates with the windows size. This growth rate increase is a function of the surface coverage by the dielectric masking material.

There is for example a standardized growth rate equal to 1.2 only for a 40% surface coverage by SiO_2 [HAR 04a]. In addition, we have highlighted (using a study in spectroscopic ellipsometry) a Ge concentration increase fluctuating between 2% and 4%, when switching from blanket substrates to large windows on patterned substrates [HAR 04a]. It is thus suitable to gauge *for each mask type* the loading effects, global as well as local. We can also be confronted with elastic strain relaxation via the formation of surface undulations when switching from blanket to patterned substrates [BAS 01, HIK 09, MEN 03].

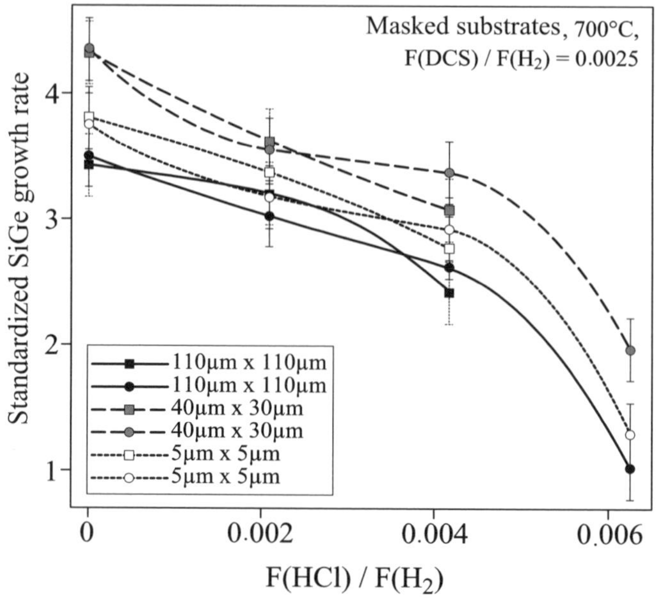

Figure 4.16. *Standardized SiGe growth rates in Si windows of variable size (110 μm x 110 μm, 40 μm x 30 μm, 5 μm x 5 μm) functions of the F(HCl) / F(H_2) mass-flow ratio for the following F (SiH_2Cl_2) / F(H_2) and F(GeH_4) / F(H_2) mass-flow ratios: 0.0025 and 1.25×10^{-4} (circles) and 0.0025 and 1.67×10^{-4} (squares). Growth temperature and pressure: 700°C and 20 Torr. Associated Ge concentrations, functions of the HCl and GeH_4 flows, were on blanket wafers in the 18% - 32% range*

To minimize the elastic relaxation, the growth temperature T should be lowered, when the Ge concentration of the layers to be deposited increases. Quantitatively, it is possible to selectively grow smooth, approximately 10 nm thick $Si_{0.85}Ge_{0.15}$ layers with $T \lesssim 750$°C. For $Si_{0.75}Ge_{0.25}$, it would be desirable that $T \leq 650$°C. For $Si_{0.65}Ge_{0.35}$, growth temperatures lower than or equal to ~ 600°C should be chosen. This is clearly illustrated in Figure 4.17, for 10 nm thick $Si_{0.65}Ge_{0.35}$ layers deposited selectively either at 650°C or at 550°C on patterned ultra-thin silicon-on-insulator (SOI) substrates [HAR 05a]. The surface of the $Si_{0.65}Ge_{0.35}$ layer deposited at 650°C is rough (root mean square (rms) roughness = 4.9 Å), while the surface of the $Si_{0.65}Ge_{0.35}$ layer deposited at 550°C is smooth (rms roughness = 0.9 Å). Lastly, once SiGe/Si stacks are satisfactorily grown, the thermal budget that will be seen by the wafers later on should be minimized. It is indeed completely possible that the elastic stresses stored in metastable stackings during the growth can be released (at least partially) by misfit dislocations emission during activation anneals after ion implantation of sources and drains, during the *in situ* H_2 bakes that are carried out prior to the growth of raised sources and drains, etc. We could also be confronted with chemical enlargements of the interfaces between Si and SiGe [HAR 03c].

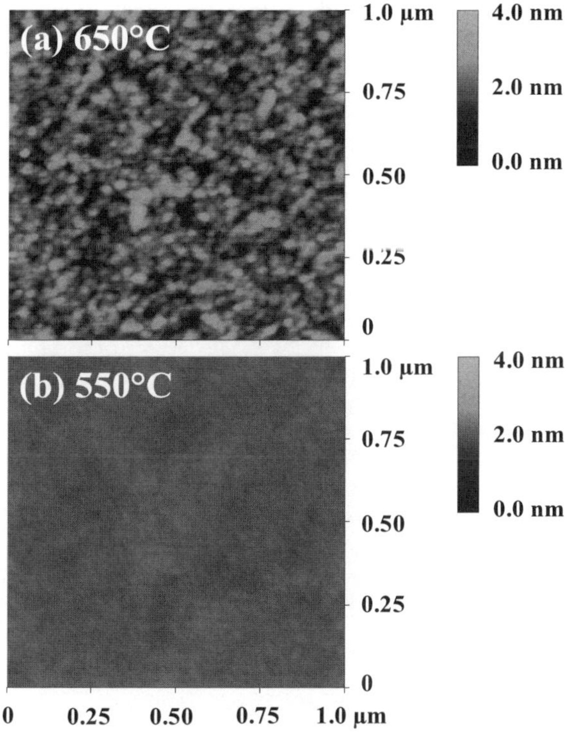

Figure 4.17. *1 µm x 1 µm AFM images of the surface of (a) a 10 nm thick $Si_{0.65}Ge_{0.35}$ layer selectively grown at 650°C and (b) an 11 nm thick $Si_{0.65}Ge_{0.35}$ layer selectively grown at 550°C inside the Si windows of patterned, ultra-thin SOI substrates*

4.2.4. *pMOS transistors incorporating SiGe layers*

These last few years, we proceeded to the integration of Si/SiGe(C)/Si stacks in short gate length pMOS transistors. Two Ge concentrations have been tested: 19% with a conventional SiO_2/Poly-Si gate stack [AND 03], and 28% with a more complex gate stack, i.e. HfO_2/TiN/W [WEB 04a, WEB 04b]. The nominal thickness for the Si buffer and cap layers were respectively about 10 nm and 3 nm, while the SiGe layers had a thickness either of 15 nm (for 19% of Ge), or of 8 nm (for 28% of Ge). We will find in Figure 4.18 cross-sectional high resolution – transmission electron microscopy (HR-TEM) images of a pMOS transistor of the second type.

First of all, we will note that the $Si_{0.72}Ge_{0.28}$ layer is planar and with a thickness close to the one aimed for [HAR 04a]. In addition, we will notice that the SiGe / HfO_2 interface is relatively abrupt. This is the result of a thorough study undertaken by Damlencourt *et al.* [DAM 04a] on the impact of the surface preparation of SiGe ("HF-last" (type A) or covered with a thin chemical oxide (type B)) on the morphology of HfO_2 layers deposited on top.

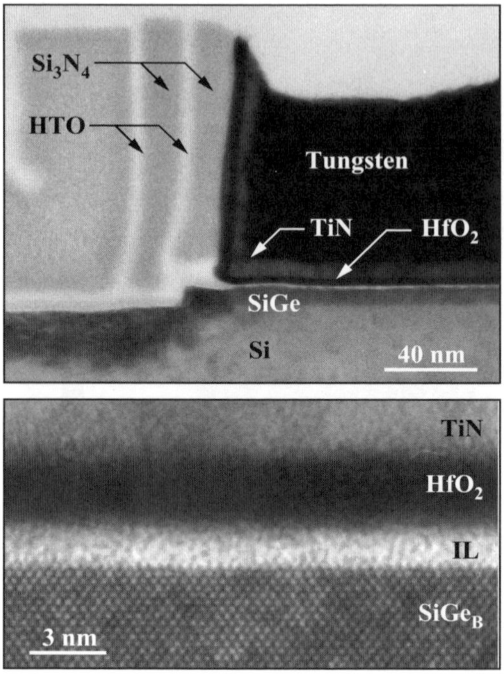

Figure 4.18. *Cross-sectional TEM images of a SiGe 28% pMOS transistor with a metal gate (damascene architecture)*

We represented in Figure 4.19 the holes mobility as a function of the effective electric field applied, for various gate stacks. Using a type B SiGeB/HfO_2 gate stack makes it possible to obtain a mobility gain of 65%, compared to the universal reference SiO_2/Si, and of 100% compared to HfO_2/Si stacks [WEB 04a]. Moreover, we can notice that an inappropriate surface preparation of SiGe prior to HfO_2 deposition (i.e. type A SiGe) generates a dramatic drop in the holes mobility. These studies also highlighted the need for reducing the thickness of the Si cap layer as much as possible, in order to avoid the appearance of a parasitic, lower mobility channel for holes at the Si/SiO_2 or Si/HfO_2 interface that would degrade the electric performances of sub-0.1 μm gate length transistors.

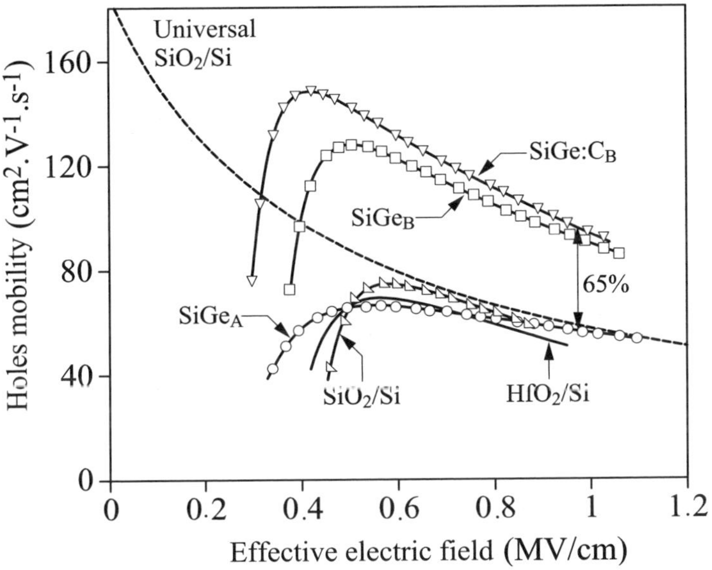

Figure 4.19. *Holes mobility as a function of the effective electric field for various stacks: reference SiO_2 / Si, HfO_2 / Si, $SiGe_A$ / HfO_2 and $SiGe(:C)_B$ / HfO_2*

4.3. Engineering of the nMOSFET transistor channel using pseudomorphic $Si_{1-y}C_y$ layers; SiGeC diffusion barriers

4.3.1. *Introduction*

Although the gain in terms of electrons mobility in buried, tensily strained $Si_{1-y}C_y$ channels still remains to be shown, it can prove to be interesting to insert $Si/Si_{1-y}C_y/Si$ pseudomorphic stacks beneath the gate of n-MOSFET transistors. Because of the tensile strain present in $Si_{1-y}C_y$ layers deposited on Si(001), there is indeed a partial lift of the degeneracy of the conduction band edge of $Si_{1-y}C_y$, with a twofold degenerated level then found below the conduction band edge of Si, and a fourfold degenerated level above it (see Figure 4.20). The edge discontinuity of the conduction band $\Delta Ec. = Ec(Si) - Ec(Si_{1-y}C_y)$, being of about 6.5*y* (in eV) [SIN 99], a $Si_{1-y}C_y$ layer inserted in Si will thus behave like a quantum well for electrons. The edge discontinuity of the valence band between Si and $Si_{1-y}C_y$ is generally regarded as negligible.

We can also profit from the fact that the C atom is much smaller than those of Si or Ge, to compensate by the addition of C the compressive strain present in SiGe layers. Indeed, let us assume that the $Si_{1-x-y}Ge_xC_y$ lattice parameter is given by the following formula (which takes into account the nonlinear dependence of the lattice parameter with the Ge concentration and which has as a hypothesis a linear dependence on the $Si_{1-y}C_y$ lattice parameter between Si and β-SiC): a $(Si_{1-x-y}Ge_xC_y) = 5.43105 + 0.2005x + 0.0263x^2 - 2.1428y$ (in Å).We will easily check that 1% of C atoms in substitutional sites compensates for the compressive stress induced by 10.5% of Ge atoms.

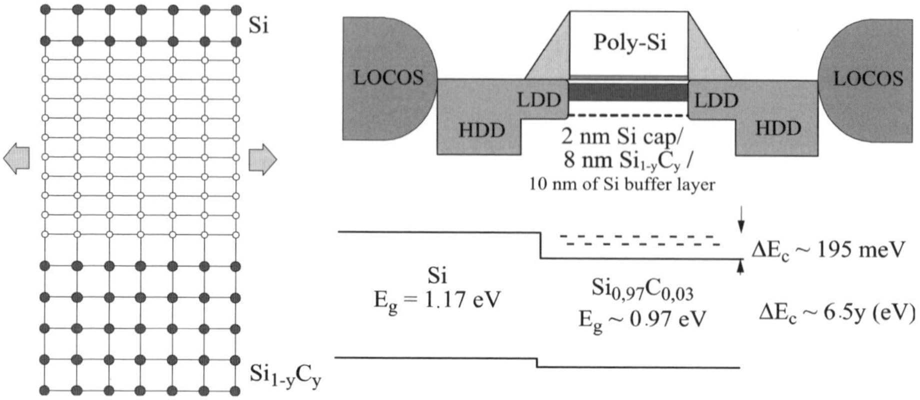

Figure 4.20. *Schematic representation of a short gate length (sub. 0.1 µm) nMOSFET transistor with in the channel region a Si / $Si_{1-y}C_y$ / Si stack (with tensily-strained $Si_{1-y}C_y$ layers) and alignment of the valence and conduction bands between Si and $Si_{0.97}C_{0.03}$*

C atoms incorporated in interstitial sites in SiGeC or $Si_{1-y}C_y$ layers otherwise block B atoms diffusion. This can be advantageously used in order to obtain retrograde doping profiles, in particular under bulk nMOS transistors gates (where multiple B ion implantation steps are used to adjust their threshold voltage) [RUC 98].

From the epitaxy point of view, it is important to know that, contrary to SiGe, miscible in all proportions, it is difficult to incorporate substitutional amounts of C atoms higher than one percent in a Si or SiGe lattice. Indeed, the C atoms will then tend to be incorporated not only in substitutional sites, but also in interstitial sites or in the form of β-SiC nano-precipitates. To reach high substitutional concentrations in C, as we will see later on, we should work with the suitable chemistry at the lowest possible temperature with still reasonable growth rates. It will also be necessary to minimize the post-epitaxy thermal budget, in particular during annealing, in order to avoid the transition of C atoms from substitutional to interstitial sites.

Finally, we will see that, in the case of $Si/Si_{1-y}C_y/Si$ stacks, it is essential to adopt an integration strategy, taking into account the non-selectivity of the deposition.

4.3.2. *Incorporation of C in Si and SiGe*

These last few years, we have worked a lot on the incorporation of C into Si and SiGe with either silane [HAR 02c, HAR 04b, HAR 05a, LOU 02, LOU 03a] or dichlorosilane as the Si gaseous precursor [LOU 03b]. We highlighted that a dichlorosilane chemistry made easier the incorporation of C in the substitutional sites of the lattice at given temperature and total quantity of C [LOU 03b]. We are however confronted, notably during $Si_{1-y}C_y$ growth, with much lower growth rates for dichlorosilane than for silane at a given temperature. We will thus have to use a silane chemistry for the growth of high sunstitutional C concentration $Si_{1-y}C_y$ and low Ge% SiGeC layers.

On the other hand, dichlorosilane chemistry will be chosen as soon as we wish to obtain high amounts of substitutional C atoms in SiGeC layers with percentages in Ge higher than or equal to 10%. First of all, we will take a look at SiGeC layers with concentrations in Ge higher than or equal to 10%. We will find in Figures 4.21 and 4.22, X-Ray Diffraction profiles and SIMS profiles, corresponding to SiGeC layers with 32% of Ge and a thickness close to 20 nm deposited at 550°C with a dichlorosilane chemistry on Si (001). We extracted from the XRD profiles, the concentrations in substitutional C atoms in our layers, by adopting a linear extrapolation between the lattice parameter of Si and that of β-SiC. First, we checked that the percentage of Ge was not altered in them, when the quantity of C atoms fluctuated. [LOU 03b]. SIMS profiles will provide us with the total concentration in C atoms (i.e. substitutional + interstitial).

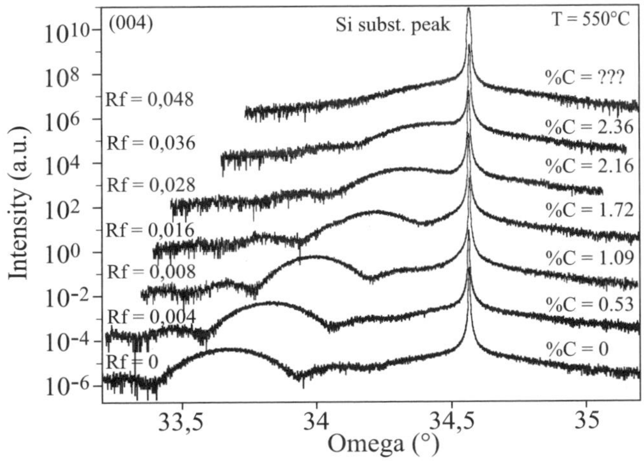

Figure 4.21. *ω -2θ profile around the (004) XRD order for SiGeC layers with 32% of Ge and variable substitutional C concentrations.* $R_f = F(SiCH_6) / (F(SiH_2Cl_2) + F(GeH_4))$

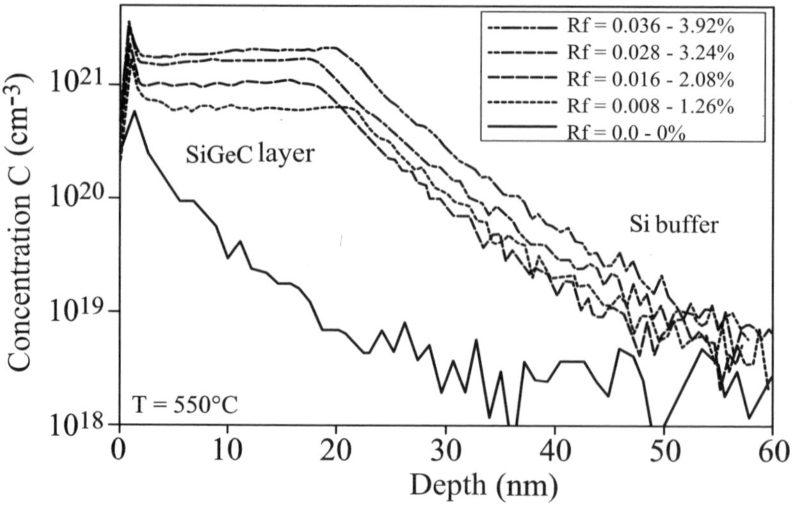

Figure 4.22. *SIMS depth profiles of the total C concentration in various SiGeC layers with 32% of Ge* $Rf = F(SiCH_6) / (F(SiH_2Cl_2) + F(GeH_4))$

We will find in Figures 4.23 and 4.24, concentrations in C atoms functions of the $Rf = F(SiCH_6) / (F(SiH_2Cl_2 \text{ or } SiH_4) + F(GeH_4))$, mass-flow ratio, with given Si gaseous precursor ($F(SiH_2Cl_2$ or $SiH_4) / F(H_2) = 0.0025$) and germane ($F(GeH_4) / F(H_2) = 1.25 \times 10^{-4}$) mass-flows and for various deposition temperatures (see Figure 4.23) or for various growth chemistry (see Figure 4.24). For low *Rf* values, almost all the C atoms are incorporated in substitutional sites. For higher *Rf* values, we have more and more interstitial C atoms, as attested by the growing discrepancy between the XRD and SIMS curves. Examining Figure 4.23, we notice that the lower the growth temperature is, the higher the quantity of C atoms in substitutional sites is. This tendency was confirmed, whatever the Ge concentration or the growth chemistry [HAR 04b, LOU 03a, LOU 03b]. Moreover, the use of a chlorinated growth chemistry leads to concentrations in substitutional C atoms which are higher than with a hydrogenated chemistry, as mentioned in the introduction and illustrated in Figure 4.24. We can thus reach for several tens of nm thick SiGeC layers, very high concentrations of substitutional C atoms: 1.42% for SiGeC 12% in Ge at 600°C, and 2.50% for SiGeC 32% at 550°C. These values are the highest reported nowadays in the literature for conventional Si precursors and growth strategies.

By comparison, Mi *et al.* have obtained (with the same hypothesis regarding the SiGeC lattice parameter) a maximum of 1.72% [MI 95]. The total quantity C_T in atoms C incorporated into the lattice almost does not depend on the growth temperature (see Figure 4.23). For silane chemistry, its dependence function of *Rf*, is given by $C_T / (1-C_T) = p[F(SiCH_6) / (F(SiH_4) + F(GeH_4))$, with $p = 1.10 - 1.34$ in the temperature range of 550°C - 650°C [LOU 03a]. For dichlorosilane chemistry, the dependence is not linear, as we can see in Figure 4.23. Experimental data are well accounted for with a $C_T^{1.4}$ / $(1-C_T)$ = q[F($SiCH_6$) / (F(SiH_2Cl_2) + F(GeH_4)) relationship, with q = 0.28 - 0.34 in the temperature range of 550°C - 650°C [LOU 03b]. Regarding the growth of high C concentration $Si_{1-y}C_y$ layers with silane chemistry, we highlighted the need to work at low-temperatures (i.e. 550°C), in order to maximize the concentration of C atoms in substitutional sites [LOU 03a]. Moreover, we showed that, at a given deposition temperature, it was extremely interesting to increase the absolute values of the silane and mono-methylsilane mass-flows, in order to increase the growth rate (from 0.4 to 1.3 nm $min.^{-1}$), and to thus maximize the concentration of substitutional C atoms [HAR 04b, HAR 05a] (see Figure 4.25). We thus could obtain 15 nm thick $Si_{1-y}C_y$ layers with 2.1% of C atoms in substitutional sites and only 0.1% of C atoms in interstitial sites. This value of 2.1% is the highest reported nowadays in the literature for the silane chemistry and this type of thickness (1.8% for Mitchell *et al.* [MIT 97]). Even higher concentrations of C atoms in substitutional sites, i.e. 2.8% [BAU 07], have recently been obtained thanks to trisilane (i.e. Si_3H_8), the liquid precursor of Si, and a cyclic {deposition/etch (with Cl_2)} growth strategy. Trisilane indeed decomposes at temperatures much lower than silane [GOU 09a].

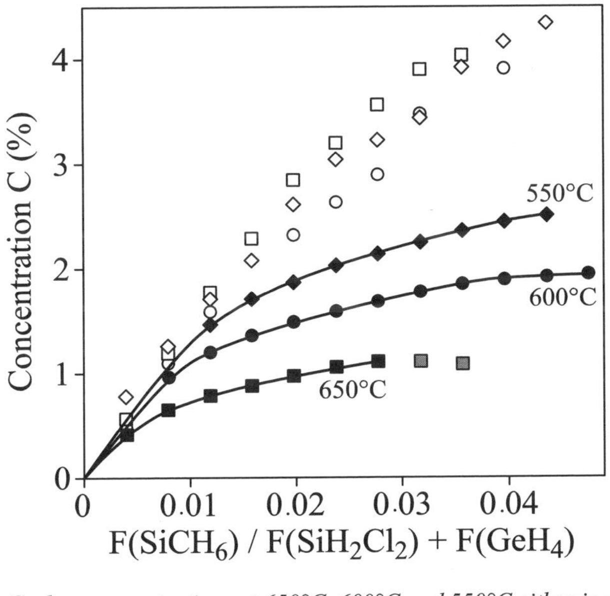

Figure 4.23. *Carbon concentrations at 650°C, 600°C and 550°C either in substitutional (XRD) or in substitutional + interstitial (SIMS) sites, for the same flows of dichlorosilane ($F(SiH_2Cl_2) / F(H_2) = 0.0025$) and germane ($F(GeH_4) / F(H_2) = 1.25x10^{-4}$)*

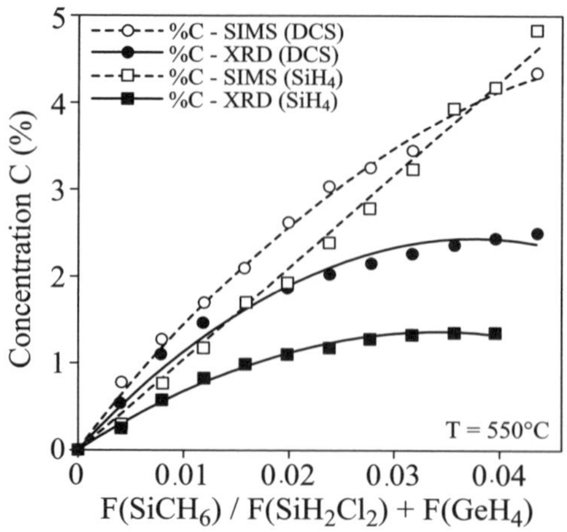

Figure 4.24. *Concentrations at 550°C of C atoms in substitutional (XRD) or substitutional + interstitial (SIMS) sites, for the same Si and Ge precursor mass-flows (i.e. $F(SiH_2Cl_2$ or $SiH_4)$ / $F(H_2)$ = 0.0025 and ($F(GeH_4)$ / $F(H_2)$ = $1.25x10^{-4}$)*

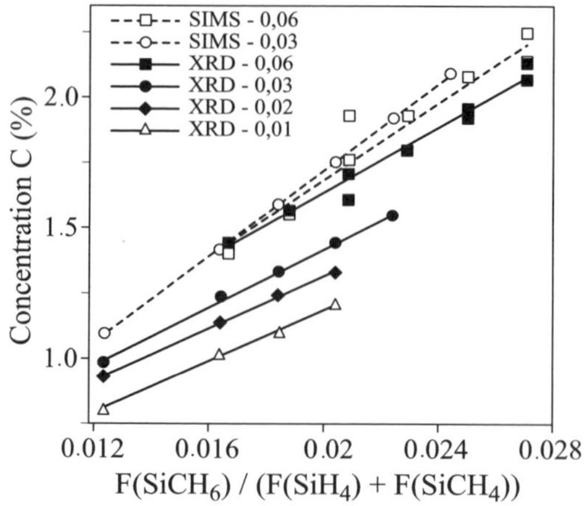

Figure 4.25. *Concentrations at 550°C of C atoms in substitutional (XRD) or substitutional + interstitial (SIMS) sites functions of the $F(SiCH_6)$ / $F(SiH_4)$ mass-flow ratio, this for four different silane flows: $F(SiH_4)$ / $F(H_2)$: 0.01, 0.02, 0.03 and 0.06*

4.3.3. *Si/Si$_{1-y}$C$_y$/Si stacks for nMOS transistors*

The aim being the integration of Si/Si$_{1-y}$C$_y$/Si stacks in nMOS transistors, we have studied their growth on patterned substrates and their stability with respect to various post-epitaxy thermal processes [HAR 02c, HAR 04b]. We will find in

Figure 4.26, an AFM image of the surface of a patterned substrate after the deposition of a {Si buffer (10 nm) / $Si_{0.984}C_{0.016}$ (7 nm) / Si cap (3 nm)} stack at 550°C with SiH_4. We note that, despite the adoption of a chlorinated chemistry for the growth of the Si buffer layer, depositing even 10 nm with silane does not make it possible to obtain selectivity with respect to SiO_2, as attested by the high roughness of the LOCOS surface. However, we can notice that the surface roughness in the Si windows of the deposited stack is very low. The corollary of this lack of selectivity is the absence of any loading effect. We will find in Figure 4.27, X-ray diffraction profiles of Si/$Si_{0.989}C_{0.011}$/Si stacks having been subjected post epitaxy to various thermal process steps. We notice that the angular position of the peak associated with the $Si_{1-y}C_y$ layer almost does not change, when a gate oxide has been formed and such stacks have been submitted to rapid thermal annealing. On the other hand, an *in situ* H_2 bake at 950°C during 2 minutes (that would be used prior to the selective epitaxial growth of raised sources and drains) generates an important angular shift of the XRD peak towards low angles, characteristic of the transition of C atoms from substitutional to interstitial sites (formation most likely of SiC nano-precipitates). It will thus be necessary, as for Si/SiGe/Si stacks, to minimize the post-epitaxy thermal budget.

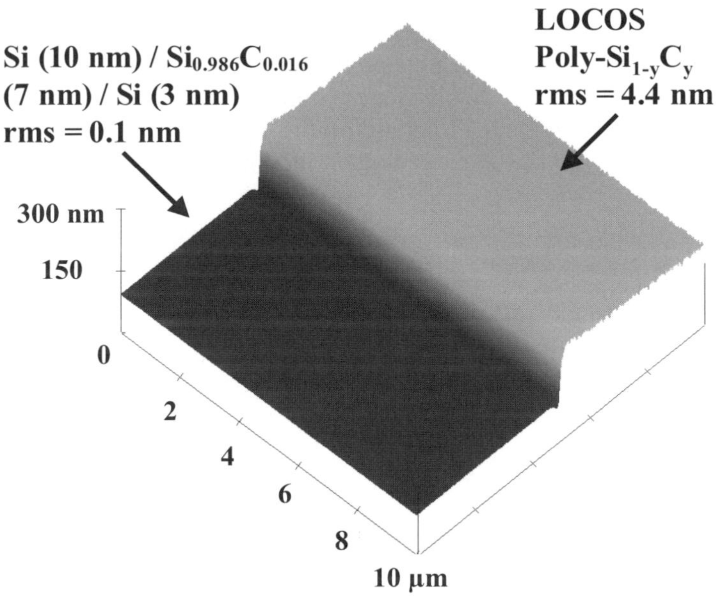

Figure 4.26. *AFM image in the proximity of a window edge of a Si/$Si_{1-y}C_y$/Si stacking, confirming the absence of selectivity with silane chemistry (conservation of the step height and LOCOS surface roughness)*

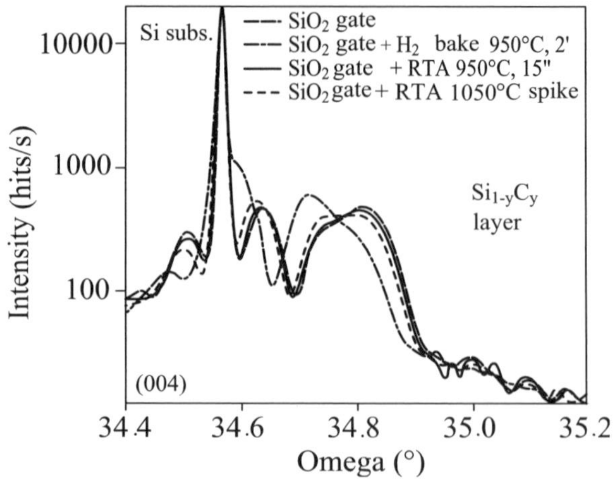

Figure 4.27. *ω -2θ scans around the (004) XRD order for Si/Si$_{0.989}$C$_{0.011}$/Si stacks after various thermal process steps*

We have also quantified the benefits of using such stacks as diffusion barriers for B. We will find in Figure 4.28, SIMS depth profiles of B in {poly-Si/SiO_2/Si/$Si_{1-y}C_y$/Si/Si substrate pre-implanted with B} stacks. Whatever the concentration in C and/or the epitaxy temperature, we obtain thanks to the presence of C atoms a more retrograde B doping profile in the channel (drop by a factor which can go up to 100 of the concentration in B in this region) [HAR 02c, HAR 04b].

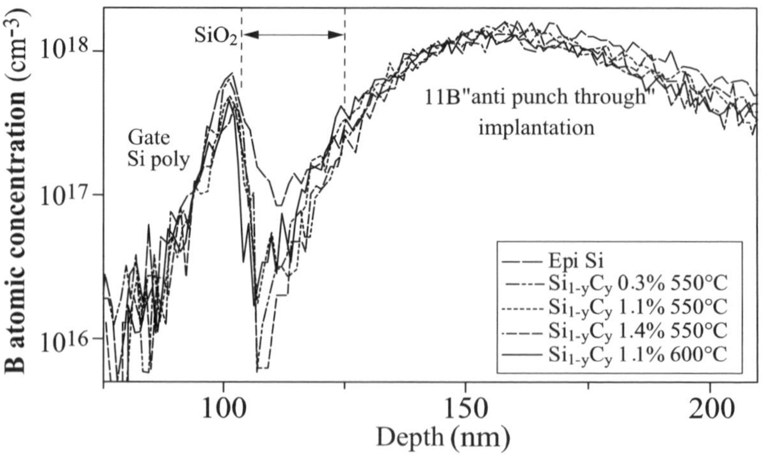

Figure 4.28. *SIMS profiles of the concentration of B for various stackings under the gate*

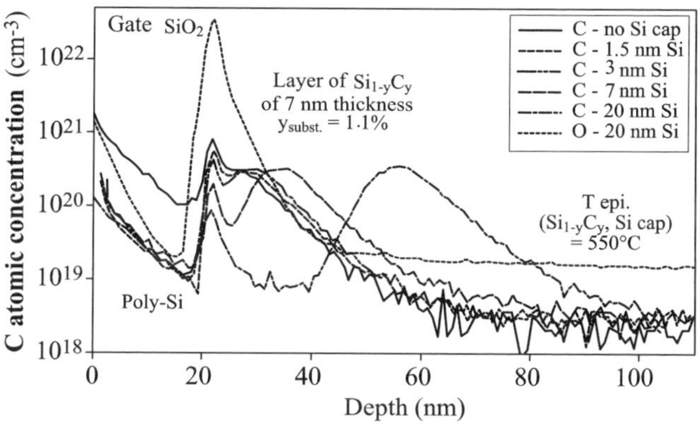

Figure 4.29. *SIMS profiles of the concentration of C for various stackings under the gate*

We have otherwise highlighted, from the electric point of view, the harmful influence of the C atoms, notably in interstitial sites, close to the gate [DUC 04a, ERN 02, ERN 03, WEB 03]. Finally, we have studied the impact of the Si cap layer thickness on the concentration of C atoms (see Figure 4.29). A drop by a factor 10 of the sheet concentration of C atoms in the SiO_2 gate dielectric occurs when the thickness of the Si cap layer goes from 0 to 20 nm [HAR 04b].

4.3.4. *nMOS transistors incorporating $Si_{1-y}C_y$ layers or SiGeC diffusion barriers*

At the end of 2004, we demonstrated the beneficial effect of double Si/SiGeC diffusion barriers on the electric performances of very short gate length MOS transistors [DUC 04b]. For that, we selectively deposited in the Si windows of patterned, bulk Si(001) substrates with STI isolation (5 nm Si buffer/ $Si_{0.88}Ge_{0.12}$:C 10 nm / Si 5 nm / $Si_{0.88}Ge_{0.12}$:C 10 nm / Si channel 5 nm) stacks. These diffusion barriers make it possible to control the doping profiles in the Si channel, as well as in the source and drain zones (presence of 0.3% of C in the SiGeC layers). Regarding pMOS transistors, a reduced junction depth and a lower sheet resistance of the sources and drains are obtained, thanks to very retrograde B implantation profiles (presence of C atoms in interstitial sites). For nMOS transistors, we clearly highlighted that the presence of Si/SiGeC multilayer reduces the diffusion coming from the implanted B pockets, and thus short channel effects due to an over-doping of the channel. We have, moreover, integrated Si 10 nm/$Si_{0.98}C_{0.02}$ 7 nm/Si 2 nm stacks right under the gate of fully-depleted silicon-on-insulator (FD-SOI) nMOS transistors with mesa-type isolation [DUC 06]. We will find in Figure 4.30 HR-TEM images of these transistors. These transistors show drain current I_d – gate voltage V_g characteristics similar to those of reference Si FD-SOI transistors (see Figure 4.31). $Si_{1-y}C_y$ curves are however shifted towards lower gate voltages than the Si ones. The

$Si_{1-y}C_y$ threshold voltage V_{th} reduction is about 0.13 ± 0.02 V, whatever the gate length (see Figure 4.32). However, we have seen above that the tensile strain in a $Si_{1-y}C_y$ layer deposited on a Si substrate, generates a lowering of the conduction band edge by about 65 meV by percent of C atoms in substitutional sites. This V_{th} reduction is in very good agreement with the expected conduction band edge shift between $Si_{0.98}C_{0.02}$ and Si (no additional V_{th} shift is induced by doping, which is nonexistent in the channel of a FD-SOI transistor). This result clearly shows how interesting it is to use a $Si_{1-y}C_y$ channel, to adjust the threshold voltage of nMOS FD-SOI transistors with a *midgap* metal gate (i.e. with a work function in the middle of the Si energy band gap). Threshold voltage tuning can then be done by varying the concentration in substitutional C atoms.

The same type of beneficial engineering of the threshold voltage (thanks to an adjustment of the alignment of the valence band between compressively-strained SiGe and Si) was highlighted in FD-SOI pMOS transistors with compressively-strained SiGe channels [AND 05b, HUT 10a]. Because of this V_{th} reduction, a gain of ~ 15% of the I_{on} on-state current was obtained in $Si_{0.98}C_{0.02}$ nMOS FD-SOI transistors with a gate length $L_g = 0.35\mu m$ at $V_g=V_d=1.2V$.

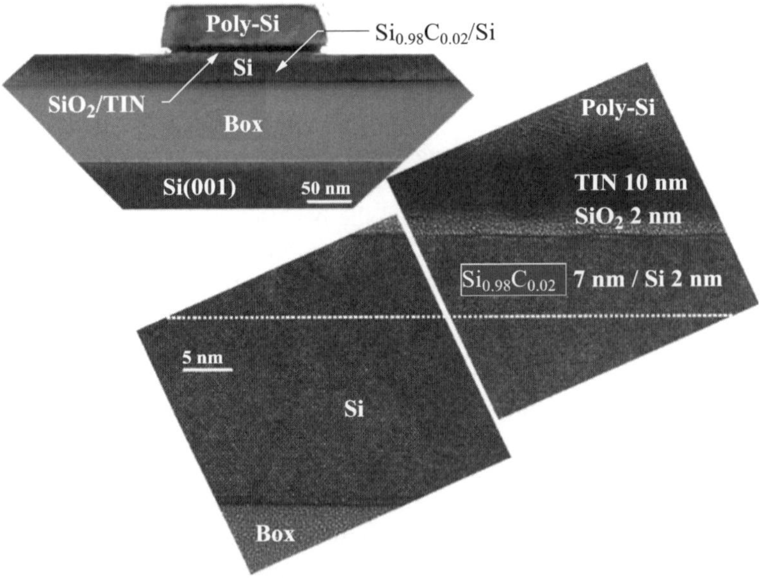

Figure 4.30. *HR-TEM images in cross-section of (from top to bottom) (i) a nMOS FD-SOI transistor with a Si/$Si_{0.98}C_{0.02}$/Si channel, (ii) the Poly-Si/TiN/SiO_2 gate stack and (iii) the Si cap/$Si_{0.98}C_{0.02}$/Si (20 nm)/buried oxide stack*

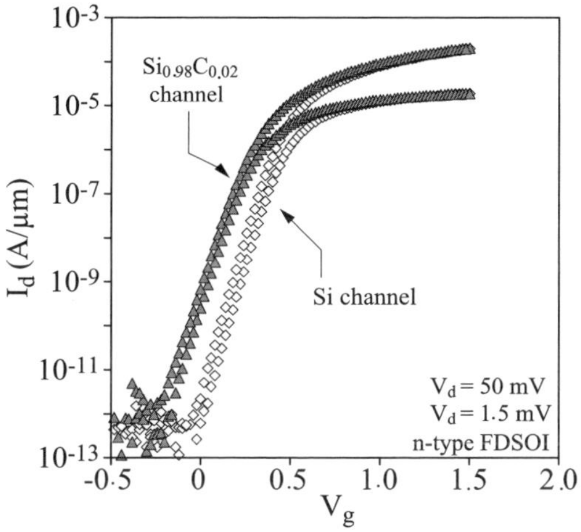

Figure 4.31. *I_d- V_g characteristics of a nMOS FD-SOI transistor with a $Si_{0.98}C_{0.02}$ channel and comparison with an FD-SOI device with Si reference channel (gate width = 0.2 μm, gate length = 0.35 μm)*

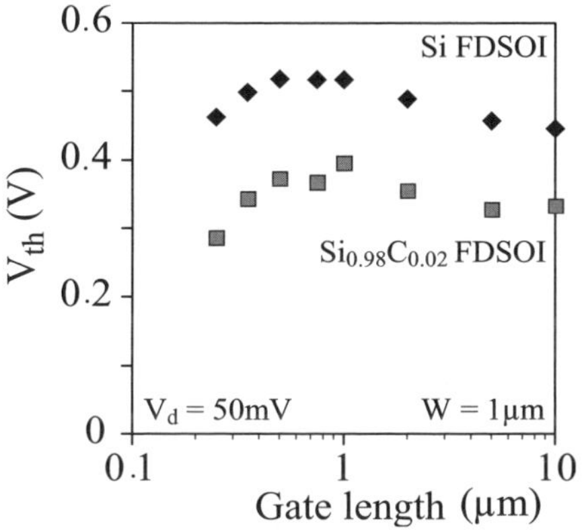

Figure 4.32. *Threshold voltage according to the gate length illustrating the constant V_{th} shift between $Si_{0.98}C_{0.02}$ and reference Si FDSOI devices due to the conduction band edge discontinuity*

4.4. Epitaxy of Si raised sources and drains on ultra-thin SOI substrates

4.4.1. *Introduction*

In the microelectronics industry, SOI substrates are increasingly used for the development of MOS transistors, because with them we can: (i) obtain very thin source/drain junctions, (ii) profit from low threshold voltages with low leakage currents (=> low voltage integrated circuits – low consumed powers), (iii) reduce the parasitic source-drain capacities for very powerful high frequency devices, but more energy-consuming, (iv) have circuits able to operate at high temperature, resistant to radiations, etc. However, the unceasing miniaturization in all dimensions of the components will require in the short term, to start from SOI substrates with much thinner Si and SiO_2 layers (i.e. e(Si) < 10 nm and e(SiO_2) < 100 nm), than those of the current commercial SOI substrates (e(Si) ~ 70 nm and e(SiO_2) ~ 145 nm).

We will find in Figure 4.33 a cross-sectional TEM image in the proximity of the gate of a MOSFET transistor fabricated on an ultra-thin SOI substrate. Under the gate, the Si film is 8-10 nm thick only; however, for transistors architecture reasons, it is necessary to grow Si in a selective way with respect to SiO_2 and Si_3N_4 (*the isolation regions between individual transistors and the sidewall spacers, respectively*) in the source and drains regions, in order to obtain in these zones final Si thicknesses of about 40 nm (for silicidation). We will then speak of Si raised sources/drains.

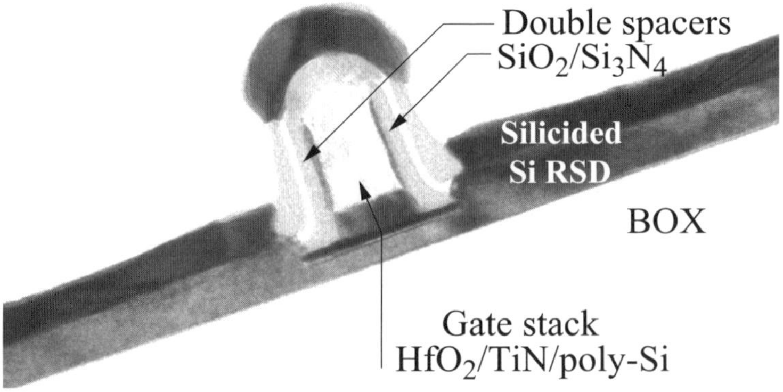

Figure 4.33. *Cross-sectional TEM image of a MOS transistor fabricated on an ultra-thin SOI substrate. The Si channel under the gate has a thickness close to 10 nm, ~ 10 nm of Si (raised extensions) then ~ 20 nm of Si (raised sources and drains) have been deposited selectively in the source and drains regions, using an optimized growth process which is selective with respect to SiO_2 and Si_3N_4 (see section 4.4.3 below)*

It can also be interesting to grow a few nm of Si on blanket ultra-thin SOI substrates, in order to obtain SOI substrates with a slightly thicker upper layer, but with much lower surface roughness and macroscopic dispersion of thickness on the wafer surface, as shown in [HAR 03a].

4.4.2. *Problems encountered on ultra-thin SOI substrates*

With a conventional Si Selective Epitaxial Growth process [HAR 03a] (i.e. an *ex-situ* "HF-last" wet cleaning in an automated wet bench followed by an *in situ* H_2 bake at 800°C for 2 minutes and a selective deposition of 40 nm of Si with a dichlorosilane + hydrochloric acid at temperatures typically around 800°C [MIY 99]) in the sources and drains regions of patterned, ultra-thin SOI wafers (i.e. Si thickness under the gate : 8-10 nm, in the sources and drains regions : 6-8 nm), we are confronted with two catastrophic problems for MOS transistors :

–Si withdrawal (i.e. "moat recess") at the boundaries with the isolation oxide;

–Si film which loses its continuity, with the formation of islands.

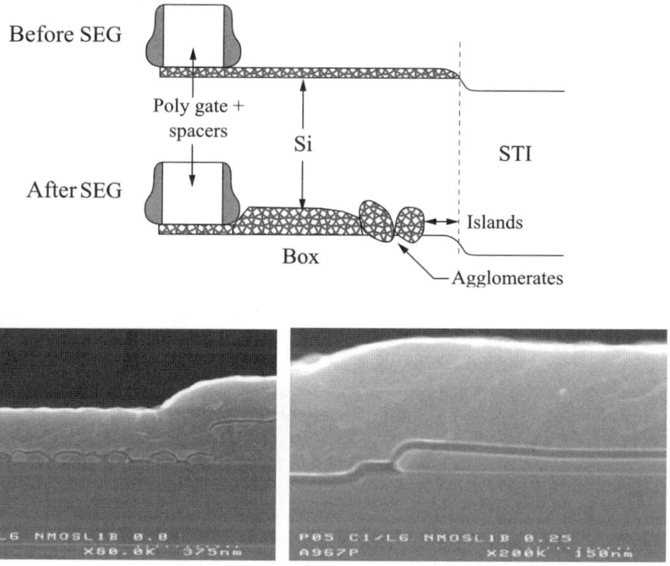

Figure 4.34. *Schematic representation and cross-sectional Scanning Electron Microscopy (SEM) images of the problems encountered with a conventional Si SEG process on an ultra-thin patterned SOI substrate (mesa isolation scheme): withdrawal ("moat recess") at the boundaries of the active regions with the isolation oxide and agglomeration or islanding of the initially continuous Si film*

This Si agglomeration phenomenon on an ultra-thin SOI substrate has already been highlighted by Ono *et al.* [ONO 95] and by Ishikagawa *et al.* [ISH 02]. It occurs at increasingly lower temperatures, as the Si thickness reduces. In addition, going from a blanket to a patterned SOI substrate definitely decreases, at a given Si thickness, the temperature above which the Si layer becomes discontinuous with the formation of islands.

4.4.3. *Method developed in response*

In order to solve these "moat recess" and islanding problems, we have optimized the steps that follow [JAH 05]:

(i) *in situ* H_2 bake before the Si growth

On these ultra-thin SOI substrates (thickness of the Si layer < 8 nm), we should use annealing temperatures lower than or equal to 700°C. Indeed, for temperatures higher than 700°C, there are indeed a withdrawal at the edge of the active zone and the formation of discontinuous islands, as illustrated in Figure 4.35.

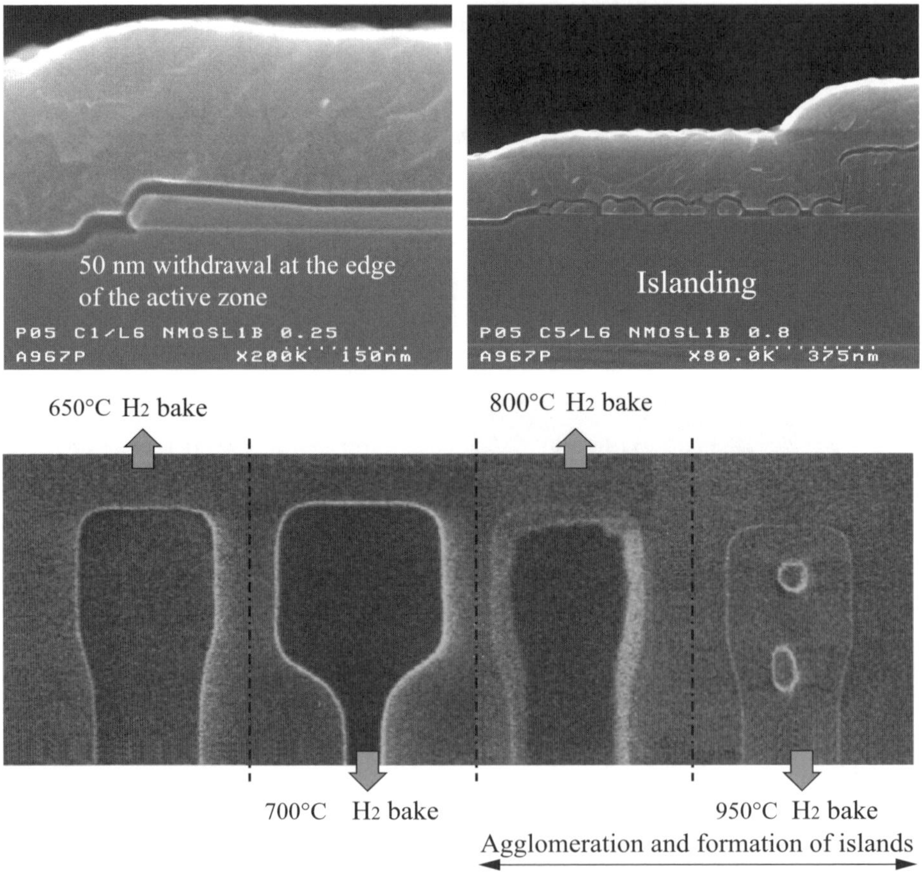

Figure 4.35. *Illustration of the effect of the H_2 bake temperature on the morphology of an ultra-thin, patterned Si layer on BOX (here: 8 nm). For two minute bakes, we see that for temperatures higher than 700°C, we have a withdrawal of the film at the edges of isolated areas, then an agglomeration in islands*

Using such H_2 bake temperatures is in opposition to the way we usually proceed, in order to prepare a Si surface before epitaxy. Indeed, in order to get rid of any trace of residual contaminants (such as C, O or F atoms) on a "HF-last" Si surface, a few minutes *in situ* H_2 bake at temperatures higher or equal to 775°C, must be used [HAR 03a]. Given that a perfect crystalline quality of Si or SiGe raised sources/drains is not essential for the technological steps used later on (i.e. ion implantation of dopants (usually As, B, P) making the layer amorphous followed by {recrystallization /electric activation of the dopants} spike anneals), some atomic contaminants at the {raised sources and drains / initial Si film} interface can be tolerated. Low temperatures H_2 bakes guarantors of a good stability of the ultra-thin Si layers can thus be used prior to SEG.

(ii) Si selective epitaxial growth

Low growth temperatures (i.e. lower than or equal to 750°C) and a chlorinated chemistry (i.e. dichlorosilane + hydrochloric acid), which is selective with respect to SiO_2 and Si_3N_4, should be used in order to thicken the Si area of ultra-thin (< 10 nm), patterned SOI wafers [HAR 04a]. We will find in Figure 4.36, an example of selective deposition of a 60 - 65 nm Si layer on an ultra-thin SOI substrate. For initial Si thicknesses of 6 nm and more, we have only little or almost no withdrawal at the edge of the active region, and no agglomeration. This method was intensively used in the last few years on standard FD-SOI electric lots, on FD-sSOI lots (with tensily-strained Si layers (see section 4.6)), on electric FD-SOI lots with not self-aligned double gates, etc. For more details concerning the obtained electrical properties, readers' should refer to [AND 05b, AND 07, BAR 07, VIN 05].

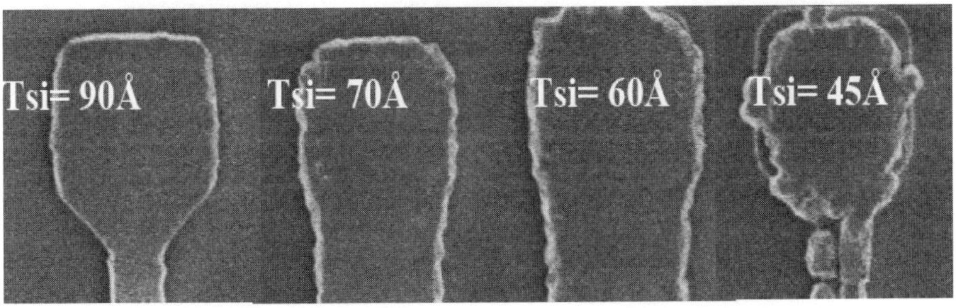

Figure 4.36. *Morphology of the Si layer after a 650°C, 2 minutes H_2 bake followed by the SEG of 60-65 nm of Si at 750°C with a chlorinated chemistry (P = 20 Torr, $F(SiH_2Cl_2)/F(H_2) = 0.01$, $F(HCl)/F(H_2) = 0.00208$). We will note that for initial thicknesses of the Si film higher than or equal to 6 nm, we have no withdrawal or agglomeration*

4.5. Epitaxy of recessed and raised SiGe:B sources and drains on ultra-thin SOI and SON substrates

4.5.1. *Introduction*

In order to boost the electrical performances of the pMOS transistors for the 65 nm and lower technological nodes, Intel proposed in 2003 to etch the sources and drains zones on both sides of the gate stack and to fill the cavities thus formed by *in situ* boron-doped SiGe. A uniaxial compressive stress is then induced (the SiGe lattice parameter is higher than the one of Si) in the Si channel for small gate lengths [ANG 05, COO 10, HUE 08, ZHA 06]. Outstanding electric performances are then obtained [HAN 06, MIS 07, THE 06, THO 06, YIN 06], justifying the adoption of this technology in production. As an example, we will find in Figure 4.37, a TEM image of a 45 nm technology node pMOS transistor of a Penryn Intel microprocessor [MIS 07].

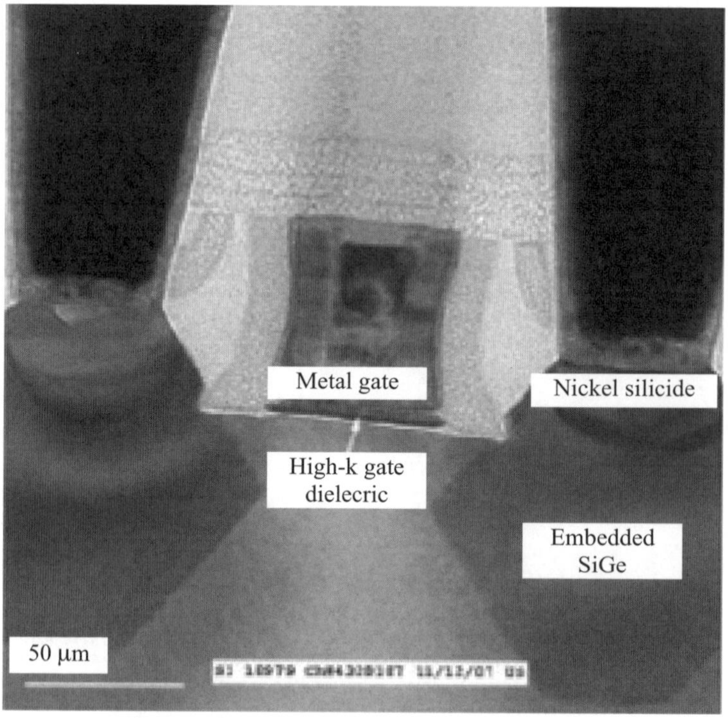

Figure 4.37. *Cross-sectional TEM image of a 45 nm technology node pMOS transistor. SiGe recessed sources and drains are present on both sides of the Si channel. The gate stack is of the "damascene" type with, as a gate dielectric, a SiON/HfO$_2$ stack*

From the epitaxy point of view, we are confronted with the following difficulties:

– surface preparation of the cavities prior to epitaxy;

– selective epitaxial growth (versus SiO_2 and Si_3N_4) of several tens, or even of a hundred nm thick heavily *in situ* boron doped (~ a few 10^{20} cm^{-3}) SiGe layers in those cavities. This might be troublesome as Ge concentrations have increased over the last few years from 20% to 35%. The appearance of defects due to plastic strain relaxation of these SiGe:B layers or at the locations where growth fronts coming from the edges of the cavities meet with the one at the bottom should be minimized or suppressed altogether.

We studied (given that our activities are centerd since 2004 on FD-SOI devices) the growth of:

– SiGe:B raised sources and drains on FD-SOI substrates and;

– recessed SiGe:B sources and drains on SON (silicon-on-nothing) substrates.

We will find below a few characteristics of the growth of SiGe:B layers on bulk, blanket Si(001) wafers, as well as morphological demonstrations of SiGe:B sources and drains on these two types of substrates.

4.5.2. *Growth kinetics and boron doping of SiGe in chlorinated chemistry*

We deposited thin, *in situ* boron doped SiGe layers on bulk, blanket Si(001) wafers slightly n-type doped (in order to have p-n junctions and thus to be able to carry out resistivity measurements) [HAR 04a, HAR 08a].

The strongly chlorinated growth chemistry used (SiH_2Cl_2 + GeH_4 + B_2H_6 + HCl), as well as the chosen growth pressure (20 Torr), will enable us, during a transposition on patterned substrates, to be selective compared to SiO_2 and Si_3N_4. The growth temperature was decreased from 700°C to 600°C, when switching from 20% to 50% of Ge, in order to remain pseudomorphic on Si and to have smooth surfaces for several tens of nm thickness.

We will find in Figure 4.38, resistivity curves associated with various Ge concentrations, 20-30 nm thick SiGe:B layers, according to the $F(B_2H_6)/(F(SiH_2Cl_2)+F(GeH_4))$ mass-flow ratio.

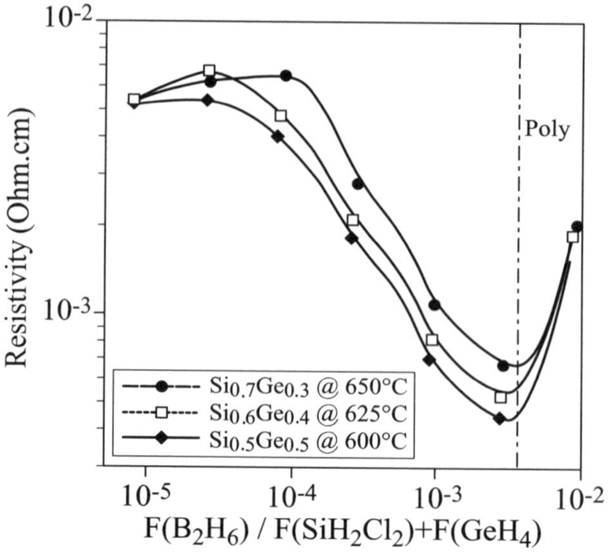

Figure 4.38. *Resistivity as a function of the $F(B_2H_6)/(F(SiH_2Cl_2)+F(GeH_4))$ mass-flow ratio, for SiGe:B layers of various Ge concentrations (see figure insert)*

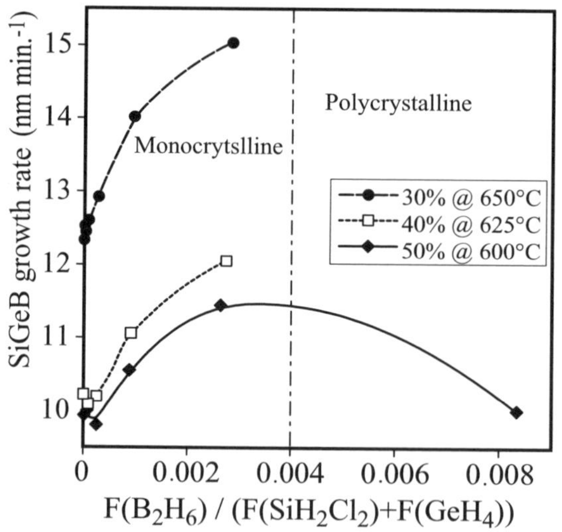

Figure 4.39. *SiGe:B growth rates functions of the $F(B_2H_6)/(F(SiH_2Cl_2)+F(GeH_4))$ mass-flow ratio, for various Ge concentrations (see figure insert: for values of %Ge and growth temperature)*

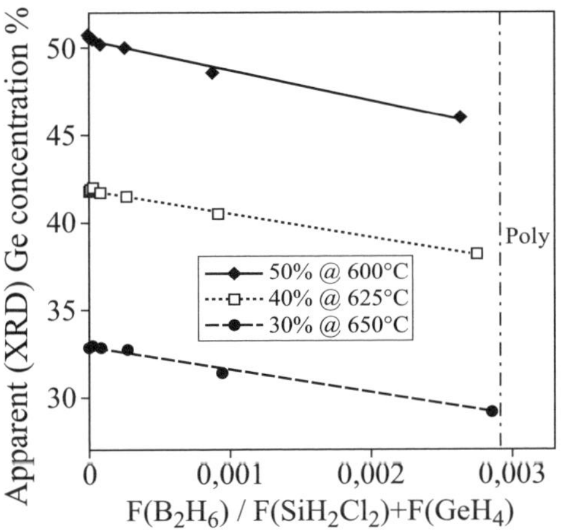

Figure 4.40. *Apparent Ge concentration functions of the $F(B_2H_6)/(F(SiH_2Cl_2)+F(GeH_4))$, for various SiGe :B layers (see insert of the figure: for values of %Ge and growth temperature)*

When the diborane flow increases, the resistivity decreases (as expected) up to values lower than 1 mOhm.cm. The minimums of resistivity of Figure 4.38 correspond to concentrations in B atoms between 2 and 3×10^{20} cm^{-3} [HAR 08a]. For more important diborane flows, SiGe:B layers become polycrystalline and rough. This explains the resistivity rise in Figure 4.38. We represented in Figures 4.39 and 4.40, the modifications induced by B doping for the growth rate and the "apparent" concentration in Ge (by supposing, during the fitting of the X-Ray Diffraction profiles that we have a binary SiGe alloy rather than a ternary SiGeB alloy). Although it is less obvious than in Si [GON 08, HAR 02a], we have a clear increase of the growth rate, related to the presence of B atoms on the surface, catalyzing at these low growth temperatures, the desorption of the H atoms, passivating the surface. The drop of the "apparent" concentration in Ge is due to high concentrations of B atoms (much smaller than those of Si or Ge [CHO 06]), inducing a contraction of the average SiGe:B lattice parameter (the real Ge content is rather steady).

4.5.3. *Recessed and raised SiGe:B sources and drains on FD-SOI and SON substrates*

We started from the data points described in section 4.5.2, in order to deposit at 650°C several tens of nm thick, heavily B doped $Si_{0.7}Ge_{0.3}$ raised sources and drains on both sides of FD-SOI transistors [HAR 08b]. We had to identify a trade-off between:

– *in situ* H_2 bake (that follows the *ex-situ* "HF-last" wet cleaning) which had to be carried out at low temperature considering the thinness of the starting Si film;

– loading effects inherent to the growth on patterned substrates (see sections 4.2 and 4.3);

– decrease of the growth rate combined with a slight increase of the Ge concentration during any growth on SOI substrates [HAR 05b, HAR 09a].

We will find in Figure 4.41 cross-sectional high-resolution TEM images of the sources and drains. We can notice on the right image, the excellent crystalline quality of the $Si_{0.7}Ge_{0.3}$:B layer, the presence of {111} facets, as well as the perfect growth selectivity, notably with respect to Si_3N_4 spacers. Given the very small quantity of SiGe under the level of the gate, we hardly expect to have holes mobility due to uniaxial compressive strain. However, efficient pMOS FD-SOI transistors have recently been obtained by IBM, thanks to the presence of faceted, raised SiGe sources and drains *in situ* B doped at very high levels (4 x 10^{20} cm^{-3}), leading to a drastic reduction of the contact resistance possible [CHE 09a]. These raised SiGe:B sources and drains are certainly useful, when we wish to strongly decrease the thermal budget to which a technological stack is subjected (3D integration) [BAT 09a] or when there is a substrate with a lattice parameter larger than the one of Si (of the extra-strained SOI (section 4.6) or germanium-on-insulator type (section 4.7)).

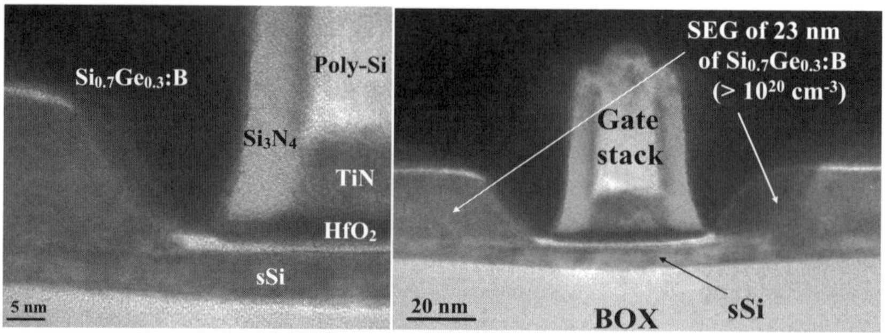

Figure 4.41. *High-resolution TEM images of raised $Si_{0.7}Ge_{0.3}$:B sources and drains with a 23 nm thickness selectively deposited on both sides of a FD-sSOI transistor*

We used the same process to grow $Si_{0.7}Ge_{0.3}$: B recessed sources and drains on both sides of SON transistors (see section 4.8 for more details). We will find in Figure 4.42 cross-sectional TEM images of these structures. We find there the SON transistor, i.e. a conventional gate stack on the thin film of Si, itself separated from the Si substrate by a $SiO_2/Si_3N_4/SiO_2$ "sandwich". On both sides of the latter, we notice the presence of a Si bridging that connects the channel to the source and drain zones. This bridging is topped by a SiGe:B layer, which was later on silicided with

Ni. Although the sequence of process steps used is not yet optimum (presence of a definite mushroom of poly-SiGe on top of the gate due to the not controlled growth on poly-Si), we can clearly note that it is a configuration where SiGe on both sides could induce a uniaxial compressive stress in the thin Si channel, where the holes are propagating.

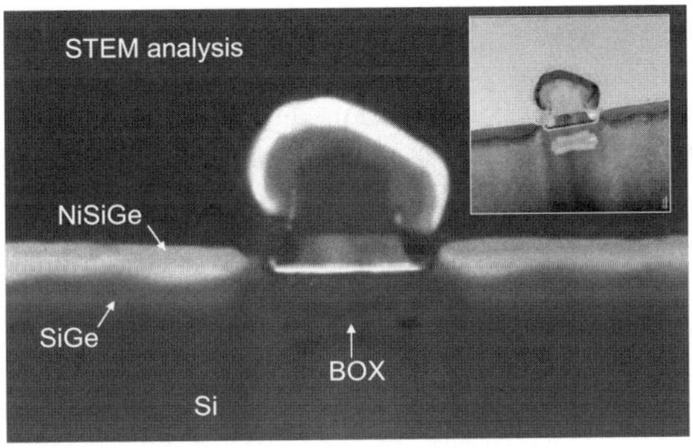

Figure 4.42. *Cross-sectional TEM image of a SON transistor with recessed $Si_{0.7}Ge_{0.3}$:B sources and drains*

To conclude this section, it is possible, thanks to $Si_{1-y}C_y$:P (y = 1.5 – 2%) recessed sources and drains with a lower lattice parameter than Si, to inject a uniaxial tensile strain in the Si channel of short gate length nMOSFETs transistors [BAU 07, GOU 09b, HUE 08]. Thanks to this strain, it is possible to significantly improve the electrons mobility and thus to obtain more efficient nMOSFETs transistors [YAN 08].

4.6. Virtual SiGe substrates: fabrication of sSOI substrates and of dual c-Ge / t-Si channels

4.6.1. *Introduction*

Biaxial tensily-strained Si layers (sSi or t-Si) grown pseudomorphically on virtual SiGe substrates are very interesting for high-efficiency CMOS devices. By SiGe virtual substrate, we mean a relatively thick SiGe layer deposited on a Si substrate, with a lattice parameter almost equal to that of a bulk SiGe substrate. There is indeed a significant increase in the electrons and holes mobility, due to the tensile strain in the Si layer. For electrons, the degeneracy of order 6 of the Δ valley of the conduction band, function of the directions <001>, is then lifted. The two valleys Δ_2 perpendicular to the plane of the sSi layer go down then by 67 meV for each 10% increment of the Ge concentration in the SiGe virtual substrate, while the four in-plane valleys Δ_2 go up by 67/2 = 33.5 meV each 10% increment of the Ge concentration (compared to un-strained Si) [MUN 08a]. The electrons' inter-valleys scatterings are then reduced, as well as their effective mass. We notice experimentally an increase in the electrons mobility in sSi for Ge concentrations higher than 10%. This mobility gain saturates at 1.8 for concentrations higher than 20% [CUR 01].

Regarding the holes, the degeneracy of the valence band in the Γ center of the Brillouin zone is lifted by the tensile strain. The light holes band (which then goes up in energy) is then preferentially occupied compared to that of the heavy holes band. As for electrons, it generates a decrease of the inter-valley scattering and of the effective holes mass. We then obtain an increase in the holes mobility in sSi for Ge concentrations higher than 25%. This gain of holes mobility saturates at 2.2 for concentrations in Ge higher than or equal to 40% [LEI 02]. These modifications are schematized in Figure 4.43.

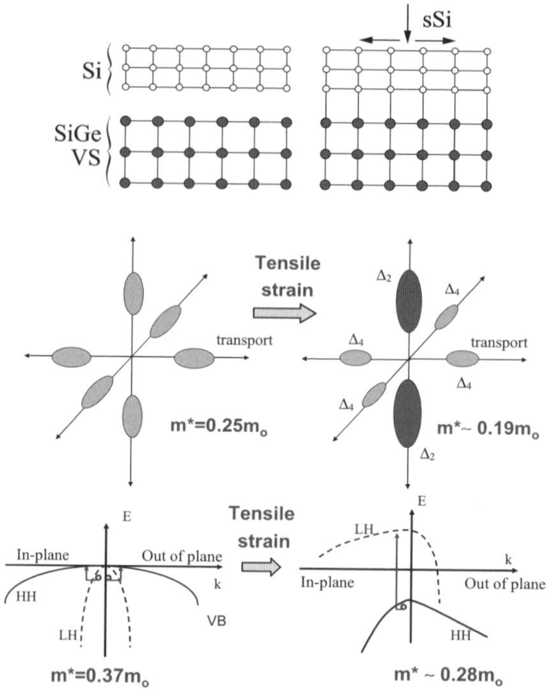

Figure 4.43. *Schematic descriptions of the structural modifications generated by the epitaxy of a thin Si layer on a virtual SiGe substrate and their impact on the Δ valleys of the minimum of the conduction band and on heavy holes/light holes sub-bands of the valence band itself (because of the tensile strain)*

Because of the lower thermal conductivity of SiGe compared to Si (harmful for CMOS devices), it can prove to be extremely interesting to transfer these tensily-strained Si layers on oxidized Si substrates, thanks to the SmartCut™ method for the fabrication of sSOI or XsSOI substrates (as we will see in more detail further on). Such substrates, with a tensily-strained Si layer on an buried insulating layer, are helpful to obtain high performance fully-depleted SOI transistors [AND 05b, AND 06, AND 07, RIM 03].

4.6.2. *Growth and structural properties of virtual SiGe substrates*

Specifications as for virtual SiGe substrates grown on Si (001) substrates are as follows: (i) a SiGe layer on the surface almost completely relaxed (i.e. with a lattice parameter the closest possible to a massive SiGe crystal); (ii) a defect density in the aforementioned layer as low as possible; and (iii) a controlled surface roughness. In order to do that, we usually insert between the surface SiGe layer (of constant composition and with a thickness typically higher than one micron) and the Si substrate, a several microns thick SiGe layer with a linear ramp of concentration in Ge of several microns thickness (where the defects resulting from the accommodation to the difference in lattice parameter between SiGe and Si will be confined).

In the past, it was shown that it was interesting in CVD to use high growth temperatures (typically 800°C - 900°C), in order to obtain virtual SiGe substrates of good crystalline quality [CHU 98]. We thus started by studying the growth kinetics of SiGe at high temperatures with a SiH_2Cl_2 + GeH_4 chemistry. Thanks to the chlorinated Si precursor, we should thus not clog the domes of the epitaxy chamber too much, despite the large thickness deposited. For that, we deposited several hundred nm thick, constant composition SiGe layers with Ge concentration discreetly increasing from one layer to another, separated with thin Si layers [BOG 05a, BOG 05b, BOG 06a, and HAR 06a]. We used X-Rays Diffraction (omega-2theta scans around the (004) and (224) orders) and Scanning Electron Microscopy, in order to determine the Ge concentration and the thickness (thus the associated growth rate) of each SiGe layer.

As an example, we will find in Figure 4.44, concentrations and growth rates obtained at 20 Torr and temperatures of 800°C, 850°C and 900°C. First of all, we will note that the Ge concentration increases in an almost linear way with the $F(GeH_4)/F(SiH_2Cl_2)$ mass-flow ratio. We used the simple relation proposed by Robbins *et al.* (describing the incorporation of Ge in silane chemistry [ROB 91]), in order to quantify the incorporation of Ge at high temperatures in dichlorosilane chemistry, i.e. $x / (1-x) = m * F(GeH4) / F(SiH2Cl2)$. We found, as an example $m(800°C) = 2.43$, $m(850°C) = 2.41$ and $m(900°C) = 2.16$ for the variable GeH_4 flow data points of Figure 4.44 (i.e. a Ge concentration decreasing when the temperature increases).

In the same way, we will see that the growth rate at these high temperatures is catalyzed only a little by the presence of Ge atoms on the surface (the fraction of surface sites which is passivated by H atoms is small). This is shown by a modest (800°C), even almost null (900°C) increase of the growth rate, when the GeH_4 flux increases (with constant SiH_2Cl_2 flux).

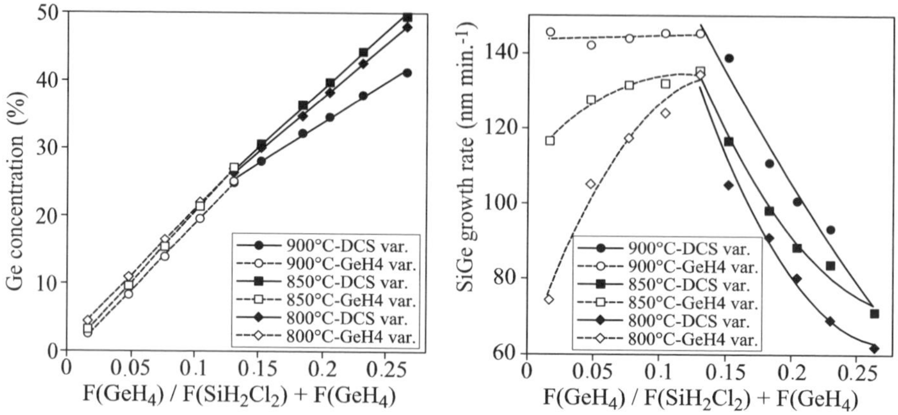

Figure 4.44. *Concentration in Ge (on the left) and growth rate (on the right) functions of the $F(GeH_4)/(F(SiH_2Cl_2) + F(GeH_4))$ mass-flow ratio (MFR), for: (i) a $F(SiH_2Cl_2)/F(H_2)$ MFR equal to 0.025 and a $F(GeH_4)/F(H_2)$ MFR increasing from $4.17x10^{-5}$ up to $3.75x10^{-4}$ (open symbols) or (ii) a $F(GeH_4)/F(H_2)$ MFR equal to $3.75x10^{-4}$ and a $F(SiH_2Cl_2)/F(H_2)$ MFR decreasing from 0.0025 to 0.00104 (full symbols). The growth temperature was equal to 800°C, 850°C or 900°C, while the growth pressure was equal to 20 Torr*

Thanks to these type of data points, we have been able to obtain virtual SiGe substrates of final concentration ranging between 20% and 50%. We will describe later on their main characteristics. In order to be exhaustive, we will however note that we have been able, on polished $Si_{0.5}Ge_{0.5}$ virtual substrates, to deposit, during a second epitaxy step, virtual SiGe substrates of final concentration between 55% and 100%. The interested reader can refer to [BOG 06a].

We will find in Figure 4.45, the SIMS profile of the Ge concentration of a virtual $Si_{0.5}Ge_{0.5}$ substrate. First of all, we deposited a layer with a Ge concentration increasing linearly from several % percent up to 50% (grading: ~ 9%Ge/μm), that we capped with a slightly more than one micron thick $Si_{0.5}Ge_{0.5}$ layer. A similar sample was pictured in TEM (see Figure 4.46). The 60° dislocations necessary to accommodate the lattice mismatch between SiGe and Si (glide planes: {111}) are very clearly restricted in the Ge concentration ramp.

No threading dislocation is visible (at least at the TEM local scale) in the constant composition SiGe layer on top.

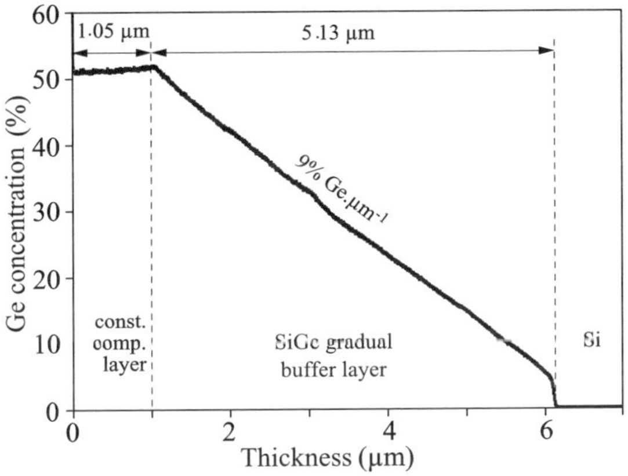

Figure 4.45. *SIMS profile of the Ge concentration in a $Si_{0.5}Ge_{0.5}$ virtual substrate*

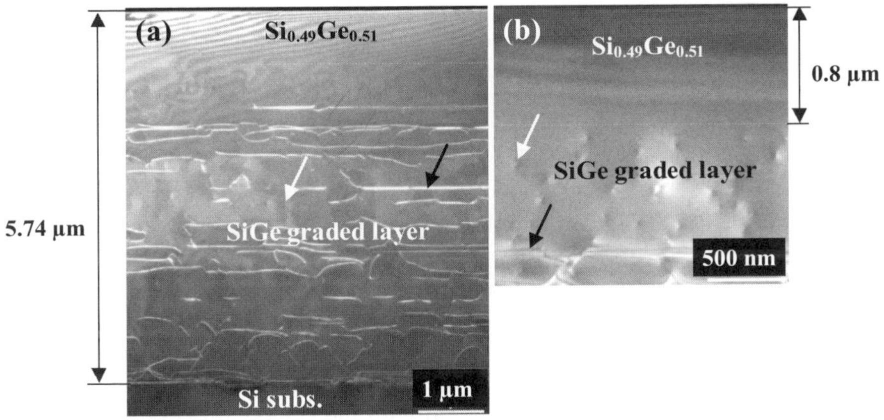

Figure 4.46. *Bright field TEM image in the <110> direction of a $Si_{0.5}Ge_{0.5}$ virtual substrate. The misfit dislocations in the plane of the TEM sample appear in the form of light gray or white lines, while those perpendicular to that plane appear as points*

We used X-ray diffraction, in order to have access to the macroscopic degree of strain relaxation of the layer of constant composition defined as $R(\%) = (a_{SiGe}^{//} - a_{Si})/(a_{SiGe}^{bulk} - a_{Si})$. $a_{SiGe}^{//}$ and a_{SiGe}^{bulk} are the in-plane and bulk (i.e. unstrained) lattice parameters of SiGe, while a_{Si} = 5.43105 Å is the lattice parameter of the Si substrate. For more information on the protocol used, readers can refer to

[BOG 05a]. We will find in Figure 4.47 the aforementioned macroscopic degree of strain relaxation functions of the germanium concentration of the constant composition layer, for relaxed substrates deposited at 850°C or 900°C.

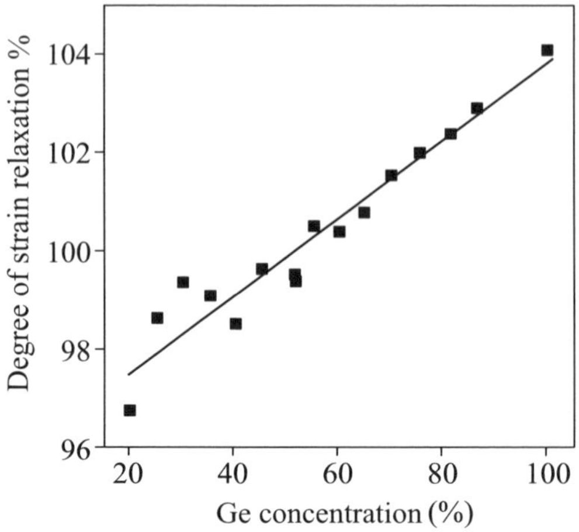

Figure 4.47. *Macroscopic degree of strain relaxation as a function of the germanium concentration of the constant composition layer*

We can see that *R* increases in a monotonous way with the Ge concentration, going from values of about 96% for $Si_{0.8}Ge_{0.2}$ to 104% for a Ge virtual substrate. Values lower than 100% correspond to a residual strain in compression, while values higher than 100% mean than the layers are slightly tensily-strained. Such a train state could seem paradoxical at first glance (since we deposit on a substrate, a material with a higher lattice parameter). This is without counting with the coefficients of thermal expansion lower for Si than for SiGe [REE 96]. The high Ge concentration SiGe layers, almost completely relaxed at the epitaxy temperature (i.e. 850°C-900°C), see their in-plane lattice parameter shrinking with constants close of those of the Si substrates underneath (because of the epitaxy relationship). They are then slightly tensily-strained.

Because of the gliding on {111} planes of the threading arms of the misfit dislocations (whose intersections with a (001) surface are the [110] and [1-10] directions) and the periodic strain fields resulting from it [CHE 02, MER 10], we have on a SiGe virtual substrate the appearance of a "*cross-hatch*" surface structuring. There is indeed two perpendicular arrays of surface undulations along the <110> directions, with amplitudes that can be up to several tens of nm and with a spatial period between 1 and 3 µm. As an example, we will have in Figure 4.48, an AFM image of the surface of a $Si_{0.5}Ge_{0.5}$ virtual substrate. In Figure 4.49, we showed the surface root-mean-square (rms) roughness as well as the amplitude in z or "Z range" (= $Z_{max.} - Z_{min.}$), functions of the Ge concentration of the constant composition SiGe top layer, for virtual substrates deposited at 850°C or 900°C. We can notice that the rms roughness, between 2 and 3 nm for 20%-30% of Ge, significantly increases for higher concentrations, with values of about 10 nm for 50% of Ge.

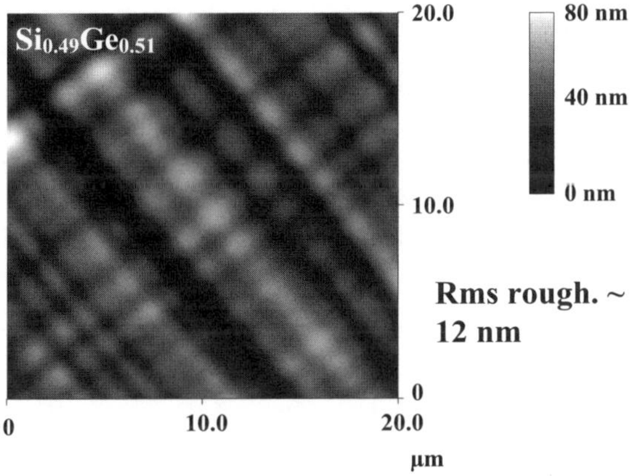

Figure 4.48. *Tapping mode AFM image of the surface of a $Si_{0.5}Ge_{0.5}$ virtual substrate. Image edges are approximately along the <100> directions*

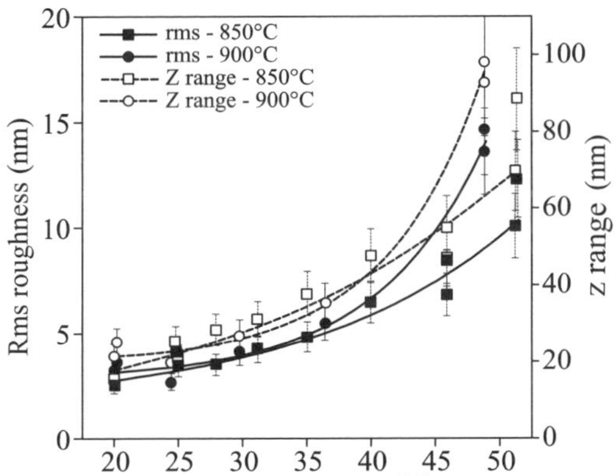

Figure 4.49. *RMS roughness and Z range (deduced from 20 µm x 20 µm AFM images of the surface of SiGe virtual substrates) according to the Ge concentration of the constant composition SiGe layer sitting on top*

We can however get rid of this surface roughness using chemical mechanical polishing (removal of approximately 0.5 µm of SiGe), followed by dedicated wet cleaning ([ABB 06, ABB 08]). We then obtain "cross-hatch"-free surfaces, with rms roughness of about 0.4 nm.

In order to conclude the characterization of these virtual substrates, the threading dislocations density in the constant composition SiGe layers on top remains to be evaluated. Those threading dislocations (either isolated from one another, called thereafter "field", or in pile-ups) will indeed be present in layers subsequently grown on top of our polished SiGe virtual substrates (for example, tensily-strained Si layers). We can either use standard Secco or Schimmel chemical solutions [ABB 07a], or profit from the fact that we have gaseous HCl in our epitaxy reactors [BOG 05c], in order: (i) to decorate the threading arms (creation of etching pits) and (ii) determine their density using optical microscopy. We will find in Figure 4.50, surface images of a $Si_{0.5}Ge_{0.5}$ virtual substrate, after HCl decoration of the threading dislocations.

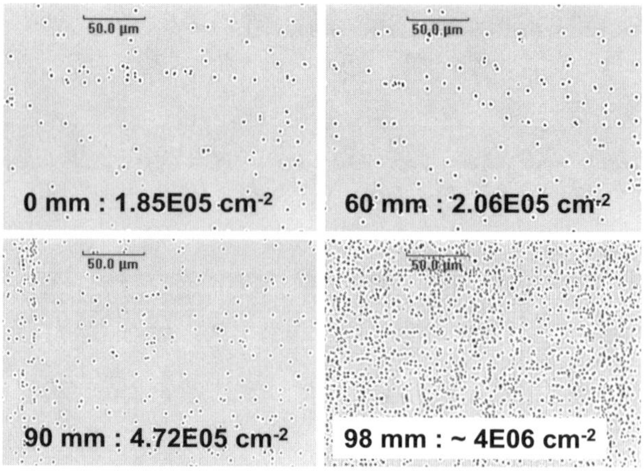

Figure 4.50. *260 µm x 195 µm bright field optical microscopy images of the surface of a $Si_{0.54}Ge_{0.46}$ virtual substrate after HCl decoration of the defects. The images were acquired along the wafer radius. The distance to the center of the 200 mm Si(001) substrate is indicated for each image, as well as the density of threading dislocations*

We will note that the density of threading dislocations is relatively constant up to the three-quarters of the wafer radius, with almost no piled-up threading dislocations. However, the density increases significantly when we approach the wafer edge. There are indeed values 20 times higher ($4x10^6$<=> $2x10^5$ cm^{-2}) 2 mm from the edge and the presence of pile-ups.

We will find in Figure 4.51, the evolution of the threading dislocation density (isolated as well as in pile-ups) close to the wafer center, depending on: (i) the growth temperature for virtual $Si_{0.8}Ge_{0.2}$ substrates and (ii) the Ge concentration of the constant composition layer (between 20% and 50%), for growth temperatures of 850°C and 900°C.

Increasing the growth temperature from 750°C to 950°C led to dramatically decreasing, for $Si_{0.8}Ge_{0.2}$, the density of threading dislocations. Moreover, we have values increasing monotonously from 10^5 to $2x10^5$ cm^{-2}, by going from 20% to 50% of Ge (at 900°C). Increasing the growth temperature further would however not allow us to drastically gain in terms of densities.

For temperatures higher than 900°C, we have indeed noticed that the surface became rougher, inhibiting the propagation of the threading arms of misfit dislocations to the wafer edges and their mutual elimination.

The data in Figure 4.51 was obtained for $R_{\%Ge/\mu m}$ concentration gradients in the gradual layer between 8% and 10%. However, it was shown that the density of threading dislocations ρ was partly predicted by the following relation: $\rho \propto R_{\%Ge/\mu m}\ GR\ exp(E_a/k_BT)$, GR being the growth rate of these virtual substrates, T the growth temperature and E_a a characteristic energy of the decrease of the dislocations density when the growth temperature increases [FIT 99, LEI 01]. We found the Ea value to be 1.43eV (see Figure 4.51), i.e. a value very close to the one found in the literature (1.38 eV) [LEI 01]. This relation was partly verified for $Si_{0.8}Ge_{0.2}$ virtual substrates. A threading dislocations density of 6 x 104 cm^{-2} only was obtained for 4% of Ge/µm and a growth rate of 140 nm min^{-1}.

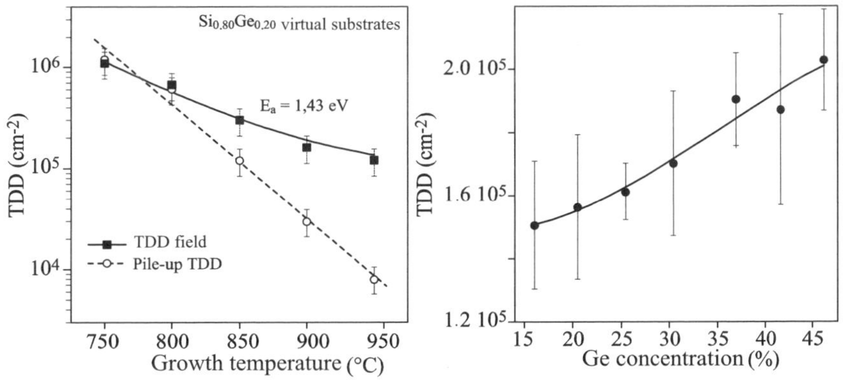

Figure 4.51. *Densities of threading dislocations present in (a) $Si_{0.8}Ge_{0.2}$ virtual substrates deposited at various temperatures and (b) SiGe virtual substrates of final concentration ranging between 20% and 50% deposited at 900°C*

Such a value is at the state of the art for $Si_{0.8}Ge_{0.2}$ virtual substrates (7 x 10^4 cm^{-2} [WES 04] and 4.5 x 10^4 cm^{-2} [ERD 05]). Reducing the Ge concentration gradient and thus thickening the gradual layer is however quickly limited, when switching to SiGe virtual substrates with a Ge concentration higher than 20% (because of the convex shape of the wafer and the prohibitive growth time).

4.6.3. *Growth and structural properties of tensily-strained Si layers on SiGe virtual substrates*

We have studied the epitaxial regrowth of thin, tensily-strained Si (sSi) layers on our polished SiGe virtual substrates. It is indeed critical to check that they are smooth, of good crystalline quality (i.e. no multiplication of the defects compared to the virtual substrates underneath), really tensily-strained, etc. We will describe here the main features of the sSi layers deposited at 700°C with dichlorosilane on SiGe virtual substrates with Ge concentrations ranging between 20% and 50%, this after a "HF-last" wet cleaning, followed by a H_2 bake at 850°C for two minutes. [HAR 07b, HAR 07c, HAR 08c]. We will let the reader take note in [HAR 08d] of the impact of the *in situ* H_2 bake conditions on the properties of sSi layers on $Si_{0.7}Ge_{0.3}$ and $Si_{0.6}Ge_{0.4}$.

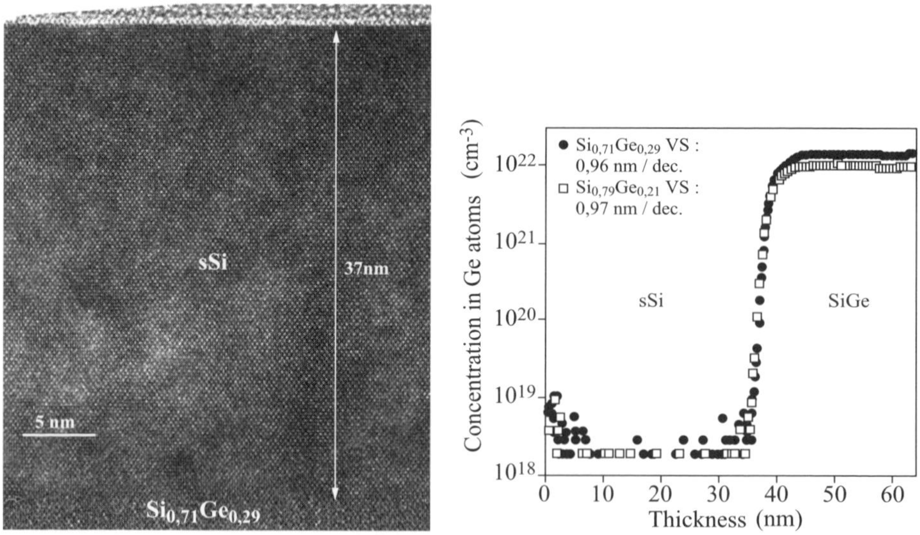

Figure 4.52. *High-resolution TEM image (on the left) and SIMS depth profiles of the Ge concentration (on the right) for 37 nm thick Si layers deposited on $Si_{0.8}Ge_{0.2}$ and $Si_{0.7}Ge_{0.3}$ virtual substrates*

We will find in Figure 4.52, a cross-sectional TEM image of a 37 nm thick sSi layer deposited on a polished $Si_{0.7}Ge_{0.3}$ virtual substrate. This layer is locally of great crystalline quality, as proved by the presence of well-ordered {111} inclined planes and the absence of extended defects. We will also find in Figure 4.52 a SIMS profile of the Ge atoms in 37 nm thick sSi layers deposited on polished $Si_{0.8}Ge_{0.2}$ and $Si_{0.7}Ge_{0.3}$ virtual substrates. The interface between sSi and SiGe is abrupt, as proved by the interface Ge concentration decay: between 0.7 and 1.1 nm/decade for Ge concentrations between 20% and 50%.

We will find in Figure 4.53, an AFM image of the surface of a 37 nm thick tensily-strained Si layer deposited on a polished $Si_{0.7}Ge_{0.3}$ virtual substrate. We will note that except a small reappearance of the "cross-hatch", related to the presence of periodic, tensile strain fluctuations in Si (see UV-Raman cartographies in [MER 10, ROU 08]), the surface remains smooth.

We have indeed obtained rms roughness ranging between 0.3 and 0.4 nm, this almost independently from the thickness of the tensily-strained Si layer or from the Ge concentration of the polished SiGe virtual substrate underneath.

We used Raman ultraviolet spectroscopy, in order to quantify the tensile strain in our sSi layers. We will see in Figure 4.54, that the strain almost does not depend on the thickness of the sSi layers. The stress levels, 1.32, 1.87, 2.58 and 3.15 GPa for sSi on $Si_{0.8}Ge_{0.2}$, $Si_{0.7}Ge_{0.3}$, $Si_{0.6}Ge_{0.4}$ and $Si_{0.5}Ge_{0.5}$, are close to those theoretically expected for completely pseudomorphic sSi layers. Similar strain values have been obtained thanks to the Geometric Phase Analysis (GPA) of high-resolution cross-sectional TEM images of these samples [HUE 07].

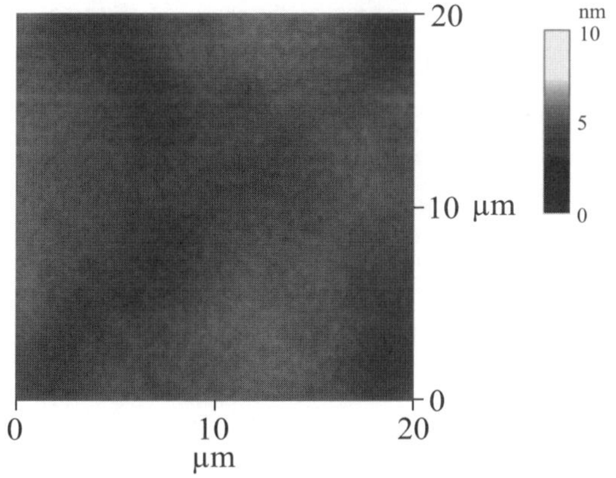

Figure 4.53. *Tapping mode, 10 µm x 10 µm AFM image of the surface of a 37 nm thick sSi layer on a polished $Si_{0.7}Ge_{0.3}$ virtual substrate. Image sides are approximately along the <100> directions*

We also studied and quantified the defects found in these sSi layers, because of their epitaxy on dislocated SiGe virtual substrates. We will find in Figure 4.55, a plane view TEM image of a 37 nm thick sSi layer on a polished $Si_{0.7}Ge_{0.3}$ virtual substrate and in Figure 4.56, an optical microscopy image of the same sample, after a Secco revelation of the defects in the sSi layer.

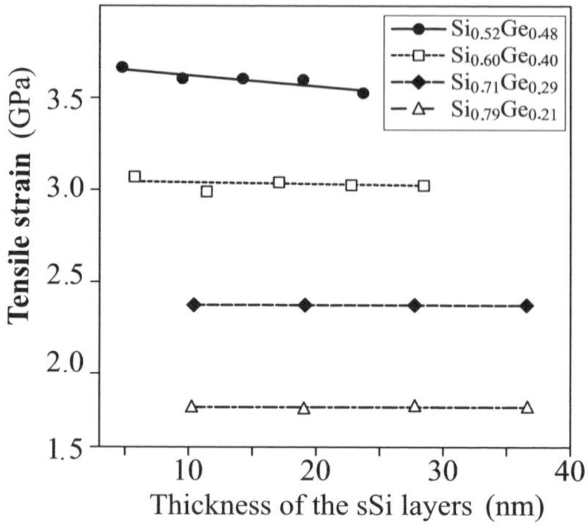

Figure 4.54. *Tensile strain (GPa), functions of the thickness of the sSi layers grown on polished SiGe virtual substrates for various Ge concentrations*

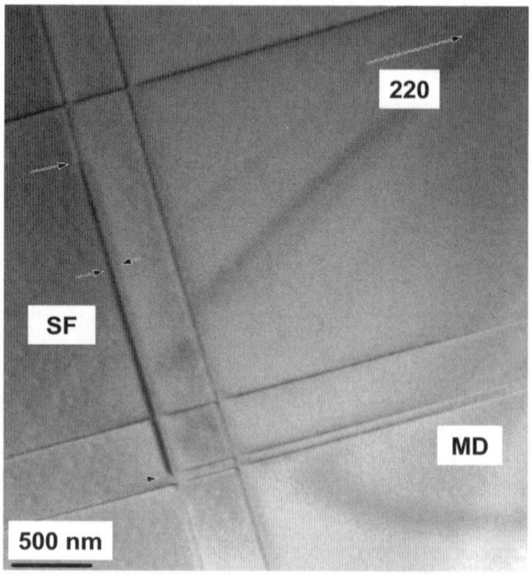

Figure 4.55. *Plane-view TEM image of a 37 nm thick sSi layer on a polished $Si_{0.7}Ge_{0.3}$ virtual substrate*

In Figure 4.56, we can notice the presence: (i) of white dots (characteristic of the presence of threading dislocations) and (ii) of lines along the <110> directions. Those are most likely due to the presence of stacking faults (in the whole sSi layer) or in-plane segments of misfit dislocations (at the sSi/SiGe interface). We expect indeed that the 60° misfit dislocations present in the SiGe virtual substrate underneath dissociate themselves (because of the tensile strain in the sSi layer) in partial 30° and 90° dislocations, propagating at different speeds in sSi, leaving in their wake stacking faults, as illustrated in Figure 4.55 [BED 04, HIR 05, KIM 06]. These stacking faults are more deleterious from the electrical transport point of view [YAN 00], than the in-plane segments (at the sSi/SiGe interface) of 60° misfit dislocations (although these act as dopant diffusion channels in bulk nMOS transistors [FIO 04]).

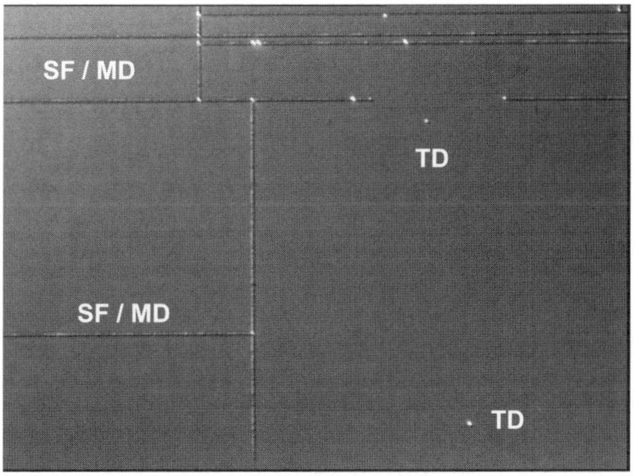

Figure 4.56. *Optical microscopy image of the surface of a 37 nm thick sSi layer on a polished $Si_{0.7}Ge_{0.3}$ virtual substrate after the Secco etching of the totality of the sSi layer and of a few nm of the SiGe layer underneath*

After Secco defect decoration, we quantified the surface density of threading dislocations (see Figure 4.57) and the linear density of <110> extended defects (see Figure 4.58), functions of the sSi thickness and of the concentration in Ge of the virtual substrate. The density of threading dislocations is close to that of virtual SiGe substrates. It increases only a little (from ~ 10^5 to $3x10^5$ cm^{-2}) when the sSi layer thickness increases.

The linear density of extended defects, almost null for thicknesses lower than 10 nm, monotonously increases with the sSi thickness, reaching values as high as 800 cm^{-1}. It seems that this linear density depends not only on the Ge concentration (the higher the concentration is, the higher the linear density will be), but also on the flatness of the polished virtual SiGe substrate [HAR 07c, HAR 08b].

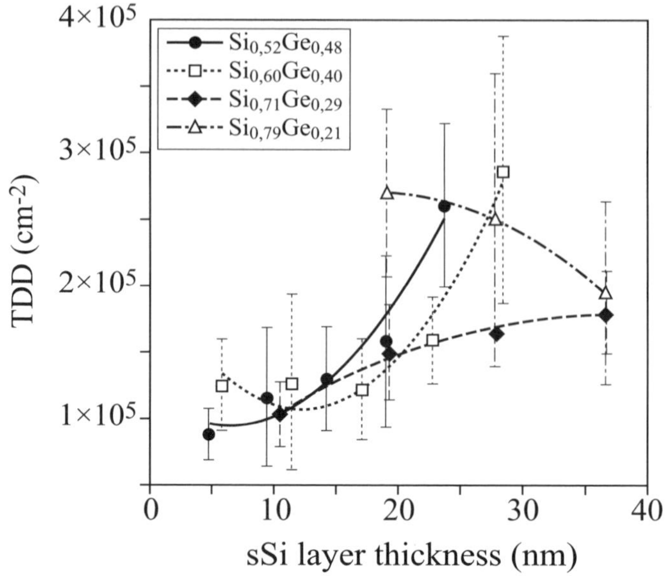

Figure 4.57. *Surface density of threading dislocations according to the thickness of sSi layers deposited on polished SiGe virtual substrates (of %Ge between 20% and 50%)*

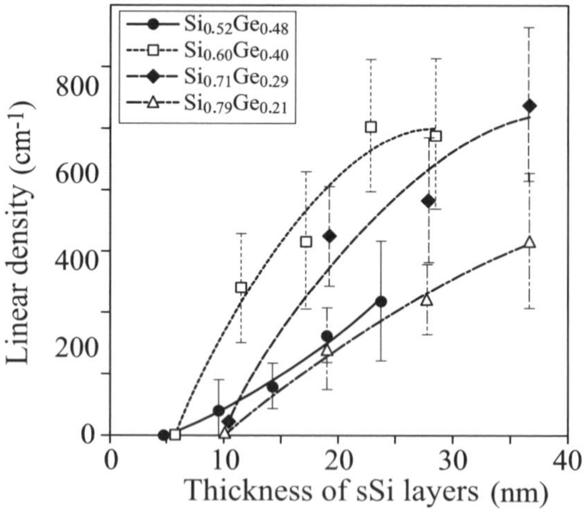

Figure 4.58. *Linear density of <110> extended defects, according to the thickness of sSi layers deposited on polished SiGe virtual substrates (of %Ge between 20% and 50%)*

As a conclusion, we highlighted that it was possible to obtain pseudomorphic sSi layers on SiGe for thicknesses more important than those expected (by reversing of course the sign of the strain), according to the plastic relaxation curves of compressively-strained SiGe on Si (see Figure 4.9 of section 4.1). We will then speak about supercritical sSi layers, which have recently been inserted in "bulk" [FIO 04, SUG 05] or partially depleted-sSOI [COL 06, THE 05a, THE 05b, THE 06, YIN 06] transistor lots, with excellent electric properties.

4.6.4. *Fabrication of sSOI and XsSOI substrates & transport properties*

sSi layers on SiGe virtual substrates can be transferred thanks to the Smartcut™ technology on oxidized Si substrates. We thus obtain substrates, known as strained SOI [GHY 04] (when Si was tensily-strained on $Si_{0.8}Ge_{0.2}$) or as eXtra-strained SOI [AKA 06a] (when Si was tensily-strained on SiGe with [Ge] > 20%).

We will find in Figure 4.59, the main process steps leading to the fabrication of these structures. After deposition of a SiGe virtual substrate that we polish, we deposit a thin sSi layer (see sections 4.6.2 and 4.6.3), that we encapsulate by several tens of nm of SiO_2, deposited at a reduced temperature. We carry out a H^+ ion implantation (high dose and implantation energy). Then, we bond the aforementioned stack on an oxidized Si substrate. Annealing leads to the agglomeration of hydrogen atoms in planar cavities, thus creating a zone of mechanical fragility, which is put to good use for the fracture. We find ourselves then with a rough SiGe layer on sSi, itself on a buried oxide. We then use a selective chemical etching, to remove the remaining SiGe, while stopping on sSi. We then obtain sSOI or XsSOI substrates which are locally of excellent crystalline quality (see Figure 4.59) and whose bonding interface has been strengthened thanks to suitable annealing [BOG 06b].

We have been able to check, thanks to UV-Raman measurements that the sSi layer kept its tensile strain [MUN 08a, ROU 08]. HCl and Secco defect revelations showed that the threading dislocations were not multiplied during the use of the Smartcut™ process [ABB 07b]. Moreover, we called upon Hall effect [BOG 04, HAR 04c] and photoluminescence measurements at 3-9K [BOU 04, MUN 08a, MUN 08b], in order to check the electronic quality of our tensily-strained Si layers. Those were either: (i) inserted in SiGe in the proximity of an *in situ* phosphorous-doped SiGe layer used as an electron reservoir (structures known as "MODFETs" [BOG 04, BOU 04, HAR 04c]) or (ii) directly on buried oxide with a thin high temperature oxide layer over it (in order to confine the charge carriers and suppress non-radiative surface recombination) [MUN 08a, MUN 08b].

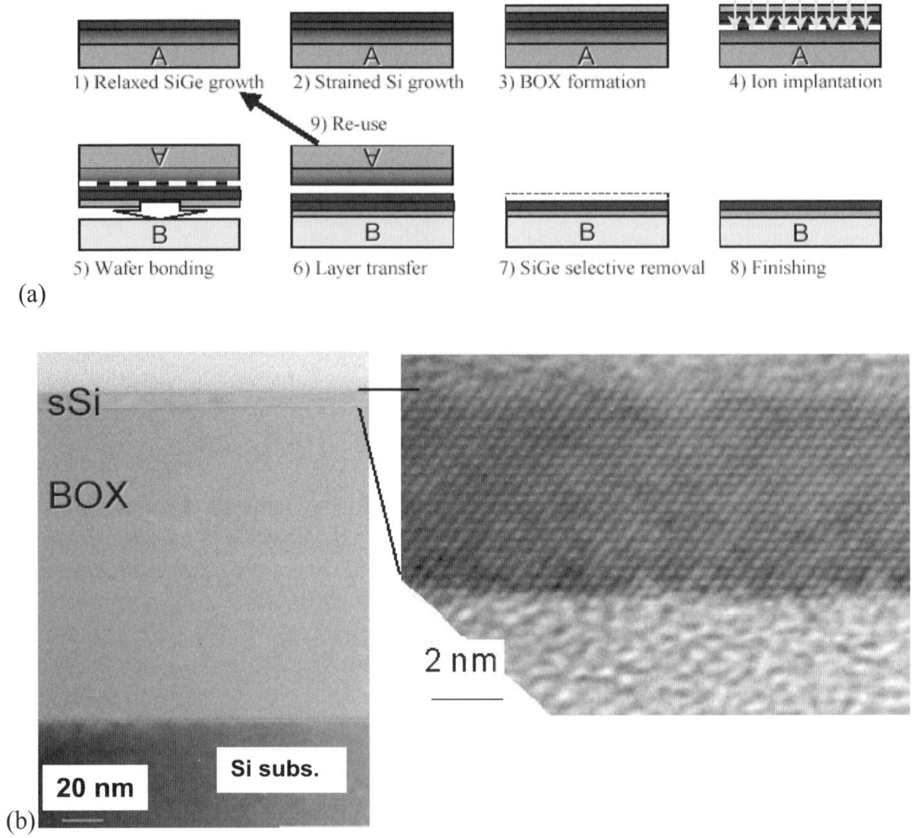

Figure 4.59. *(a) Main process steps in order to obtain sSOI or XsSOI substrates. (b) Cross-sectional TEM images of a 8 nm thick XsSOI 40% substrate*

We have been able to confirm (thanks to Shubnikov-De Haas oscillations of the longitudinal resistivity and to the Hall plateaux of the transverse resistivity) the presence at liquid helium temperature of 2D electron gases in MODFETs structures. The mobilities obtained (212,000 $cm^2\ V^{-1}\ s^{-1}$ for a $5.4x10^{11}\ cm^{-2}$ electrons sheet density) are comparable to those obtained using dedicated tools in academic laboratories. In the same way, we have been able to clearly highlight the presence in photoluminescence of NP and TO lines, related to the radiative combination, assisted by phonons (TO) or not (NP) of electron-hole pairs in sSi layers. Their energy position depends not only on the value of the tensile strain [MUN 08a], but also on the sSi layer thickness [MUN 08b].

We thus used those 1.3 GPa sSOI and 1.9 GPa XsSOI substrates (sSi on $Si_{0.8}Ge_{0.2}$ and sSi on $Si_{0.7}Ge_{0.3}$, respectively) to obtain high- performance FD-SOI nMOS and pMOS transistors [AND 05b, AND 06, AND 07]. We will find in Figure 4.60, a TEM image of one of the short gate length transistors, and in Figure 4.61, the obtained carrier mobilities (for long transistors). As expected, we have an electron mobility gain by a factor of 2. The gain is however much more modest regarding the mobility of the holes: +10% by going from SOI to XsSOI substrates.

Figure 4.60. *Cross-sectional TEM image of a short gate length (25 nm) FD-sSOI transistor*

An elegant CMOS integration solution would then be to start from patterned sSOI substrates and to selectively deposit {compressively-strained SiGe / thin Si cap} bi-layers on the pMOS active areas. With that, we would gain simultaneously in electron and hole mobilities, this on the same wafer. We will see in Figure 4.62, a short gate length pMOS transistor fabricated on a sSOI substrate with a 20 nm thick $Si_{0.6}Ge_{0.4}$ channel for the holes capped with a few nm of sSi (in order to have a good electrical quality gate dielectric).

Changing the material and the strain state of the channel (c-SiGe instead of sSi) is extremely beneficial for the holes mobility, as illustrated in Figure 4.63: it is indeed multiplied by two!

Given the obtained hole mobility gains, we could push the reasoning a little further and move over to pMOS transistors with, as conduction channels, highly compressively strained Ge layers. That is what has been done, as shown in section 4.6.5.

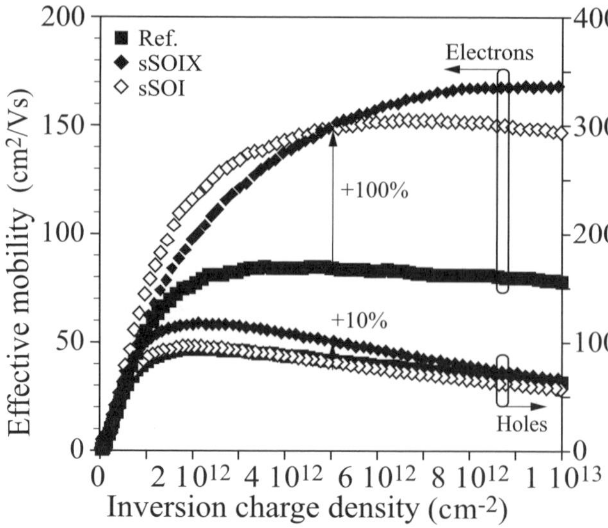

Figure 4.61. *Effective mobility of the electrons and holes, according to the inversion charge density in conventional SOI, sSOI and XsSOI substrates*

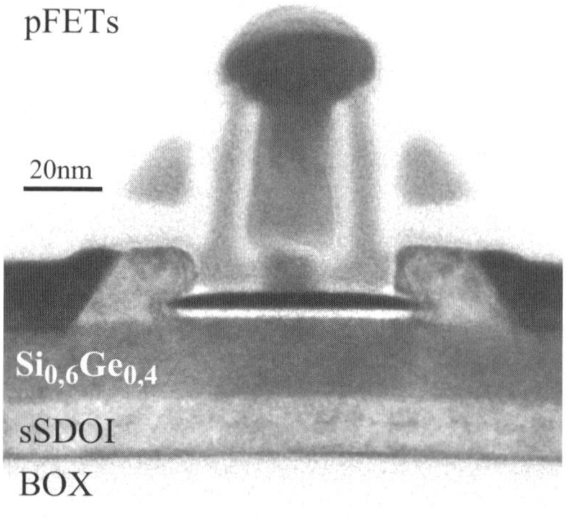

Figure 4.62. *Cross-sectional TEM image of a pMOS transistor on sSOI with, as the conduction channel for the holes, a $Si_{0.6}Ge_{0.4}$ layer*

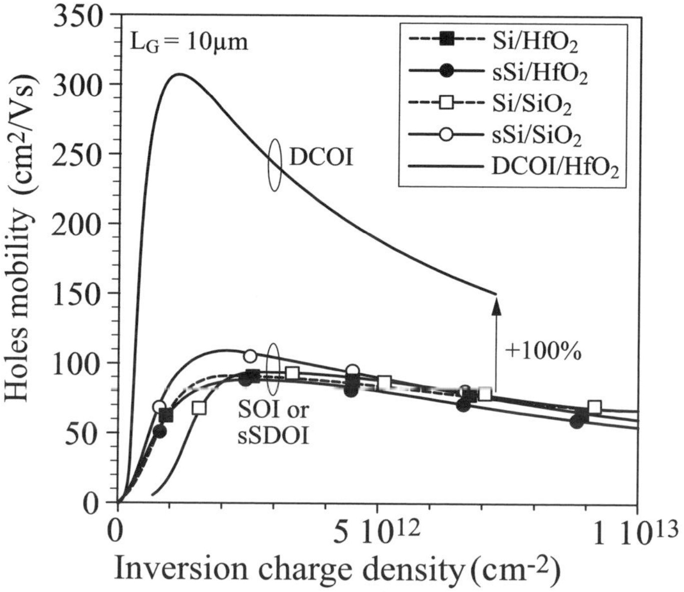

Figure 4.63. *Effective hole mobility functions of the inversion charge density in conventional SOI substrates, sSOI substrates and BOX/sSi/$Si_{0.6}Ge_{0.4}$/sSi stacks (known as dual channel on insulator – DCOI)*

4.6.5. *c-Ge/t-Si dual channels on $Si_{0.5}Ge_{0.5}$ virtual substrates*

We started from polished $Si_{0.5}Ge_{0.5}$ virtual substrates, in order to make symmetrical – as much as possible – the strain in the {compressively-strained Ge / tensily-strained Si (noted hereafter c-Ge / t-Si)} dual channels that we wish to have. Given the thicknesses aimed for (c-Ge : ~ 8 nm, t-Si : ~ 5 nm), it seemed vital to (in agreement with literature [LEE 04]):

(i) significantly lower the epitaxy temperature of the c-Ge layer and not to exceed 8-9 nm, in order to avoid any elastic (undulations) or plastic (misfit dislocations) relaxation of the compressive stress;

(ii) optimize the Si encapsulation process of this Ge layer.

The deposition temperature of the c-Ge layer, using GeH_4, 350°C, is close to the lower detection and piloting limit of our infra-red pyrometers (~ 330°C). We nevertheless obtain a growth rate of about 2.8 nm min.$^{-1}$. As for its encapsulation by Si, we first of all carried out a Si passivation of the c-Ge layer at 350°C with SiH_2Cl_2 chemistry (the growth rate, initially non-null because of the presence of Ge atoms on the surface, becomes non-existent as soon as the surface is entirely covered with Si atoms, i.e. after ~ 5 Å of deposition). Then, we slowly go (1°C/s) under SiH_2Cl_2 up to 650°C, temperature at which we thickened the aforementioned t-Si layer (growth rate: ~ 0.5 nm min.$^{-1}$) [BOG 05d].

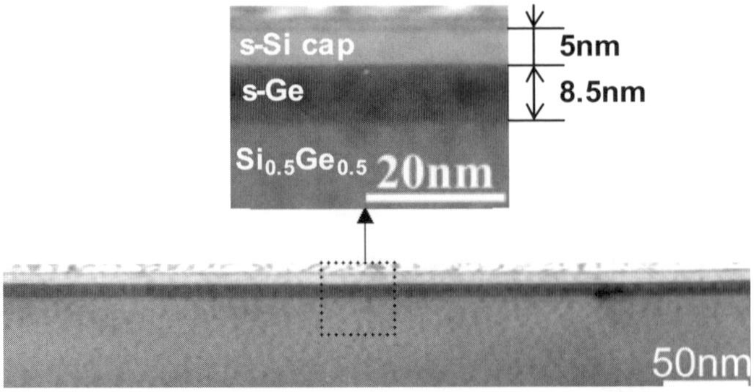

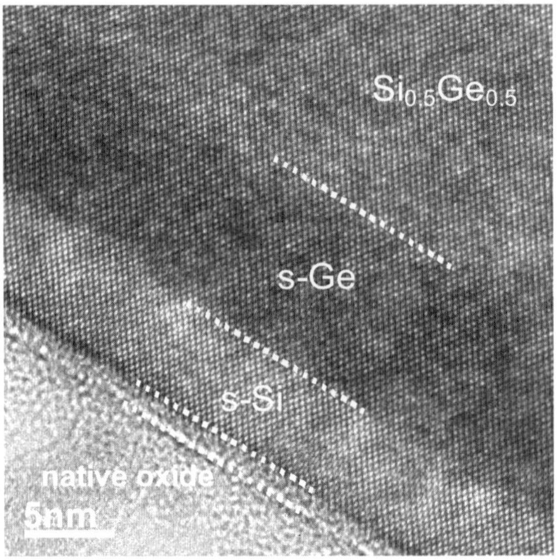

Figure 4.64. *(On the left) dark field and (on the right) high-resolution cross-sectional TEM images of a c-Ge/t-Si dual channel on a $Si_{0.5}Ge_{0.5}$ virtual substrate*

We can note in Figure 4.64, the planarity of the surface and of the interfaces, as well as the absence of extended defects. We checked, thanks once again to the Geometric Phase Analysis of high-resolution TEM images, that these Ge and Si layers were indeed pseudomorphic on $Si_{0.5}Ge_{0.5}$, with the expected levels of compressive and tensile strains [CHE 06].

We thus proceeded with the integration of these dual channels in pMOS transistors and of tensily strained Si layers in nMOS transistors, the dimensions of the devices being relaxed and the integration scheme being simplified [WEB 05]. We will find in Figure 4.65 cross-sectional TEM images of one of these dual channel transistors. The interface between t-Si and c-Ge is well defined and abrupt, just as those of the HfO_2/TiN/Poly-Si gate stack. We showed in Figure 4.66, the mobility of charge carriers (electrons in Si or t-Si, holes in Si or in c-Ge/t-Si). As observed before with this type of gate stack, there is a drop in the un-strained Si of the mobility, compared to the universal mobility in bulk Si.

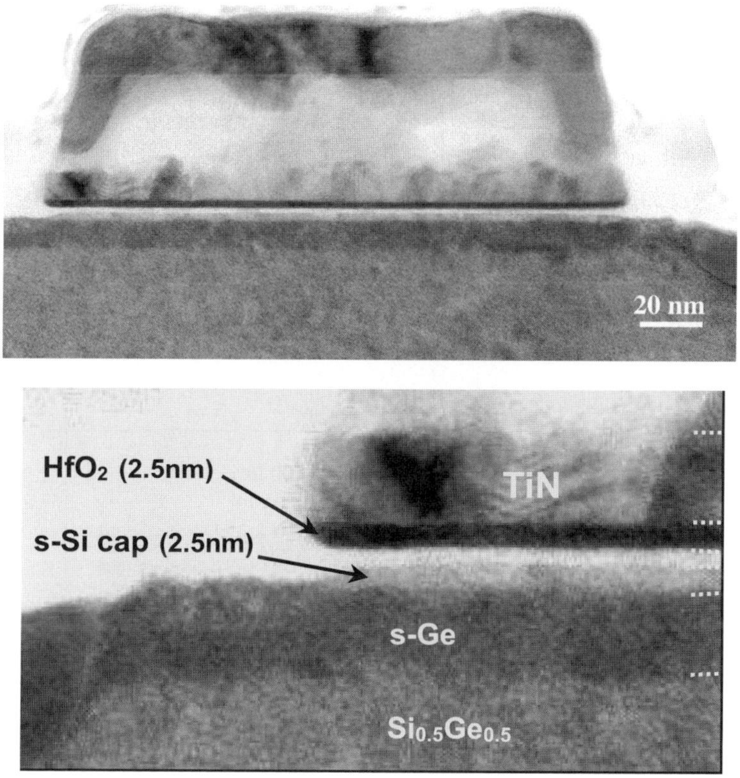

Figure 4.65. *Cross-sectional TEM images of a pMOS transistor on a c-Ge/t-Si dual channel*

As expected, we find a gain of 65% of the electrons mobility in t-Si. The mobility gain of the holes is however extremely high: a factor of 9! Thanks to this gain, we can balance the on-state characteristics of nMOS and pMOS transistors, as illustrated in Figure 4.67. We will however note that the I_{ON}/I_{OFF} (on-state current/off-state current) compromise is only 10^3 in this type of stacking, a value much lower than the one usually obtained for Si transistors (likely due to band-gap lowering in compressively strained Ge).

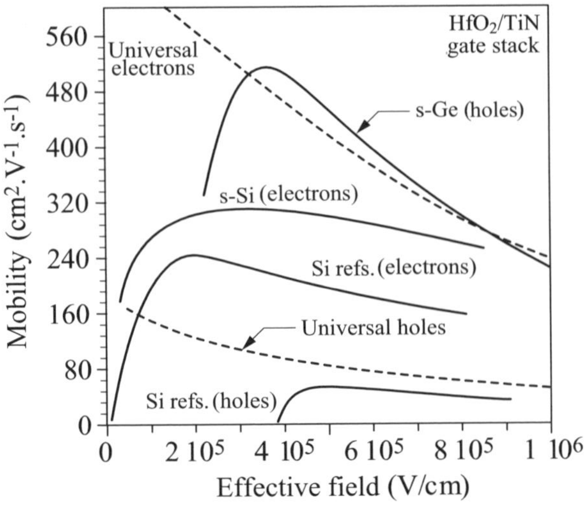

Figure 4.66. *Mobility of the charge carriers in: (i) bulk Si, (ii) tensily-strained Si and (iii) a c-Ge/t-Si dual channel, functions of the effective field*

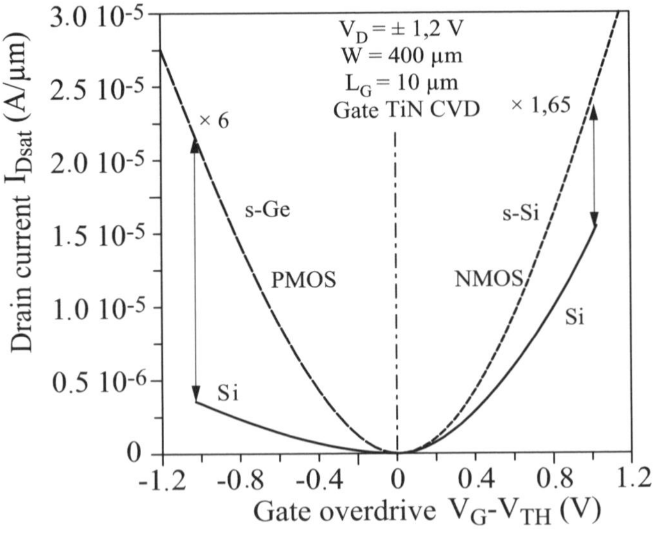

Figure 4.67. *On-state drain current functions of the gate overdrive V_G-V_{TH} (threshold voltage) in: (i) bulk Si, (ii) tensily-strained Si and (iii) a c-Ge / t-Si dual channel*

4.7. Thin or thick layers of pure Ge on Si for nano and opto-electronics

4.7.1. *Introduction*

Thin or thick Ge layers deposited on Si are very interesting because of their intrinsic properties. First of all, the energy band gap of Ge is much lower than the one of Si (E_g (Ge) = 0.67 eV ⇔ E_g (Si) = 1.11 eV at room temperature). This allows its use as a light emitting or detecting material in the near infra-red, i.e. at the wavelengths used in optical waveguides (1.3–1.55 μm). We will favour pseudomorphic heterostructures such as multiple planes of Ge quantum dots in Si for electroluminescence [BRU 00, DAM 04b, HAR 05c]. Several hundreds of nm thick pure Ge layers selectively grown in Si cavities at the end of optical waveguide (see Figure 4.68) lead to the efficient conversion of a light signal into electric impulses (i.e. light detection) [HAR 04d, HAR 04e, OSM 09, ROU 06, VIV 07, VIV 09].

Multiple planes of Ge quantum dots in Si or high Ge concentration Si/SiGe superlattices can also act as active material for photo- detection in these domains of wavelength, with however a lesser absorption efficiency [KUR 02a, KUR 02b, MAS 03].

Moreover, we can take advantage from the fact that the lattice parameter of GaAs (sphalerite structure) is very close to the one of Ge (a_{Ge} = 5.65785 Å ⇔ a_{GaAs} = 5.6533 Å) to deposit GaAs-based heterostructures such as solar cells [AND 05a], laser diodes [KWO 06], bipolar or high electronic mobility transistors [LIU 09, LUB 08], etc., on Ge (see Figure 4.68). In that case, vicinal surfaces (typically misorientated by 6° in one of the <110> directions) will be used, so that there are biatomic steps on the surface and thus to minimize the appearance of anti-phase boundaries harmful to the electronic properties of active structures [HAR 09b, TAN 06].

With a low electric field, electrons and holes mobilities in bulk Ge are respectively twice and four times higher than those in Si (see Figure 4.68). Significant gains in performances are thus expected in MOSFETs transistors fabricated on Ge, provided that there is an efficient gate stack (with, as a gate dielectric, a high-K material such as HfO_2 (considering the strong instability of Ge oxide)). GeOI (Ge-on-insulator) substrates fabricated thanks to the SmartCut™ process with as donor wafers bulk Ge substrates or thick Ge layers on Si [AKA 06b, AUG 09, CLA 06] lead to a better electrostatic control (i.e. less short channel effects, smaller junctions capacities, etc.). A gain of a factor of 2 of the holes mobility was evidenced in pMOSFETs transistors on GeOI, as we will see later on [POU 08, POU 09, ROY 08, VIN 10].

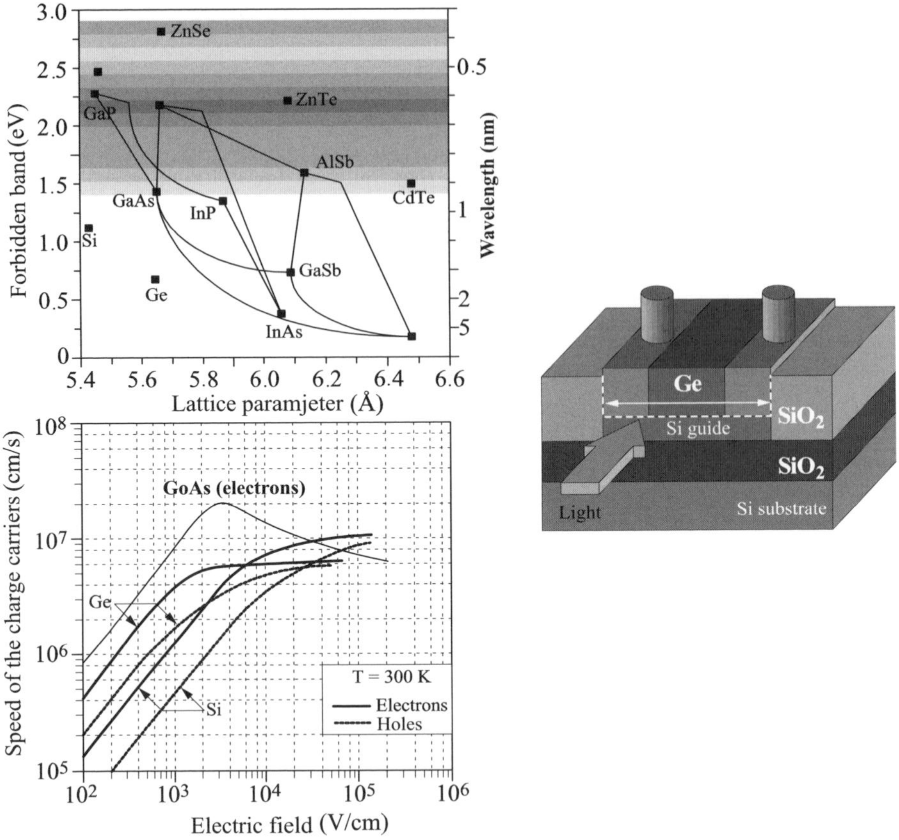

Figure 4.68. *Schematic description of a Ge photo-detector on SOI; (on the top left) energy band gap functions of the sphalerite lattice parameter for various II-VI, III-V and IV-IV semiconductors; (on the bottom left) charge carriers velocity functions of the electric field in Si, Ge and GaAs*

In the following sections, we will focus: (i) on the structural properties of thick Ge layers grown on Si(001) and of the GeOI substrates resulting from it, (ii) on the electric properties of pMOSFETs transistors fabricated on GeOI and of near-infra red photo-detectors based on pure Ge and (iii) on the growth and optical properties of Ge quantum dots inserted in Si.

4.7.2. *Structural properties of thick layers of Ge on Si (001) and of GeOI*

It is difficult, because of the important discrepancy of the lattice parameter (4.2%), to obtain on Si(001), Ge layers with the sought properties, i.e. a flatness compatible with advanced lithography or layer transfer process steps, a defect density minimized for efficient devices, etc. Several different approaches can lead to such films. Ge virtual substrates (i.e. with a Ge concentration grading from 0% to 100% in ~ 10 µm) are characterized by a macroscopic degree of strain relaxation slightly higher than 100% and a rather low threading dislocations density (~ 10^6 cm^{-2}). They are nevertheless rough (appearance of a strong cross-hatch) [BOG 06a, CUR 98, THO 03] and, in the case of a deposition on one face only of the Si substrate, severely bowed (bow of about 250 µm) [HAR 10a]. Chemical Mechanical Polishing is then mandatory, in order to recover a smooth surface. The approach thus chosen was proposed in 1998 by Colace *et al.* [COL 98], closely followed by Hernandez *et al.* [HER 03]. It consists of depositing at low-temperature a thin Ge "seed" layer, followed by the deposition at high temperature of a thick Ge layer [EAG 91, HAR 04d, HAR 04e, HAR 05d, HAR 08e, HAR 09b, LUA 99, SAK 97]. The low temperature (330°C-400°C) chosen for the first Ge layer leads to a nearly complete plastic relaxation of the elastic stresses stored without too many surface undulations. The high temperature (600°C-850°C) used to deposit the second Ge layer leads to a drastic drop in the density of threading dislocations, as well as a reduction of the growth time. Later on, we can use a thermal cycling (typically between 750°C and 875-900°C), in order to further reduce the density of crystalline defects (thanks to the thermally assisted propagation of the threading arms of misfit dislocations, their mutual annihilation and their elimination at the edge of the substrate) [HAR 04d, HAR 05e, HAR 09b, LUA 99]. Such a method is however time consuming (although a recent study showed that a reduced number of short cycles gave the same densities of emergent defects as long cycles in greater number [HAR 10b]).

From the growth protocol point of view, we proceeded as follows. First of all, we deposited on Si(001) 130 nm of Ge at low-temperature (400°C) and at a pressure of 100 Torr, in order to profit from a relatively high growth rate: 14 nm $min.^{-1}$ for a $F(GeH_4)/F(H_2)$ mass-flow ratio equal to 7.917 x 10^{-4} [HAR 08e].We then slowly ramped-up the temperature under GeH_4 from 400°C to 750°C (1°C/s), while reducing the pressure from 100 Torr to 20 Torr.~ 85 nm of Ge is deposited during this ramp. We then thickened this Ge layer at 750°C, 20 Torr, with the same $F(GeH_4)/F(H_2)$ mass-flow ratio as previously (growth rate: 47 nm $min.^{-1}$). The following thermal cycling: 890°C, 5 min + 3x{750°C, 5 min/890°C, 5 min} can be implemented on those thick Ge layers (2570 nm here),so as to minimize the threading dislocations density.

We will find in Figure 4.69 AFM images of the surface of these Ge layers functions of the thickness deposited. Just after the growth of the 130 nm of Ge at 400°C, the surface is rather rough, with low spatial wavelength undulations (a few hundred nm). After the temperature rise up to 750°C under germane, these undulations coalesced, giving rise to higher spatial wavelength undulations

(~ 1 µm). These undulations continue to stretch in the growth plane during the thickening at 750°C (images corresponding to 1,395 and 2,570 nm of deposited Ge). For such layers, the surface has a morphology which a mix between low amplitude "hills" and a "cross-hatch", i.e. an array of undulations along the <110> directions, as for SiGe virtual substrates. There is otherwise the presence of "hills" flanks of large terraces separated from one another by bi-atomic steps [HAR 10b]. The subsequent use of a thermal cycling leads to a surface morphology closer to the one of SiGe virtual substrates, i.e. a "cross-hatch" defined better at the expense of the "hills".

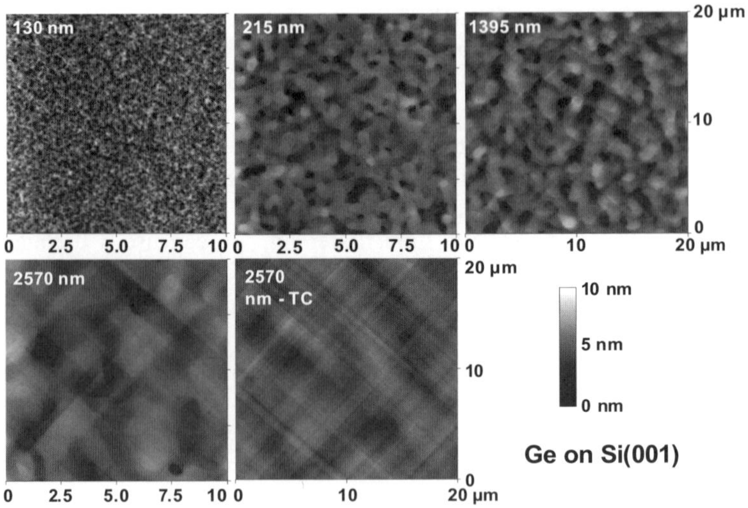

Figure 4.69. *Tapping mode AFM images of the surface of Ge layers deposited on Si(001), for ~ 130 nm of Ge at 400°C, a 130 nm thick Ge "seed layer" at 400°C + 85 nm of Ge deposited during the temperature ramping-up from 400°C to 750°C (215 nm), after the thickening of this layer with either 1180 or 2355 nm of Ge at 750°C (1395 and 2570 nm), and after thermal cycling of a 2570 nm thick Ge layer*

In Figure 4.70, we showed the evolution of the surface rms roughness and Z range (i.e. $Z_{max.} - Z_{min.}$), functions of the thickness of the Ge layer deposited on Si(001). A monotonous decrease of these two parameters is clearly highlighted, as soon as we thicken at 750°C the Ge "seed" layer deposited at 400°C. Surface rms roughness and Z ranges of about 0.7 nm and 4.5 nm are obtained for 2.5 µm thick Ge layers, thermally cycled or not. Although these values are remarkably low given the large lattice parameter mismatch between Ge and Si, it is possible to still reduce the surface roughness by using chemical mechanical polishing steps, followed by dedicated chemical wet cleanings and epitaxial re-growth process of Ge thin layers [ABB 06, ABB 08, CLA 06, HAR 04d].

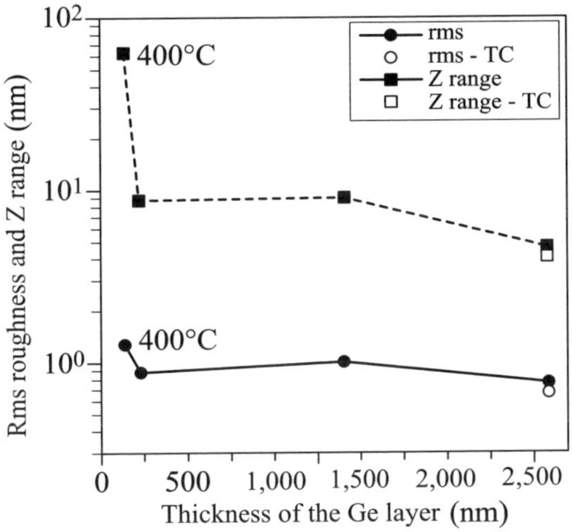

Figure 4.70. *Rms roughness and Z range according to the thickness of the Ge layer deposited on Si (001)*

We used X-Ray Diffraction, in order to determine the macroscopic degree of strain relaxation (defined by $R(\%) = \left(a_{Ge}^{//} - a_{Si}\right)/\left(a_{Ge} - a_{Si}\right)$, $a_{Ge}^{//}$, a_{Ge} = 5.65785 Å and a_{Si} = 5.43105 Å, being respectively the in-plane lattice parameter of the epitaxial Ge layer and the bulk lattice parameters of Ge and Si). As an example, we will find in Figure 4.71, the omega-2theta scans around the (004) XRD order for the Ge layers imaged in Figure 4.69. The peak associated with the Ge layer shifts towards high incidence angles, and becomes narrower and more intense, when the thickness increases. We will note the strong asymmetry of the peak, when a thermal cycling is used. This asymmetry is due to an inter-diffusion between Si and Ge at the interface.

The macroscopic degrees of strain relaxation are represented in Figure 4.72, according to the thickness of the deposited Ge layer. The 130 nm Ge "seed" layer deposited at 400°C is 95% relaxed. On the other hand, as soon as we go up to 750°C and we thicken the Ge layer, the latter is tensily-strained (R ~ 104-105%). This is at first paradoxical. We would indeed expect a compression state. This is however without the different coefficients of thermal expansion between Si and Ge [REE 96]. If we assume that a 2.5 µm Ge layer is completely relaxed at 750°C (i.e. R = 100%), and that the in-plane shrinking of the Ge lattice parameter occurs with the thermal expansion coefficient of the much thicker Si substrate underneath, we are at room temperature with $R_{theor.}$ = 105.4%, a value very close to those experimentally obtained [HAR 08e].

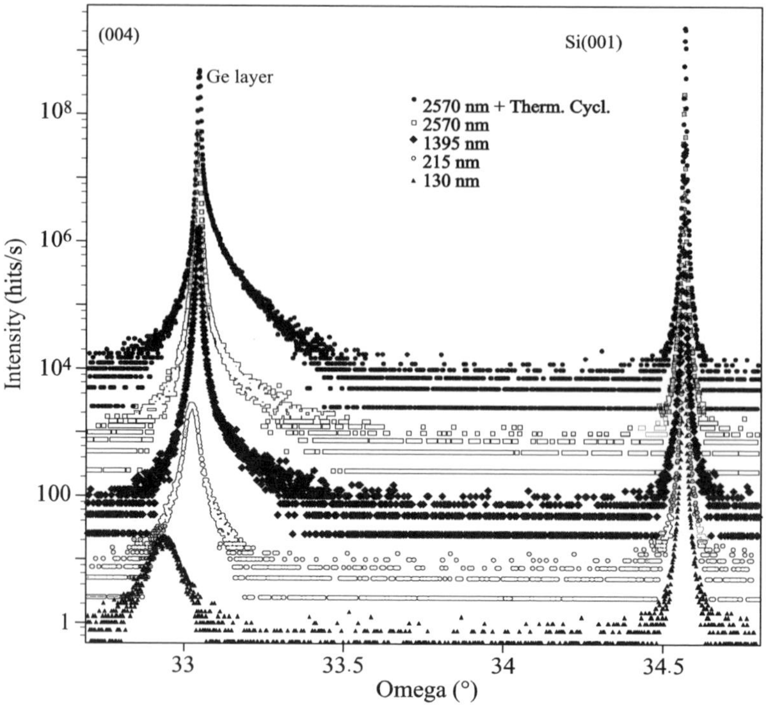

Figure 4.71. *Omega-2 theta profiles around the (004) XRD order for various thicknesses of Ge layers on Si (001)*

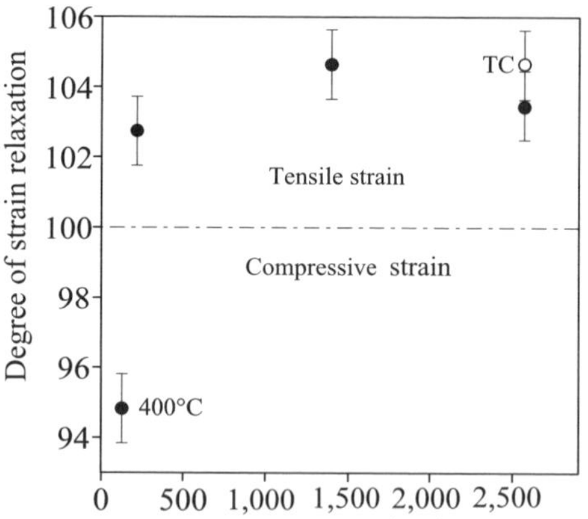

Figure 4.72. *Macroscopic degree of strain relaxation functions of the thickness of deposited Ge*

We imaged a thermally cycled, 2.8 μm thick Ge layer on Si(001) in Transmission Electron Microscopy (see Figure 4.73). There is a very important density of defects in the first 30 nm of Ge, as well as a significant number of misfit dislocations in the first micron. The top 1.8 μm of Ge is almost exempt of defects at the TEM scale, which likely means that the threading dislocations density is lower than 10^8 cm^{-2} [HAR 09b]. We thus carried wet chemical decorations (using Secco, Schimmel and chromium-free solutions) of the threading dislocations in these layers and we counted the etch pits using an optical microscope. Threading dislocations densities of about 10^7 and $5x10^7$ cm^{-2} have been obtained in 2.5 μm thick Ge layers repectively with and without thermal cycling [HAR 09b]. We can also use *in situ* HCl defect decoration, in order to gain access to these densities.

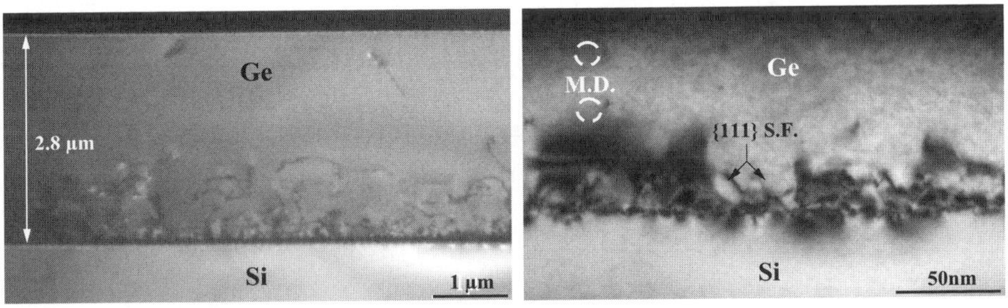

Figure 4.73. *Cross-sectional TEM image of the totality of a 2.8 μm thick, thermally cycled Ge layer on Si(001) (on the left) and zoom close to the Ge/Si interface (on the right)*

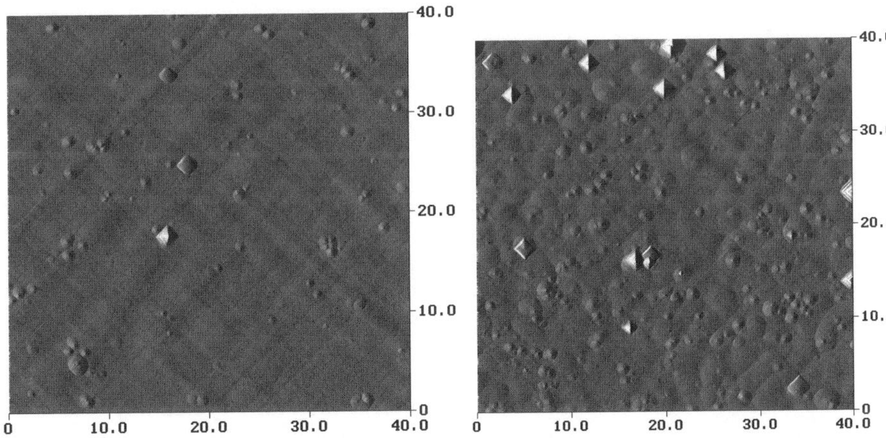

Figure 4.74. *Amplitude mode AFM images of the surface of 2.5 μm thick Ge layers without (on the left) or with (on the right) thermal cycling after HCl decoration of the threading dislocations*

We will find in Figure 4.74 amplitude mode AFM images of the surface of 2.5 µm thick Ge layers thermally cycled or not, after the HCl etch of ~ 0.3 µm of Ge. The etch pit density is three times lower after thermal cycling: $7x10^6$ cm^{-2} ⇔ $2x10^7$ cm^{-2}. The fact that ~ 1 µm of Ge was consumed during wet chemical defect decorations probably explains why higher threading dislocations density values found: we are indeed closer to the dislocated area [HAR 09b].

Finally, we used secondary ions mass spectrometry, in order to have access to the Si and Ge depth profiles in these thick layers (see Figure 4.75). A significant penetration of the Si atoms (coming from the substrate) in the Ge layer is highlighted after an "aggressive" thermal cycling (in this case, 10 x {750°C, 10 min/900°C, 10 min}). This is harmful for the absorption properties of Ge in a photo-detector, but has *a priori* only little impact from the electric transport point of view.

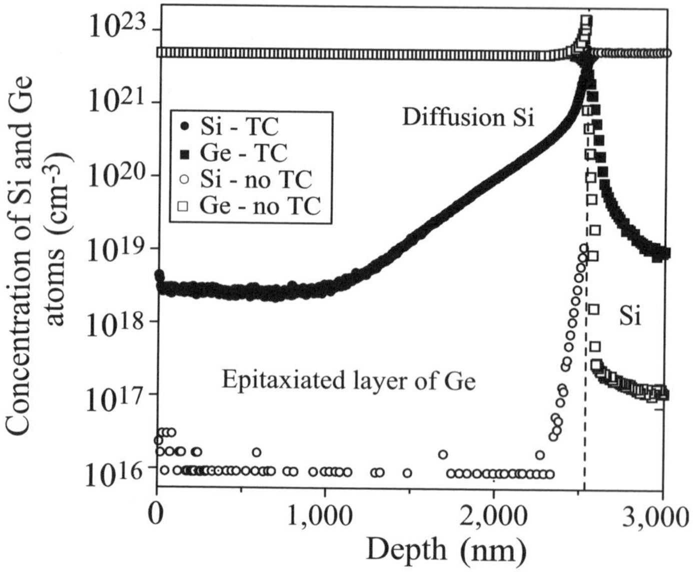

Figure 4.75. *SIMS profile of the concentration in Si and Ge atoms in thick Ge layers thermally cycled or not deposited on Si (001). The interface is symbolized by a dotted line*

The first hundred nm (from the surface) of these thermally cycled thick Ge layers can be transferred thanks to the SmartCut^{TM} process on oxidized Si substrates [AUG 09, CLA 06]. We thus obtain GeOI substrates of excellent crystalline quality, as shown in Figure 4.76. The density of threading dislocations in these GeOI substrates is indeed about the same, as the one of the Ge starting layers [ABB 07c].

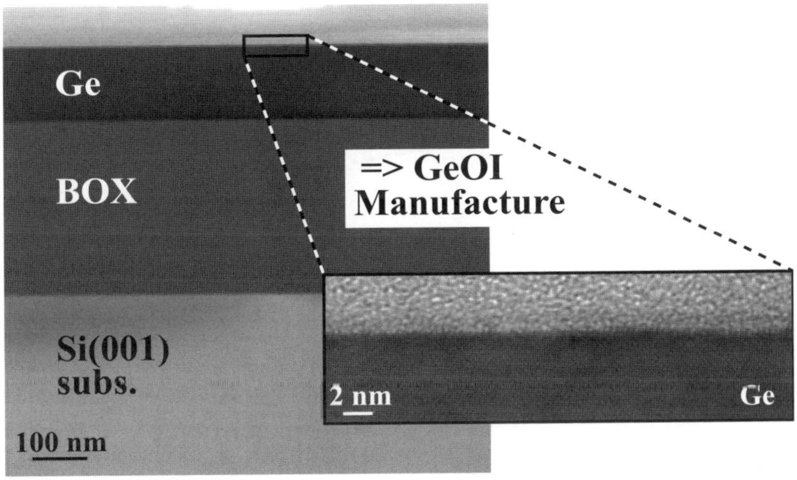

Figure 4.76. *Cross-sectional TEM images of a GeOI substrate fabricated with a thick layer of Ge deposited on Si(001)*

4.7.3. *Optical and transport properties of thick Ge layers on Si (001) and of GeOI substrates*

First of all, we will take a look at the optical properties of thick Ge layers on Si(001). In Figure 4.77, we will find the absorption coefficient of Ge layers of various thicknesses, functions of the wavelength. Despite the fact that these layers are dislocated, very high absorption coefficients are obtained at 1.3 µm (10,000 cm^{-1}), as well as at 1.55 µm (4,500 cm^{-1}). The slight tensile strain present in these Ge layers (see Figure 4.72) generates a decrease of about 30 meV of their direct energy bandgap (which determines the absorption threshold). We go indeed from 0.81 eV for bulk Ge to 0.78 eV for epitaxial Ge on Si, which is most interesting for photo-detection at 1.55 µm. We thus could obtain high-speed Metal-Semiconductor-Metal [VIV 07] and vertical p-i-n [VIV 09] Ge photo-detectors, at 1.31 as well as at 1.55 µm. To that end, we selectively grew several hundred of nm up to more than one micron thick Ge layers in Si cavities at the end of SOI waveguides, thanks to a low /high temperature growth strategy, without subsequent thermal cycling.

The latter is indeed harmful for thin Ge layers: the GeSi alloy, then formed, (from the Ge/Si interdiffusion) absorbs light in the near infra-red far less efficiently than pure Ge [HAR 10b]. In the case of p-i-n Ge photo-detectors, Ti/TiN/AlCu electrodes have then been deposited and etched, in order to contact the higher and lower *in situ* doped layers (see Figure 4.78). A cut-off frequency at -3dB, able to go up to 42 GHz at 1.53 µm, was then obtained (see Figure 4.79).

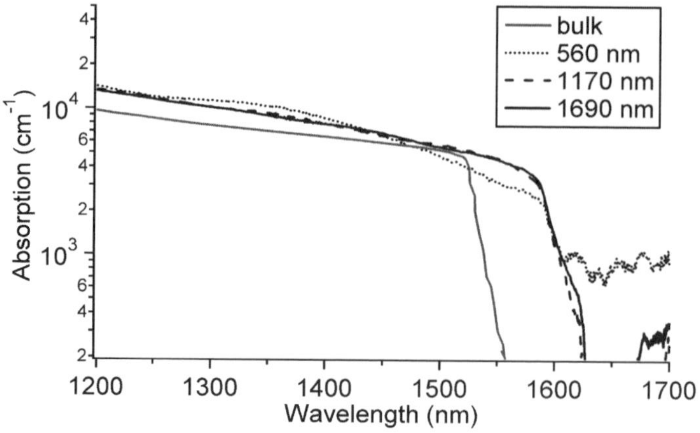

Figure 4.77. *Absorption coefficient of bulk Ge and of various thickness Ge layers deposited on Si(001), functions of the wavelength*

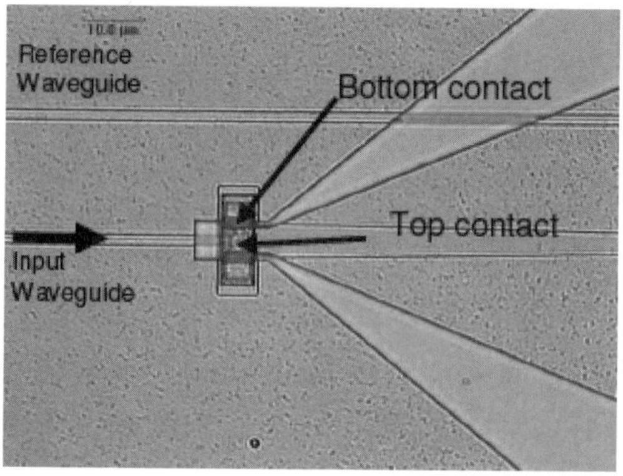

Figure 4.78. *Top view optical microscopy image of a Ge photo-detector inserted at the end of a ridge optical waveguide*

It is possible, when combining the tensile strain present in Ge layers deposited on Si(001) and n-type phosphorous ion implantation, to obtain an intense emission of light at a wavelength of about 1.6 µm, at room temperature [SUN 09]. The electrons given by the n-type impurities to the Ge crystalline lattice then occupy the L valleys of the conduction band, exalting the direct transitions between the Γ valley of the conduction band and the heavy and light holes sub-bands of the valence band. A room temperature laser has thus been obtained [LIU 10], letting us believe for the near future in the monolithic integration of IV-IV light emitters, optical waveguides, modulators [MAR 09] and photo-detectors on Si substrates.

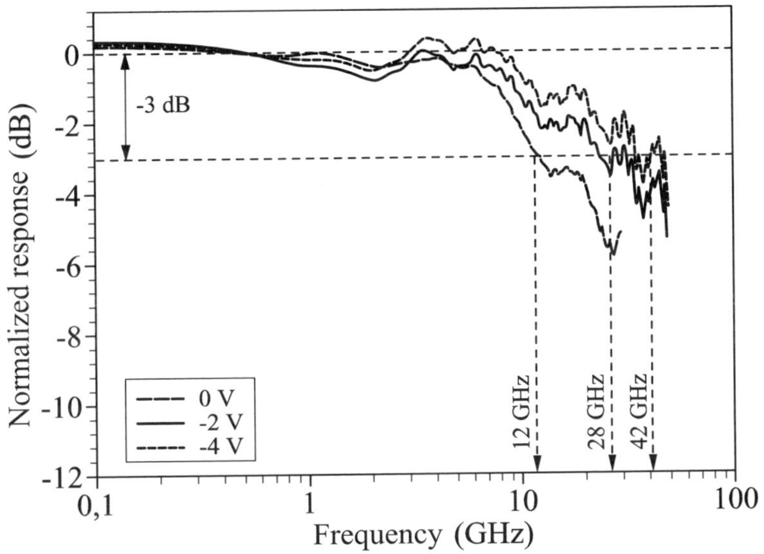

Figure 4.79. *Normalized optical functions of the frequency for a Ge pin diode integrated at the end of a ridge SOI optical waveguide for 0, -2.0V and -4.0 V applied voltages, this for a 1.53 μm wavelength. The length of the Ge photo-detector is 15 μm and the width of the ridge is 3 μm*

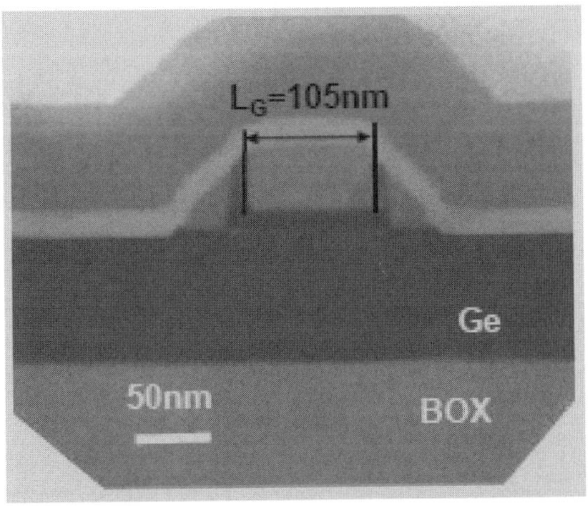

Figure 4.80. *Cross-sectional TEM image of a pMOSFET transistor on GeOI*

We will now briefly show the electric properties of pMOSFETs transistors fabricated on GeOI SmartCut^{TM} substrates (see Figure 4.80) [POU 08, POU 09, ROY 08, VIN 10]. The integration scheme is as follows: after the definition of mesas on the GeOI substrate, a surface passivation of Ge by 1 to 2 nm of epitaxial Si is carried out [HAR 09b, HAR 10b, MIT 09]. This Si layer, partially chemically oxidized, is then covered with a HfO_2/TiN/poly-Si gate stack.

The gates are then patterned using ultra-violet lithography, the extensions are implanted using BF_2, the HfO_2/SiN spacers are etched and the sources and drains are BF_2 implanted. The thermal budgets of the post implantation activation anneals and those of the various deposition steps, have been lowered. Those pMOSFETs are indeed fabricated on GeOI substrates, and not SOI substrates (temperature of solid-liquid transition for Ge: 928°C ⇔ 1,414°C for Si). From the device point of view, we have been confronted to the appearance of a parasitic channel at the interface between Ge and the buried oxide generating a very important off-state current and thus a degraded I_{ON}/I_{OFF} trade-off. The use of a high backside voltage V_{BG} = 60 V (for an intrinsic GeOI substrate), the insertion of a thin Si passivation layer between Ge and the buried oxide [ROM 09] or a counter-doping using phosphorus atoms, yielded good pMOSFET characteristics, however (i.e. a negative threshold voltage and a I_{ON}/I_{OFF} trade-off higher than 10^4), as illustrated in Figure 4.81.

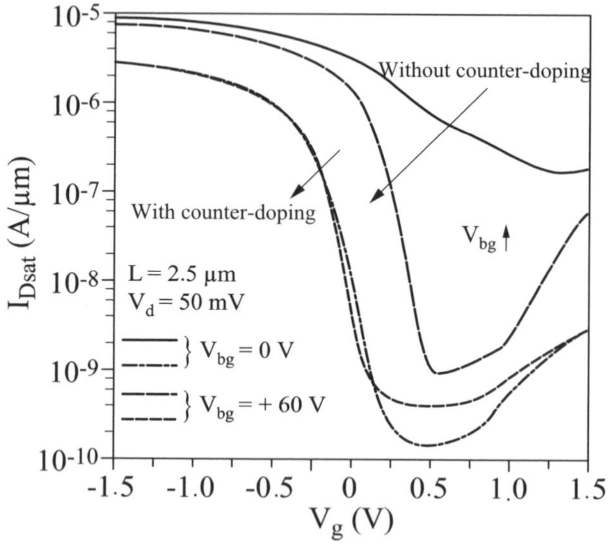

Figure 4.81. *Low drain voltage I_D - V_G characteristics, with and without P counter-doping, for 0 and 60 V backside voltages*

Holes mobilities in these pMOSFET GeOI transistors are twice higher with or without channel counter-doping, than those in FD-SOI transistors. With implanted pockets (control of short channel effects), they are 1.5 times higher (see Figure 4.82). The horizontal or vertical co-integration with SOI (or sSOI) on the nMOS side and GeOI on the pMOS side can consequently be used, in order to obtain high efficiency complementary MOS components [BAT 09b].

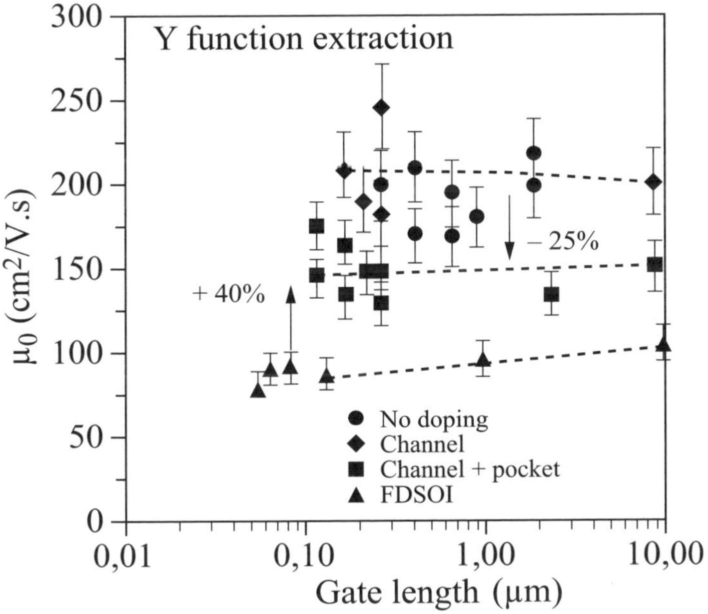

Figure 4.82. *Holes mobility according to the gate length in pMOSFET GeOI transistors with and without counter-doping and in pMOSFETs FD-SOI transistors*

In order to be exhaustive, it is possible, thanks to the Ge enrichment method, to obtain GeOI substrates without using the SmartCut™ method. In this approach, we start from thinned down SOI substrates, on which low Ge concentration SiGe layers are deposited. Thanks to high temperature dry oxidation steps, it is then possible to selectively oxidize the Si atoms and to push, thanks to a "snow-plough" effect, the Ge atoms towards the buried oxide. We thus obtain very high Ge concentrations, thin (~ 10 nm) SiGe layers, on which pure Ge layers of pure Ge are then deposited [NGU 07]. Thanks to various device integration improvements, a I_{ON}/I_{OFF} trade-off exceeding 10^5 has recently been obtained in pMOSFETs transistors fabricated on this type of substrate [HUT 10b].

4.7.4. *Structural and optical properties of Ge islands on Si (001)*

In order to conclude section 4.7, we will take a look at the growth of Ge islands on Si (001) and at the optical properties obtained when those are encapsulated by Si [HAR 05c, DAM 04b]. When we deposit Ge on Si at relatively high temperatures (about 600°C-700°C in RP-CVD, lower in molecular beam epitaxy), we have, because of the compressive strain stored in the Ge layer, a growth mode called Stranski-Krastanov, i.e. an initially 2D growth mode, which becomes 3D (formation of islands) when it is energetically favorable that the surface undulates (elastic relaxation of the built-in strain).

These Ge islands have intensively been studied these last 20 years, with the objective of benefitting (post-encapsulation by Si) of quantum effects in IV-IV semiconductor heterostructures. Their sizes, shapes and surface density depend on the amount of Ge sent on the surface, on the temperature and on the growth pressure, etc., [HER 99, KAM 97, KUR 02a, KUR 02b, KER 04, LOO 01, MO 90]. Low growth temperatures are indeed recommended, in order to obtain high densities of pyramidal Ge islands with {105} facets. With higher growth temperatures, we will probably have islands nucleation in the shape of domes, with higher slopes.

In Figure 4.83, we will find 1 µm x 1 µm AFM images of the surface for various numbers of Ge monolayers sent at 600°C, 20 Torr on Si(001). The growth rate is about 1.5 nm $min.^{-1}$ for a $F(GeH_4)/F(H_2) = 4.17x10^{-5}$ mass-flow ratio. The surface is smooth for 3.9 atomic monolayer of Ge (i.e. 5.7 Å). On the other hand, for 5.1 atomic monolayers of Ge, there is an appearance of a substantial number of {105} pyramids. When the number of atomic monolayers goes from 5.1 to 15.3, these pyramids grow in size and gradually evolve into domes.

We extracted from such AFM images the surface density of the islands, as well as their average diameter and height functions of the number of atomic monolayers of Ge deposited at 600°C or 650°C on Si (001) (see Figure 4.84). The total density of islands (pyramids + domes) is much lower at 650°C than at 600°C: $9.4x10^9$ cm^{-2}⇔ $2.4x10^{10}$ cm^{-2}. We have an increase in the surface density of domes, at the expense of one of the pyramids, when the number of atomic monolayers of deposited Ge increases.

In addition, let us notice that the diameters of the pyramids and domes are relatively independent from the number of Ge monolayers deposited, but depend however on the growth temperature: the higher it is, the larger the diameter will be (between 60 and 90 nm).

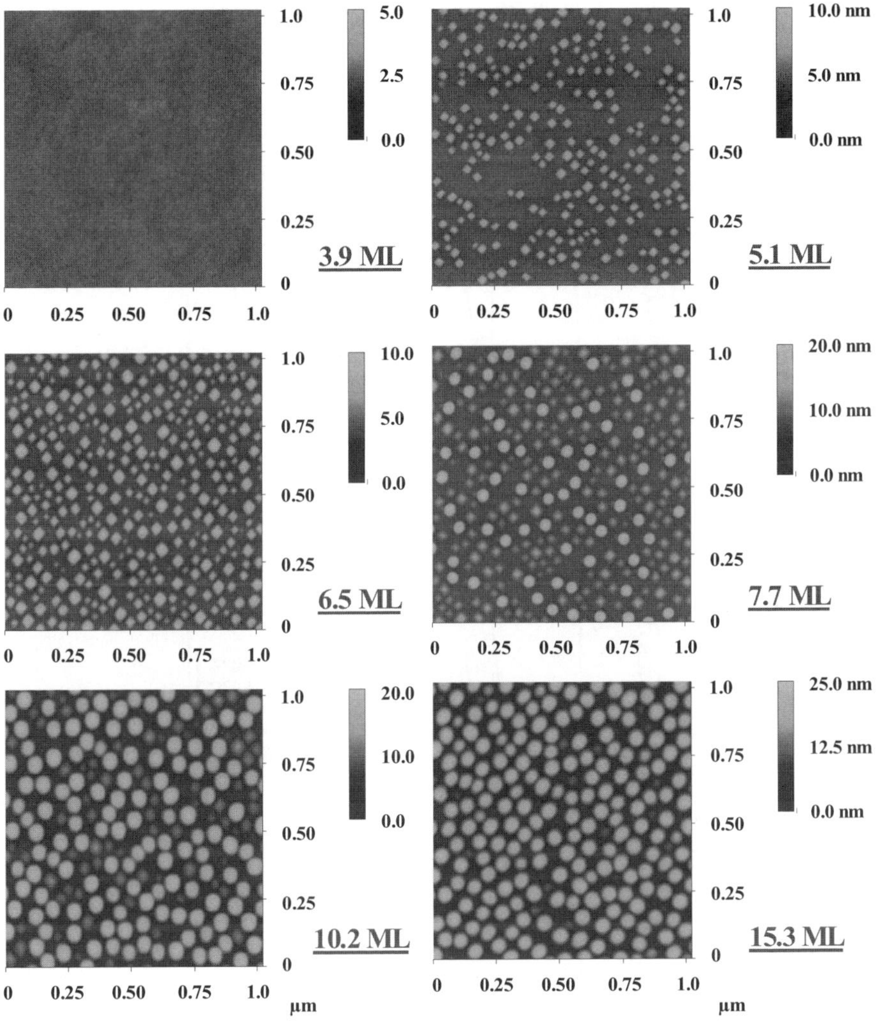

Figure 4.83. *Evolution of the surface morphology functions of the number of atomic monolayers of Ge deposited at 600°C on Si (001)*

The height of the pyramids is, as expected, constant at ~ 5 nm, given the relative stability of their diameter ({105} facets). On the other hand, the height of the domes definitely increases with the growth temperature and the number of Ge atomic monolayers deposited. It reaches values higher than 20 nm for 15 atomic monolayers.

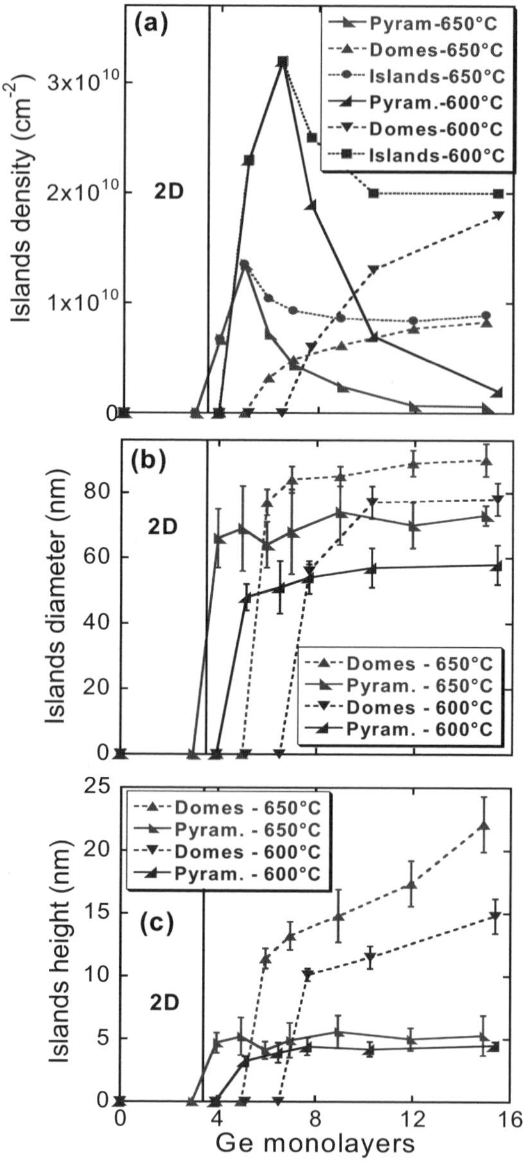

Figure 4.84. *Evolution (from left to right, from top to bottom) of the density, of the diameter and of height of the Ge pyramids or domes, functions of the number of atomic Ge monolayers deposited at 600°C, as well as 650°C on Si (001). The diameters of the islands are most likely over-estimated because of the convolution between the shape of the AFM tip and the actual surface*

A single plane of Ge islands deposited at 600°C on Si (001) has been encapsulated by 150 nm of Si and has been studied at T = 10 K in photoluminescence. We will find the spectra obtained in Figure 4.85. The impact of the number of Ge atomic monolayers on the energy position of no-phonon (NP) lines and of their replica, assisted by an optical transverse phonon (TO) is illustrated in Figure 4.86. These lines shift towards low energies, i.e. towards large wavelengths, when the number of Ge atomic monolayers increases. It is thus possible to tune the properties of luminescence of these Ge quantum islands, functions of the applications aimed for (emission at 1.3 or 1.55 μm).

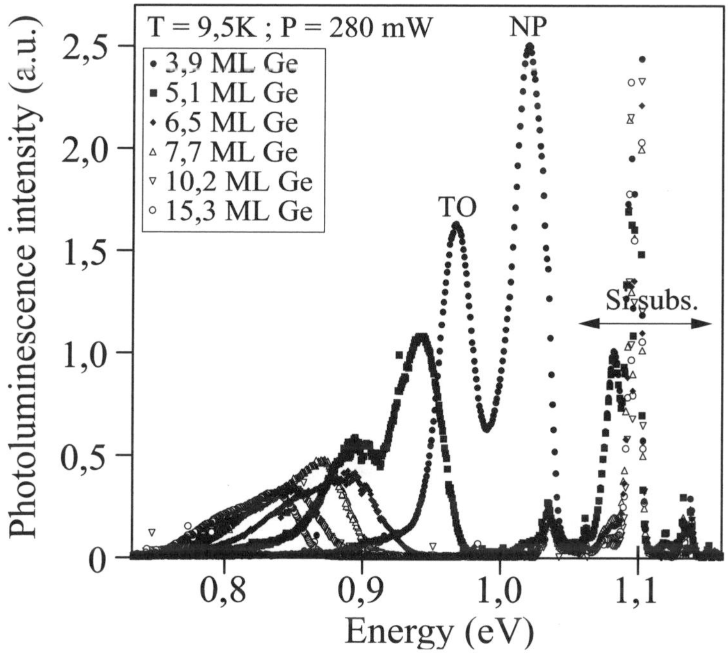

Figure 4.85. *Photoluminescence spectra associated with a single plane of Ge islands capped with Si, according to the number of Ge atomic monolayers deposited*

Multiple planes of Ge islands (with tens of nm thick Si spacers) can be stacked one upon the other in order to exalt the luminescence output per surface unit. Depending on the thicknesses of the Si spacer layers, there will be either a vertical alignment of the islands one above the other, or a "A-B-A-B" stacking (presence of stress fields in these stacks) [DAM 04b, KER 04]. Rather interesting photo- or electro-luminescence properties at 0.8 eV, i.e. 1.55 μm, have then been obtained.

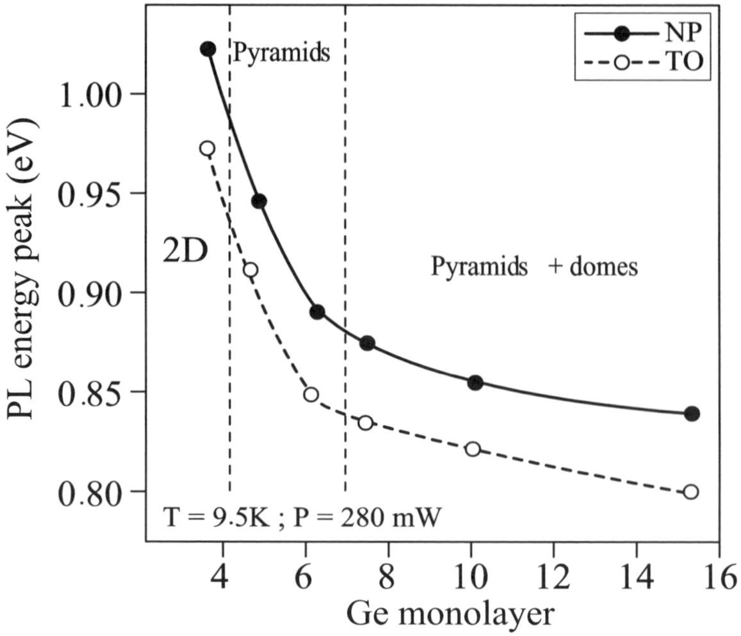

Figure 4.86. *Energy positions of the NP and TO lines functions of the number of Ge atomic monolayers deposited*

4.8. Devices based on sacrificial layers of SiGe

4.8.1. *Introduction*

To be able to etch several nanometers thick mono-crystalline layers selectively compared to others is mandatory in order to obtain localized SOI or silicon-on-nothing devices [MON 07, MON 10], multi-wires [DUP 08a, DUP 08b, TAC 09] or multi-channels [BER 08, BER 09a, BER 09b] transistors, etc. In the following figures, we will show a schematic description of the technological steps leading to some of these structures.

All these devices rely on the fact that, in pseudomorphic SiGe/Si stacking, it is possible to laterally etch SiGe layers with a Ge concentration higher or equal to 20%, selectively compared to Si. This obviously presupposes having lateral access to said SiGe layers, thanks to anisotropic etching steps (definition of mesas) or to more complex sequences of technological steps [DUT 07]. Several techniques are available for these selective etchings. A plasma etching at low pressure and room temperature with CF_4 as a precursor gas is normally used [BAR 08, BOR 06]. We can also call upon wet chemical solutions, such as $HF{:}HNO_3{:}CH_3COOH$ [HOL 10,

SAL 08]. In the first case, we will have etching selectivity between 60 and 100 (depending on the Ge concentration of the layer). In the second case, the selectivity will be about 200 [SAL 08]. The etching will be perfectly isotropic in both cases.

The last few years, we explored a third option: thermal HCl etching directly in the epitaxy chamber [BOG 06c, DES 08a, DES 08b]. This technique will be presented in section 4.8.2.

4.8.2. *Selective HCl etching of SiGe selectively compared to Si*

In 2004, we profited from the fact that gaseous hydrochloric acid cannot etch SiO_2 or Si_3N_4, in order to etch on patterned substrates the Si windows surrounded by oxide (shallow trench isolation or STI). These cavities were filled afterwards with several hundred nm of $Si_{0.8}Ge_{0.2}$, the idea being to locally have thin SiGe virtual substrates, on which tensily-strained Si layers can then be deposited [BOG 05e, DEL 04]. That is why we studied the etching kinetics of Si, of $Si_{0.65}Ge_{0.35}$, of $Si_{0.5}Ge_{0.5}$ and of Ge. We found ourselves in the presence of (i) a high temperature regime, where the etch rate dependence on the Ge concentration is low, and (ii) a low-temperature regime, where the etch rate dramatically increases with the Ge concentration. From this came the idea to use HCl, in order to laterally etch SiGe selectively compared to Si, idea whose feasibility we proved in 2006 [BOG 06c]. Relatively high etching temperatures (typically 700°C) were then necessary (because of the low total pressure (20 Torr) and the low partial pressure of HCl (0.208 Torr)), in order to have reasonable etching rates. We thus increased the total pressure, as well as the HCl partial pressure in considerable proportions: 300 Torr and 180 Torr, respectively.

We thus could increase by a factor of 10 (at given temperature) the etch rate of Si and have substantial SiGe etch rates at rather low temperatures (see Figure 4.87) [DES 08a]. In Figure 4.88, we can see the theoretical selectivities expected (by dividing at a given temperature, the vertical etch rate rate of SiGe by the (extrapolated) vertical etch rate of Si). Those are about 30 for $Si_{0.8}Ge_{0.2}$, 200 for $Si_{0.7}Ge_{0.3}$ and 5,000 for $Si_{0.6}Ge_{0.4}$!

What about the lateral selectivity of the etching of SiGe compared to Si? In order to evaluate it, we deposited SiGe 20 nm/Si stacks with various Ge concentrations, that we capped with deposited oxide. Optical lithography and anisotropic etching (all the way to the Si substrate underneath) was then used in order to define to have access laterally to the Si and SiGe layers (various size mesas with <110> sidewalls). In Figure 4.89, we will find a cross-sectional scanning electron microscopy image of such a stack, after an HCl etching at 650°C for twenty minutes.

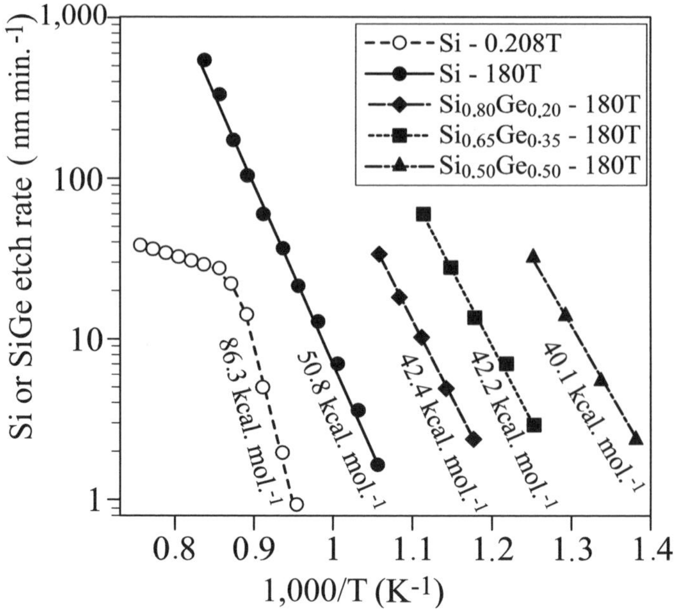

Figure 4.87. *Etch rate of Si (for partial pressures of HCl of 0.208 and 180 Torr) and of SiGe (partial pressure of HCl equal to 180 Torr), functions of the opposite of the absolute etching temperature*

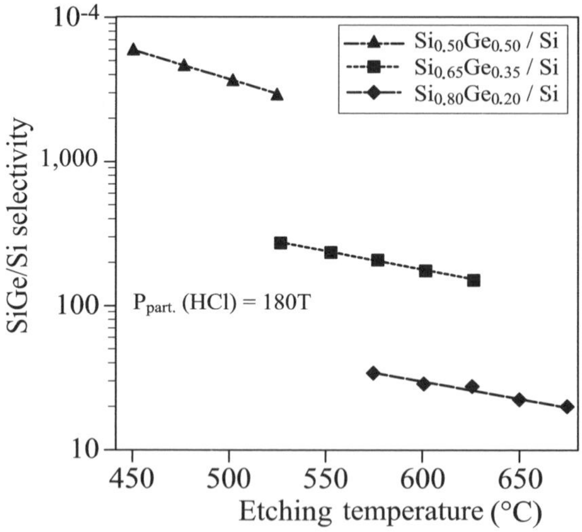

Figure 4.88. *Theoretical selectivities expected functions of the etching temperature, for Ge concentrations of 20%, 30% and 40%*

We can notice that, as expected, the depth of the channels dramatically increases with the Ge concentration: it goes from 20 nm for $Si_{0.8}Ge_{0.2}$ to 270 nm for $Si_{0.7}Ge_{0.3}$ to 740 nm for $Si_{0.6}Ge_{0.4}$. Selectivities of 100 for $Si_{0.7}Ge_{0.3}$ and of 130 for $Si_{0.6}Ge_{0.4}$, are associated with this high HCl partial pressure lateral selective etching [SAL 08].

However, we did not succeed in selectively etching $Si_{0.8}Ge_{0.2}$ layers compared to Si, even by playing with the etching parameters. This is one of the restrictions of HCl etching, compared to the dry plasma and wet etching techniques. We used cross-sectional transmission electron microscopy, in order to visualize more precisely these channels (see Figure 4.90). Floors and ceilings of tunnels are smooth. <111> facets are otherwise present at the end of the tunnels. Moreover, we can note, in Figures 4.89 and 4.90, that the SiGe layers, located at the bottom of stackings, are slightly more etched than those at the top. With given process parameters, <110> lateral SiGe etch rates are, for 20 nm thick SiGe layers, approximately ten times lower than the vertical etching rates (along [001]) of Figure 4.87.

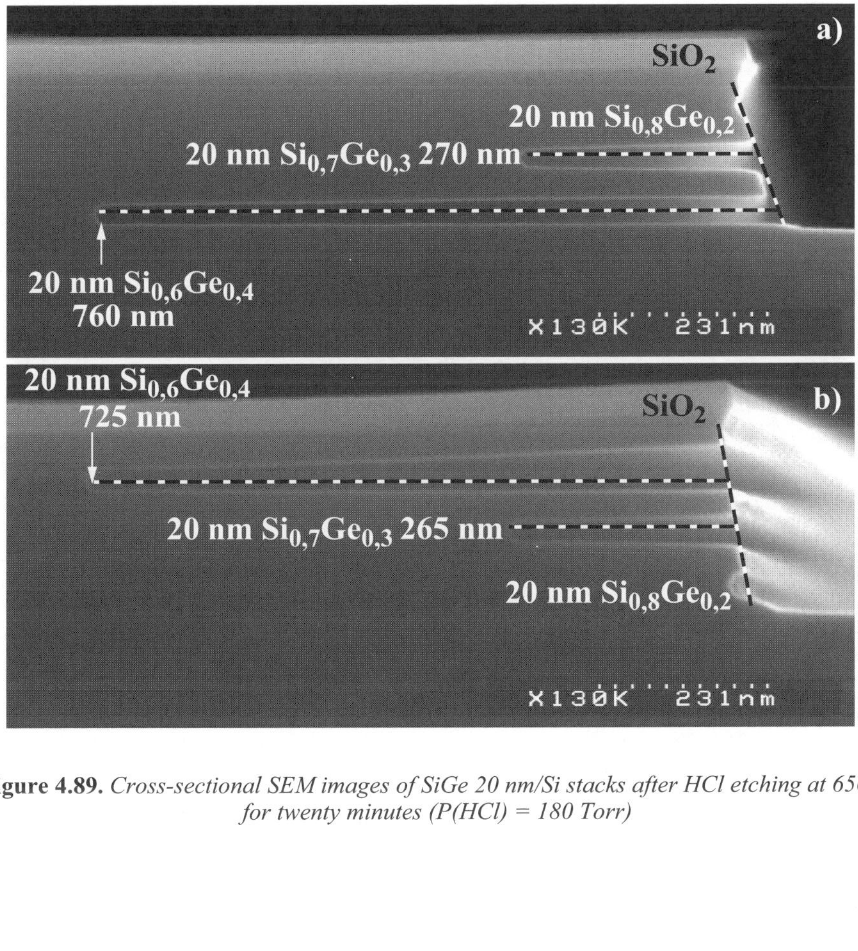

Figure 4.89. *Cross-sectional SEM images of SiGe 20 nm/Si stacks after HCl etching at 650°C for twenty minutes (P(HCl) = 180 Torr)*

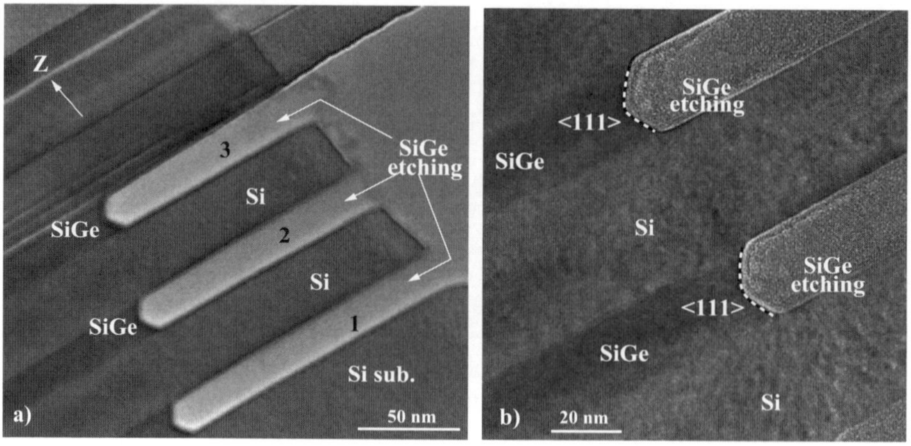

Figure 4.90. *Cross-sectional TEM images of a $Si_{0.6}Ge_{0.4}$ 20 nm/Si multilayer after HCl etching at 610°C for twenty minutes (P (HCl) = 180 Torr)*

The isotropy of the HCl etching method is thus questionable. A top view SEM image of patterned SiGe/Si stacks after HCl etching can be found in Figure 4.91 (mesa edges are along the <110> directions). The regions where SiGe has been removed are clearly seen (electron transparency differences with regions where SiGe is still present). We clearly have at the corners of these <110> elongated lines <113> facets, related to a etch rate which is much higher in the [100] and [010] directions than in the <110> directions. The HCl etching method is thus definitely anisotropic, contrary to the other two.

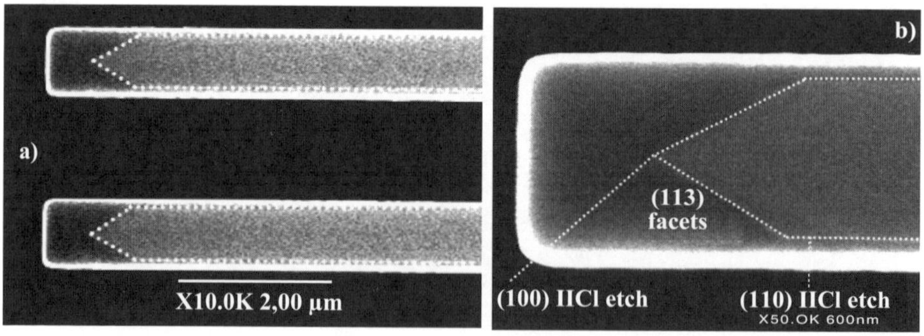

Figure 4.91. *Top-view SEM images of several tens of µm long mesa lines etched anisotropically in a (20 nm $Si_{0.6}Ge_{0.4}$/40 nm Si) x3 superlattices, which was laterally etched later on at 600°C for twenty minutes (P (HCl) = 180 Torr)*

We can add that with the HCl method, lateral etch rates definitely depend on the thickness of the SiGe layer. An increase by 18 of the etch rate has indeed been noticed, by going from 5 to 20 nm thick $Si_{0.6}Ge_{0.4}$ layers [DES 08b].

Because of these restrictions, we did not use HCl selective etching of SiGe compared to Si on the 200 mm localized-SOI and SON electrical lots processed in CEA-LETI. Our set of lithography masks impose the presence of anchors on both sides of the active central zone, which will be hollowed out and then filled by a dielectric, so that the whole pattern (in shape of a bone) does not collapse. However, it seems that these anchors are preferentially consumed compared to the active zones during anisotropic HCl etching [DES 08a]. On the other hand, 300 mm L-SOI and SON technologies do not require anchors: the small area Si slabs on top are kept rigid by the gate during selective etching. This justifies the adoption of HCl etching at STMicroelectronics Crolles [MON 07, MON 10].

4.8.3. *Localized SOI devices and SON*

We will find in Figure 4.92, a schematic description of the main technological steps used for the fabrication of localized SOI devices [MON 07]. We start from patterned Si (001) substrates with STI isolation. After a {"HF-last" wet cleaning / *in situ* H_2 bake at 850°C} surface preparation, the active zones of Si are partially recessed (consumption of approximately 20-30 nm) using HCl [DES 07, HAR 06b]. After that, a {sacrificial layer of $Si_{0.7}Ge_{0.3}$/Si channel} stack is selectively deposited compared to SiO_2 using chlorinated chemistry, by taking into account obviously the loading effects inherent in any SiGe epitaxy on a patterned substrate (see section 4.2). The desired thicknesses of $Si_{0.7}Ge_{0.3}$ and of Si are respectively about 15-20 nm and 10-20 nm.

Having hollowed out the active zones prior to epitaxy, we can thus avoid the formation of facets at the {active zone – STI} boundaries [COL 03, DUT 07]. Indeed, the flanks of the STI are then entering, instead of being open (see Figure 4.93).

A (HfO_2/TiN/poly-Si) gate stack is then deposited and etched and nitride spacers are fabricated. Si raised sources and drains (typically 20 nm thick) are then selectively deposited (thanks to the method described in section 4.4) and ions implanted. STI zones are then lowered thanks to a de-oxidation in a diluted HF bath, in order to have lateral access to the SiGe sacrificial layers. Those are selectively etched compared to Si, thanks to the CF_4 plasma etching technique (in 200 mm technology) or with a HCl etching (in 300 mm). We have then at this stage a suspended Si channel under the gate.

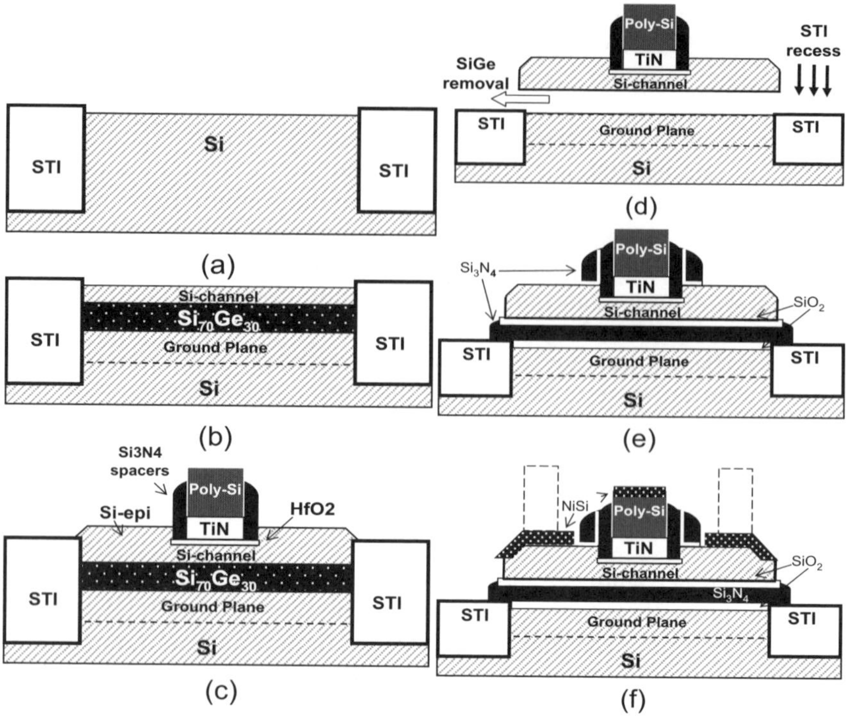

Figure 4.92. *Schematic description of the main technological steps used to fabricate L-SOI devices*

The space previously occupied by SiGe is then filled in a conformal fashion by a SiO_2/Si_3N_4 stacking. The sources and drains are then Ni silicided, prior to the so-called "back-end" process steps, where the metal interconnections are fabricated between the individual transistors. The perfectly single-crystal Si channel under the gate (thickness greater than or equal to 5 nm), is above a thin buried dielectric. That is why it is called localized SOI (we indeed started from massive Si substrates and not from SOI substrates).

In the literature, we can find a variant of the localized SOI method, called Bulk+ [MON 10]. The differences compared to the localized SOI method described in Figure 4.92, are as follows: after definition of the gate stack (see Figure 4.92c), the source and drains zones are anisotropically etched until the Si substrate is reached. We then have lateral access to the SiGe sacrificial layer under the gate, without needing to lower the STI zones. This layer is selectively etched compared to Si and the space left vacant is filled (as for the localized SOI) by a SiO_2/Si_3N_4 stacking.

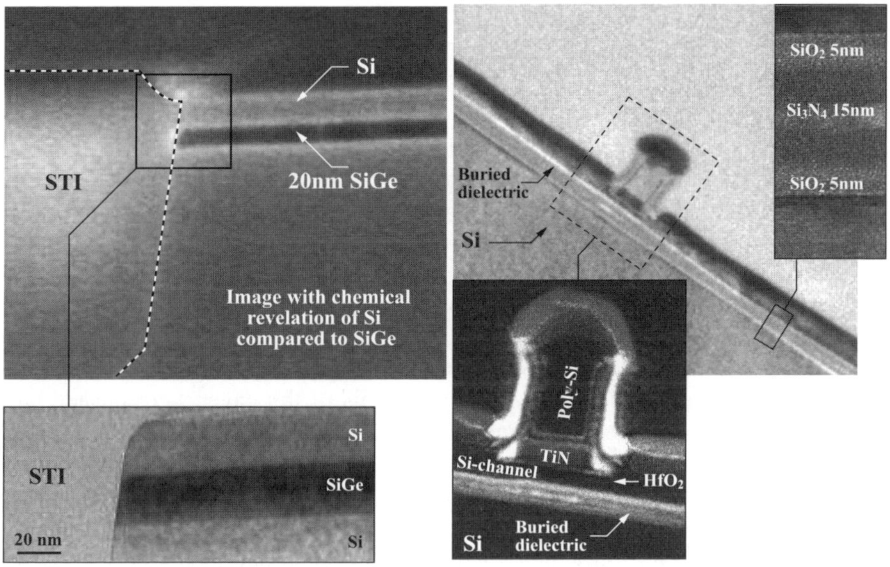

Figure 4.93. *Cross-sectional SEM or TEM images of the SiGe/Si stack (b) and of a L-SOI transistor (f)*

We then proceed to a selective epitaxy of about 30-40 nm of Si in the sources and drains areas, in order to connect the Si channel under the gate. We will find in Figure 4.94, a cross-sectional TEM image of a Bulk + transistor, for the 32 nm technological node obtained at ST Microelectronics Crolles [BID 08, BID 09a].

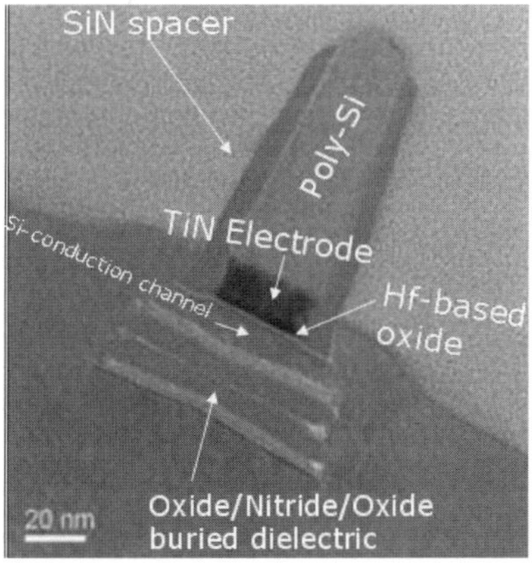

Figure 4.94. *Cross-sectional TEM image of a Bulk + 32 nm transistor with HfO$_2$/TiN/poly-Si gate stack*

Electric performances relatively similar to those of FD-SOI transistors, have been obtained with Bulk+ and localized SOI devices [MON 10]. We thus have a low cost co-integration strategy of "SOI-like" and "bulk" devices on the same bulk Si wafer (instead of starting from a much more expensive SOI substrate). Nothing indeed prevents us from selectively depositing SiGe/Si stacks in part of the Si windows of a patterned substrate, (which will be the core of upcoming Bulk+ or localized SOI transistors), the remaining windows being masked (sites of upcoming "bulk" transistors) [MON 10].

A certain number of recent developments were undertaken in order to improve the electric performances of localized SOI pMOS transistors. We notably carried out, after the etching of the SiGe sacrificial layer, a selective epitaxy of approximately 30 nm of $Si_{0.65}Ge_{0.35}$. The Ge atoms of this layer were pushed back towards the SiO_2/Si_3N_4 stacking, thanks to the Ge enrichment technique (i.e. a high temperature oxidation of the SiGe layer generating: (i) the formation of SiO_2 and (ii) the movement by a "snow-plough" effect of the Ge atoms towards the buried dielectric [NGU 07]). We thus locally obtained a dozen nm thick, very high Ge concentration GeSi layer, on which a gate was deposited and then etched. We then speak of germanium-on-nothing or GeON devices [BAT 07, BAT 08].

A second integration option has been developed, in order to have pMOS GeON transistors. A {2.3 nm pure Ge/1.3 nm Si} stack has therefore been selectively deposited at low-temperature on the Si surface layer (after etching the SiGe sacrificial layer underneath and filling the cavity thus formed by dielectrics). With: (i) the Ge/Si interdiffusion occurring during the epitaxy step itself and (ii) the thermal budget (even carefully minimized) of the subsequent technological steps, we obtain a GeSi channel (and not pure Ge) under the gate of our pMOS transistors. A 60% gain in the mobility of the holes in this channel (compared with Si) was nevertheless highlighted, stressing the interest of the approach [BAT 08].

Another way of gaining hole mobility would be to go from standard (001) to (011) crystalline planes [BID 09b, OKA 07, SHA 05, SHE 05]. These gains would be even higher by going from Si (011) to SiGe (011) [JOS 07, PAN 06]. The final objective being to fabricate high mobility localized SOI (011) pMOSFETs, we thus studied on blanket substrates the impact of the crystalline orientation on: (i) the efficiency of the H_2 bake that follows "HF-last" wet cleanings, (ii) the Si etch kinetics with HCl and (iii) the growth kinetics of Si and SiGe (the precursor gases being SiH_2Cl_2 and GeH_4) [DES 08c, DES 10, HAR 06c]. Briefly, a H_2 bake temperature $\geq$ 800°C is mandatory on Si(001) and Si(011) surfaces, in order to be free from any C or O residual contamination. Si HCl etch rates are (between 750°C and 950°C) four times higher on (011) than on (001). The associated activation energies are however close to one another (57 kcal mol.$^{-1}$ on (100) ⇔ 59 kcal mol.$^{-1}$ on (100)). The SiGe (011) growth rate at 600°C and 650°C increases parabolically then saturates when the GeH_4 flux increases. It however increases in an almost linear way on (001), reaching at 650°C, values three times higher than those on (011) for high GeH_4 flows. The parabolic relation provided in section 4.2 satisfactorily describes on (001) the dependence of the Ge concentration (between 10% and 37%) with the $F(GeH_4)/F(SiH_2Cl_2)$ mass-flow ratio. This relation is on the other hand linear on (011). The ratio between the Si(011) and the Si(001) growth rates, ~ 0.74, is close to the ratio of dangling bonds densities on the surface, i.e. 0.71. Activation energies of 63 and 65 kcal.mol.$^{-1}$ are associated with the increase of the Si growth rate with the temperature on (001) and (011).

With such data, we have been able to manufacture on patterned Si(011) substrates with STI isolation, high electric performances L-SOI transistors. Hole mobility gains as high as 250% were demonstrated [DES 10].

4.8.4. *Devices based on multi-wires and on multi-channels*

In order to gain in terms of current density per surface area, it could be interesting to stack conduction channels onto one another. A gate-all-around configuration them would lead to a better control of short channel effects for very small gate lengths. This is possible with architectures, known as "multi-channels" and "multi-wires".

We will find in Figure 4.95, a schematic description of the main technological steps used for the fabrication of multi-channels transistors. We start by depositing a Si/SiGe multi-layer on a blanket SOI substrate (step 1). Then, the active zones are defined ("mesa" isolation between those, i.e. we etch the Si/SiGe stack until the buried oxide). The gates are then defined thanks to the use of a hard mask, of a hybrid (UV/e-beam) lithography and of an anisotropic etching, stopping on the Si layer of the starting SOI substrate (step 2). A Si selective epitaxy in the source and drains zones on both sides of the gate is then carried out, in order to connect the piled up Si channels (step 3), followed by multiple ion implantation steps (to dope the S/Ds).

Then, the SiGe sacrificial layers are selectively etched, compared to those of Si, thanks to isotropic plasma etching (step 4). The cavities thus formed are filled in a conformal fashion by a HfO_2 / TiN / poly-Si gate stack. After the deposition of spacers, the source and drain zones are silicided (step 6). We thus obtain a multi-channels transistor, as illustrated by the cross-sectional TEM image in Figure 4.95.

We will find in Figure 4.96 a schematic description of the main technological steps leading to, this time, multi-wires transistors. We find the epitaxy of a Si/SiGe multi-layer on SOI (step 1), the anisotropic etching of an array of "fins" between the sources and drains blocks (step 2), the selective etching of the SiGe sacrificial layers (step 3), the conformal deposition in the cavities thus formed, as well as between the "fins" of the same type of gate stack as for multi-channels (step 4), the anisotropic etching that follows leading to the obtainment of either etched gates (step 5a), or "plugged" gates (step 5b), followed by multiple ion implantation steps and silicidation of sources and drains (step 6) and then by metal interconnections (step 7).

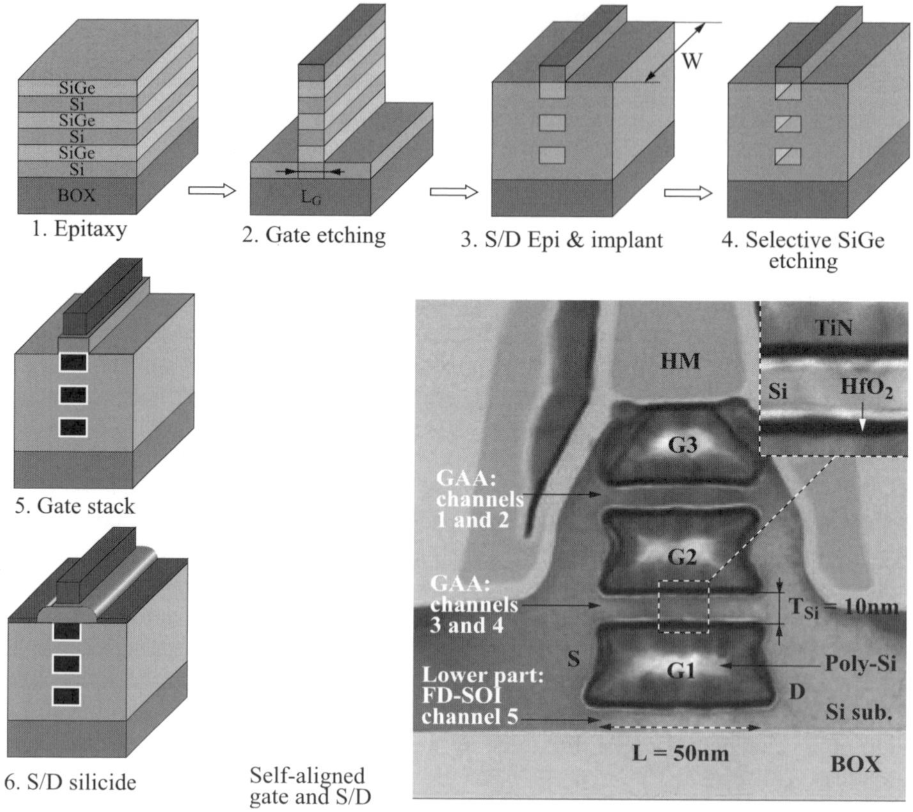

Figure 4.95. *Description of the various technological steps used in order to fabricate multi-channels transistors and cross-sectional TEM image of a transistor with 5 channels*

We notice that, for the two integration schemes, it is essential to obtain on SOI substrates Si/SiGe stacks which are in fact regular super-lattices. The individual thicknesses of the Si and SiGe layers will indeed dictate the dimensions according to z of the Si wires and of the spaces between those. And yet, the presence on SOI substrates of a buried oxide will significantly influence the growth kinetics of Si and SiGe, according to the deposited thickness, as we will see below [HAR 09a].

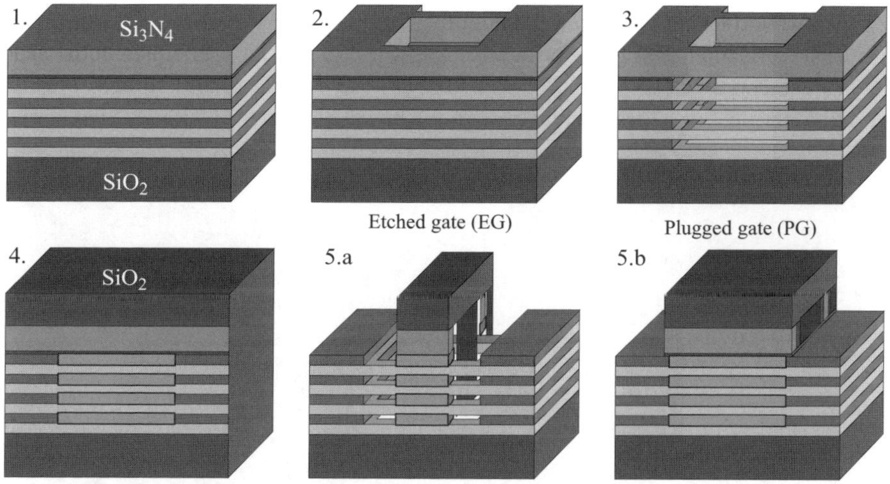

1. Epitaxy of Si/SiGe superlattices; deposition of a Si_3N_4 hard mask
2. (RIE) etching of the multilayer
3. Selective isotropic etching of the SiGe
4. Conformal deposition of the HfO_2/TiN/poly-Si gate stack; chemical mechanical polishing of poly-Si
5. Gate etching (anisotropic + isotropic); etched a) or "plugged" b) gate
6. Ion implantation, activation annealing and sources and drains silicidation
7. "Back end"

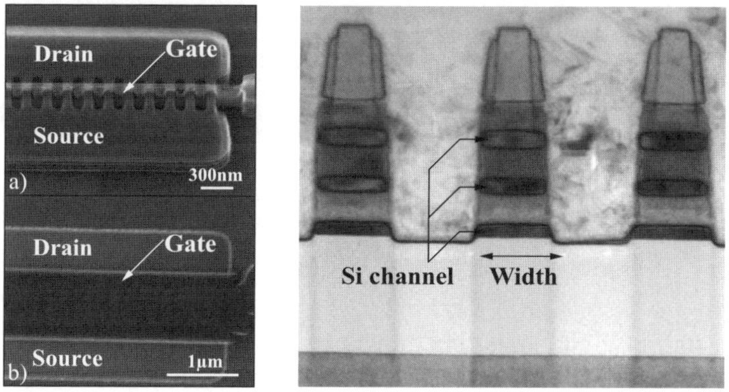

Figure 4.96. *Description of the various technological steps used for multi-wire transistors fabrication, top-view SEM images of multi-wires transistors with etched gate and "plugged" gate and cross-sectional TEM image of a transistor with 3 wires (perpendicular to the transistor)*

We will find in Figure 4.97 cross-sectional TEM images of two 19 periods $Si/Si_{0.8}Ge_{0.2}$ superlattices deposited at 650°C, 20 Torr on a bulk Si(001) substrate and on a {20 nm Si/145 nm buried oxide} SOI substrate, this using the *same epitaxy recipe*. Individual thicknesses of SiGe and Si layers are, as expected, constant on a bulk Si(001) substrate. Those on a SOI substrate are almost equal to those on a massive substrate for the first period. They however drop dramatically for the following periods, and then increase again when the deposited thickness increases.

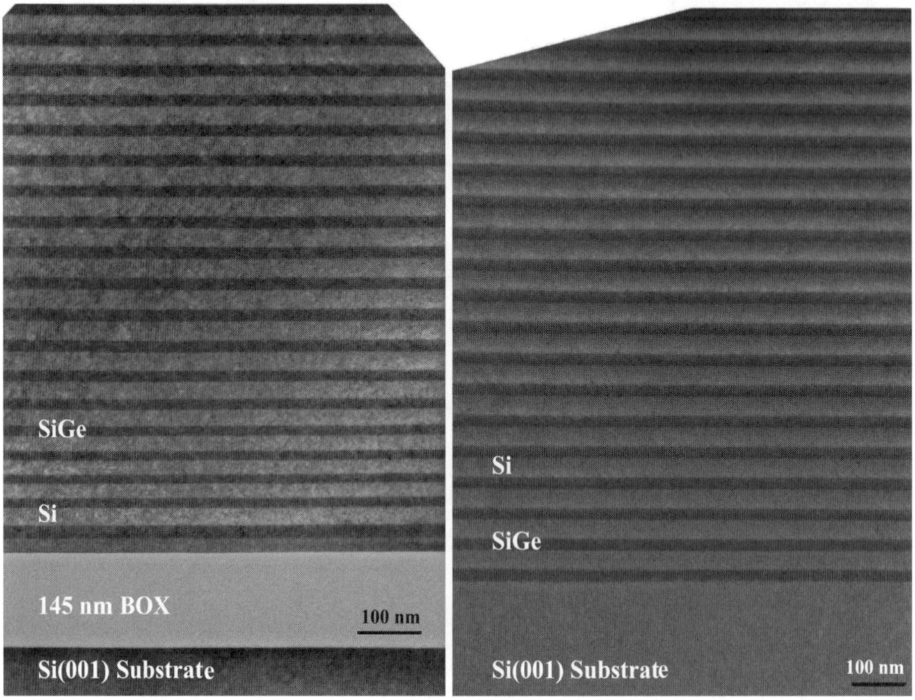

Figure 4.97. *Cross-sectional TEM images of two 19 periods $Si/Si_{0.8}Ge_{0.2}$ superlattices deposited at 650°C, 20 Torr on a {20 nm Si/145 nm buried oxide} SOI substrate (left) and a bulk Si(001) substrate (right). Same epitaxy recipe used in both cases*

We will find in Figure 4.98, the growth rates associated with the Si and SiGe layers, and in Figure 4.99 the Ge concentration according to the thickness deposited on these two types of substrates (values obtained from SIMS profiles of the concentration in Ge atoms in the stackings above).

We find there the important drop, then the increase (followed by a small decrease) on the SOI substrate, of the growth rates of Si and SiGe. This evolution occurs at the same time as a small increase, followed by a decrease, and then again by a small increase in the Ge concentration. This is due to the surface temperature fluctuations, because of the presence of a buried oxide.

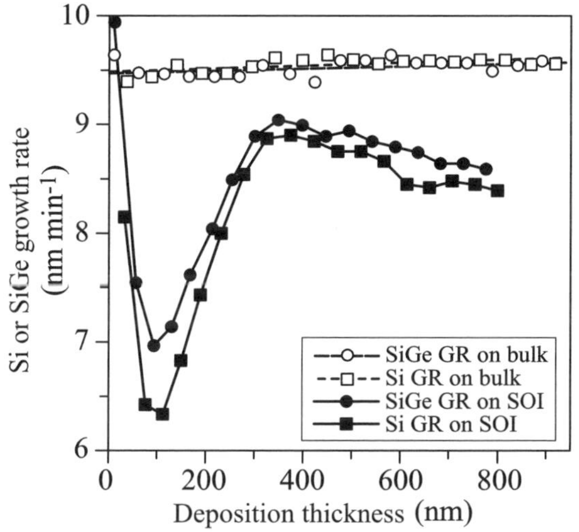

Figure 4.98. *Si and SiGe growth rates according to the thickness deposited on a bulk Si substrate and on a SOI substrate*

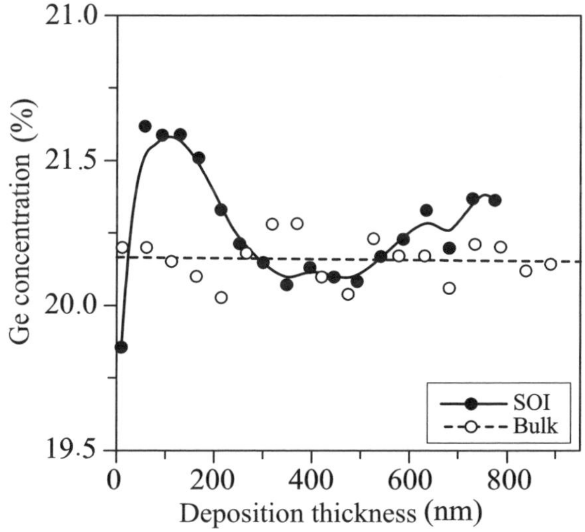

Figure 4.99. *Ge concentration according to the thickness deposited on a bulk Si substrate and on a SOI substrate*

The surface temperature, about 650°C for the starting SOI substrate (i.e. 20 nm of Si on top of 145 nm of BOX), drops to approximately 638°C for 100 nm deposited, goes back up to approximately 648°C for 350 nm deposited, and then slowly decreases to approximately 646°C for 800 nm deposited. Thanks to calibration curves such as those in Figure 4.98, we were able in the past to grow well controlled SiGe/Si stacks on SOI substrates and thus obtain regular multi-channels or multi-wires structures [HAR 05b, HAR 09a]. We will note that the thickness of the buried oxide also has an impact on the growth kinetics that we must consider if we change the type of SOI substrate [HAR 04a].

Once the Si nanowires are released, it is possible to alter their shape, using *in situ* H_2 annealing in the epitaxy reactor [DOR 07]. We can indeed go from Si wires of rectangular section (after selective etching of SiGe versus Si) to wires of circular section, while preserving an excellent crystalline quality, as shown in Figure 4.100. It is however necessary not to use too important thermal budgets, or else the nanowires' junctions with the sources and drains blocks will get thinner and then break.

Once their section is rounded off, it is then possible to drastically reduce their diameter, using {thermal oxidation (peripheral Si converted into SiO_2) / wet chemical de-oxidation} cycles [HUB 08]. We thus obtained arrays of Si nanowires with a diameter as low as 8 nm (as illustrated in Figure 4.101), or even below [ERN 09].

From the integration point of view, we also can take advantage of the possibility that we have to selectively etch (with a dry plasma process) Si compared to SiGe [CAU 06]. We thus obtain an array of SiGe nanowires (and not of Si). Those wires, once selectively encapsulated at 650°C by a few nm of Si [HAR 09b], have been integrated in pMOS transistors, with a gain of 50% regarding the holes mobility [TAC 09]. These SiGe wires could also be used as starting structures for the formation of almost pure Ge nanowires of lower dimensions, thanks to the Ge enrichment technique previously mentioned [NGU 07, SAR 09].

We also have to use, during the manufacture of multi-channels transistors, a relatively complex process in order to selectively thicken the source and drains zones (selective epitaxy of about a hundred nm or even more of intrinsic or *in situ* doped Si: step 3 of the Figure 4.95). The standard epitaxy recipe which is used to selectively grow 20 nm thick Si raised sources and drains on both sides of FD-SOI transistors [JAH 05] will not yield (with a growth duration change, of course) the selectivity desired for this type of thickness and architecture. We thus implemented a cyclic {selective deposition (SiH_2Cl_2+HCl) / etching (HCl)} process in order to gradually thicken at 750°C or 800°C, 20 Torr, the source and drains regions while suppressing the poly-Si nuclei being formed on the Si_3N_4 internal spacers and hard mask of the multi-channels transistors [HAR 10c]. As an example, we will find in Figure 4.102 examples of cyclic deposition/etch processes optimized or not (~ 10 nm of intrinsic or *in situ* doped Si deposited per cycle).

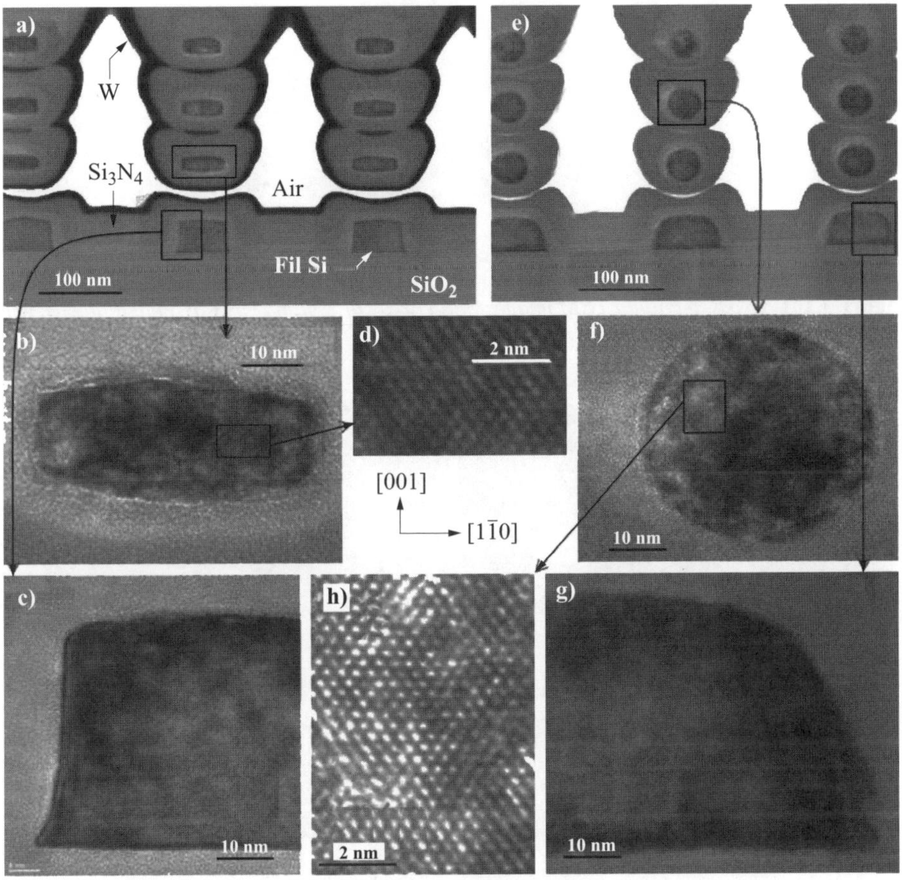

Figure 4.100. *Cross-sectional TEM images of arrays of Si nanowires just after selective etching (a), (b), (c), (d) and after H_2 annealing at 800°C for two minutes (e), (f), (g), (h)*

It should be noted that the combination of an *in situ* doping of these Si raised sources and drains with subsequent ion implantation steps leads to a significant gain in terms of electric performances for such transistors, compared to intrinsic Si raised sources and drains implanted later on [TAC 10].

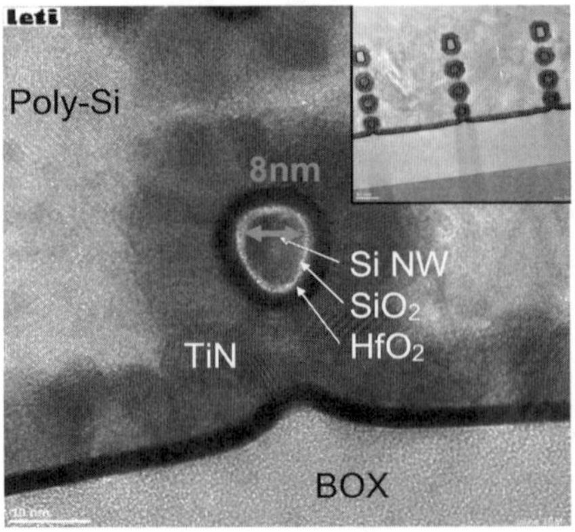

Figure 4.101. *Cross-sectional TEM images of an array of Si nanowires of 8 nm diameter with a conformal encapsulation with a HfO_2 / TiN / poly-Si gate stack*

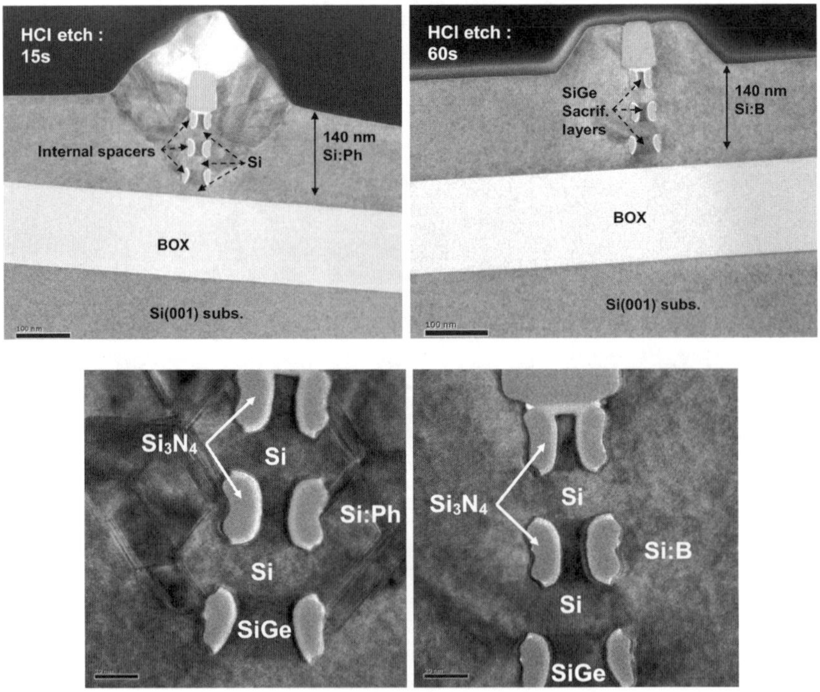

Figure 4.102. *Cross-sectional TEM images of multi-channel transistors with Si_3N_4 internal spacers after the selective deposition of ~ 140 nm of Si:P or Si:B thanks to cyclic deposition/etching processes optimized (on the right) or not (on the left)*

As a conclusion of this section 4.8, we will briefly show the electric characteristics associated with these multi-channel and multi-wire transistors. We will find in Figure 4.103 the drain currents associated with multi-channels nMOS and pMOS transistors with 50 nm effective gate length (see Figure 4.95) functions of the gate voltage. Exceptional on-state current / off-state current trade-offs (nMOS: I_{ON} = 2.27 mA /µm ⇔ I_{OFF} = 16.4 pA /µm, pMOS: I_{ON} = 1.32 mA /µm ⇔ I_{OFF} = 16.75 pA /µm) are obtained thanks to this architecture and the HfO_2/TiN/poly-Si gate stack [BER 08].

We have compared in Figure 4.104, the drain currents associated with these five channels nMOS and pMOS transistors and with standard FD-SOI transistors with 50 nm effective gate length according to the difference between the gate voltage and the threshold voltage. A gain of a factor 4 – 5.5 (as geometrically expected) is obtained while moving over from single channel (i.e. FD-SOI) transistors to transistors with five channels piled up one above the other (MC-FET).

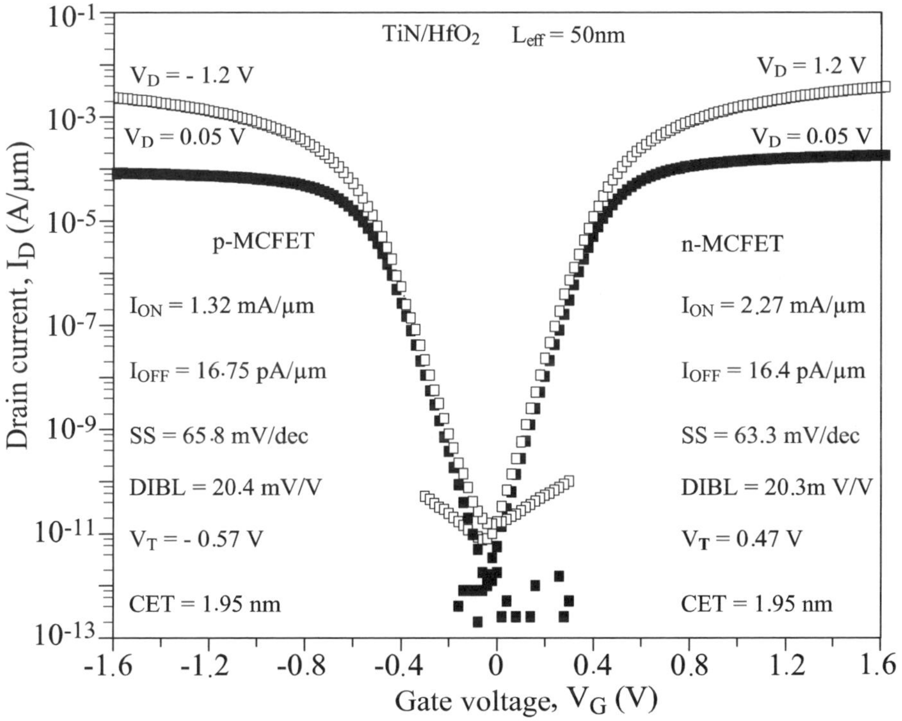

Figure 4.103. *Drain current I_D according to the gate voltage V_G (drain voltages V_D =± 0.05 V or ± 1.2V), this for five channels MC-FET transistors with 50 nm gate length*

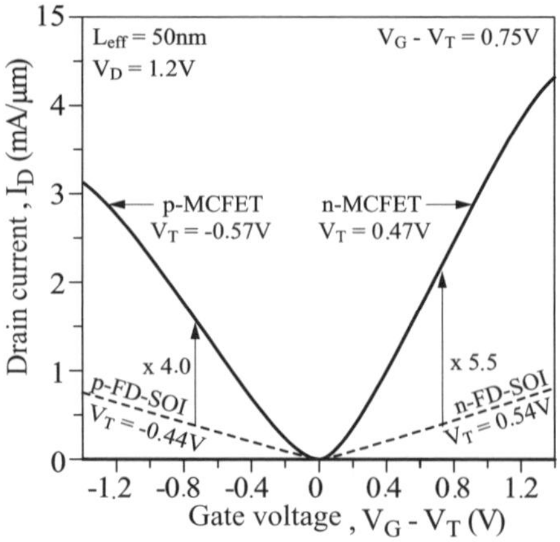

Figure 4.104. *Drain current I_D functions of the gate voltage V_G - V_T, (drain voltage V_D =± 1.2V), for 50 nm gate length transistors with either one (FD-SOI) or five channels (MC-FET)*

We will better realize the obtained electric performances by comparing in Figure 4.105, the I_{ON}/I_{OFF} trade-offs of our multi-channels nMOS and pMOS transistors with those associated with other transistors from the literature (with simple planar gates, of the "FinFET" type or even with multi-channels, but with another gate stack (SiO_2/poly-Si)).

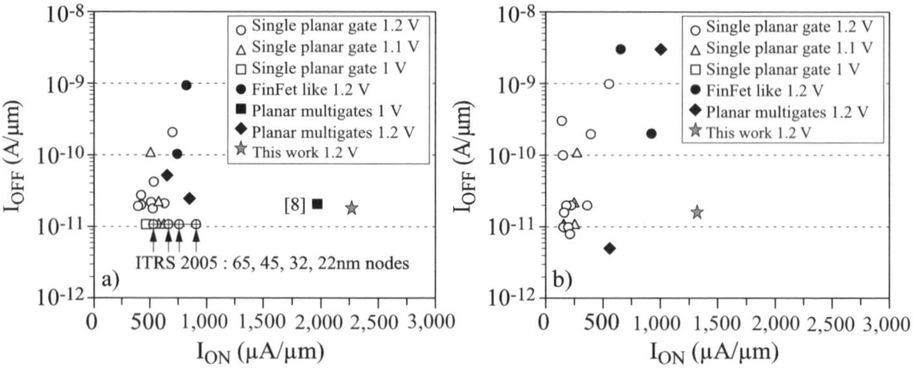

Figure 4.105. *Comparison of the I_{ON} /I_{OFF} trade-offs of our multi-channels nMOS (on the left) and pMOS (on the right) transistors with those of other nMOS and pMOS transistors, aiming for Low Stand-by Power applications. Each point corresponds to a recent article of the literature. The currents are standardized by the gate width seen from the top*

Extremely high on-state currents have recently been obtained for transistors based, this time, on ~ 15 nm diameter Si stacked nanowires (see Figure 4.106), at the detriment however of off-state currents which are more important than in multi-channel transistors: (although the gate stack is the same): nMOS: I_{ON} = 6.5 mA /µm ⇔ I_{OFF} = 27 nA/µm, pMOS: I_{ON} = 3.3 mA /µm ⇔ I_{OFF} = 0.5 nA/µm [DUP 08b]. These I_{ON} values, the highest published at this date, fully show the potentialities of this 3D architecture.

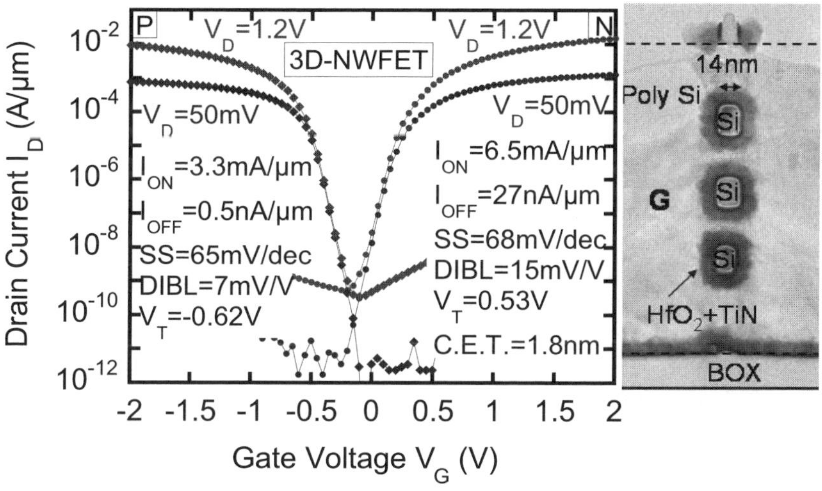

Figure 4.106. *Drain current I_D according to the gate voltage V_G (drain voltages V_D =± 0.05 V or ±1.2V), for transistors with three stacked Si nanowires. The current is standardized compared to the diameter of the wires, approximately 15 nm (see the cross-sectional TEM image perpendicular to those Si NWs)*

4.9. Conclusions and prospects

4.9.1. *General conclusion*

We briefly presented in section 4.1, the Reduced Pressure – Chemical Vapor Deposition tool used, as well as some epitaxy concepts essential to the understanding of the following of the chapter.

We described in sections 4.2 and 4.3, the specificities of the low-temperature epitaxy of thin SiGe (C) layers for the engineering of the channel of MOS transistors. The difficulties inherent in the selective epitaxy of SiGe in the Si windows of patterned substrates were highlighted (loading effects and surface roughness). We have also shown the strategy to adopt, in order to obtain SiGeC or $Si_{1-y}C_y$ layers with record concentrations in substitutional C atoms (3.2% and 2.1%):

a dichlorosilane (for SiGeC) or silane (for $Si_{1-y}C_y$) chemistry, a lowest possible growth temperature and a highest possible growth rate. Thanks to these studies, bulk, short gate length pMOS transistors with SiGe channels have been fabricated with hole mobilities twice as important and a lower threshold voltage (in absolute value) than in bulk Si. Low C concentration Si/SiGeC multi-layers inserted under the gate of bulk transistors lead to the reduction of: (i) the exo-diffusion of the atoms of B of the anti punch-through layer (pMOS) and (ii) the implantation depth of the As atoms in the sources and drains (nMOS), therefore to gain in terms of short channel effects. The situation is more contrasted for nMOS transistors with $Si_{1-y}C_y$ channels: the electrons mobility is indeed lower or equal to that of bulk Si, but lower threshold voltages have been obtained from the alignment of the conduction band (because of the tensile strain).

Sections 4.4 and 4.5 were devoted to *recessed or raised Si or SiGe:B sources and drains* selectively grown on both sides of the gate of FD-SOI or SON transistors. Lowering on ultra-thin SOI substrates the *in situ* H_2 bake temperature (650°C), as well as the epitaxy temperature (750°C, SiH_2Cl_2 + HCl) is mandatory, in order to avoid any withdrawal at the edge of active areas or islanding for initially less than 10 nm thick Si films. The impact of *in situ* B doping on (i) the resulting SiGe:B layers' structural and electrical properties and (ii) the SiGe growth kinetics with a chlorinated chemistry has been described. Two demonstrations of Selective Epitaxial Growth of SiGe:B raised (FD-SOI) or recessed (SON) sources and drains have otherwise been presented.

In section 4.6, we took a look at the structural properties of *SiGe virtual substrates and of the tensily-strained Si layers*, which we can deposit on top. Thanks to the adoption of high growth temperatures (850-900°C), threading dislocations densities at the state of the art, i.e. 10^5 cm^{-2}, and almost fully relaxed constant composition SiGe top layers have been obtained. By changing the Ge concentration between 20% and 50%, we could change the tensile strain in the sSi layers between 1.3 and 3.15 GPa, while preserving a good crystalline quality and smooth layers (thanks to the use of chemical-mechanical polishing steps of our virtual substrates, which are strongly undulated (i.e. "cross-hatched") right after epitaxy). These sSi layers, transferred on oxidized Si substrates, form the core of sSOI substrates. Gains of a factor of 2 of the electrons mobility (compared to bulk Si) have been obtained in nMOS transistors fabricated on tensily-strained Si layers. Gains of a factor 2 of holes mobility have been similarly been obtained, by inserting between these sSOI substrates and the gate stack of the pMOS transistor a thin SiGe layer in compression, encapsulated by a few nm of Si. Thanks to an epitaxy on $Si_{0.5}Ge_{0.5}$ virtual substrates of a {compressively-strained Ge channel / tensily-strained Si cap} bi-layer, we have otherwise demonstrated a gain of a factor of 9 of the holes mobility and of a factor of 6 of the on-state current (compared to bulk Si), and have thus balanced the performances of nMOS and pMOS transistors.

Section 4.7 shows the interest to deposit *thin or thick Ge layers directly onto Si substrates*. A low-temperature/high temperature growth strategy for Ge, followed by a thermal cycling yield thick, smooth Ge layers, that are slightly tensily-strained and with a relatively low threading dislocations density, i.e. 10^7 cm^{-2}. Thanks to the SmartCut^{TM} process, good structural and electrical quality GeOI substrates were thus obtained, with a hole mobility gain of a factor of 2 in pMOSFET transistors fabricated on top. Superior performance photo-detectors operating in the near infra-red (1.3 - 1.55 μm) have moreover been developed, thanks to the Selective Epitaxial Growth in Si windows at the end of optical waveguides (on SOI substrates) of several hundreds of nm thick pure Ge layers. Measurements in photoluminescence of single or multiple planes of Ge islands encapsulated Si have finally illustrated the possibility to modulate the emission wavelength of these islands, by modifying the number of Ge atomic monolayers, the thickness of the Si spacers between Ge islands planes, etc.

Lastly, section 4.8 was devoted to *devices based on sacrificial SiGe layers*. Multiple techniques are indeed available in order to laterally etch SiGe layers selectively compared to Si (plasma, wet and thermal (HCl) etching). Those are used in order to obtain localized-SOI or Silicon-On-Nothing devices. The space previously occupied by the SiGe layer is then filled in a conformal fashion by buried dielectric or the gate stacks. We thus obtain on bulk Si substrates, L-SOI and SON transistors functioning in the fully-depleted mode, without suffering from the costs associated with the use of SOI substrates. Thanks notably to the deposition on SOI substrates of SiGe/Si superlattices, it is possible thanks to a relatively complex integration scheme to obtain multi-channel or multi-wire transistors. Superior On-state currents as well as exceptional I_{ON}/I_{OFF} trade-offs per surface unit were thus obtained, because of the stacking according to z of the charge carrier channels in these type of structures.

4.9.2. *Prospects*

In the previous chapters, we have been able to realize the degree of smoothness and of electronic quality of the Si/SiGe heterostructures that can be obtained thanks to reduced pressure – chemical vapor deposition. The most eloquent illustration according to us is Ge islands (see section 4.7), whose density, shape and size (and thus the photoluminescence properties) can be modified, thanks to a control at the fraction of an atomic monolayer of the quantity of deposited Ge. We will find in Figure 4.107 another example of this control. It is a planar, multiple delta-doping of Si by B atoms. The principle is as follows: we deposit Si (thanks to SiH_4), we stop the growth, we expose the surface to a B_2H_6 flux, we stop the flux and then we encapsulate the adsorbed B atoms by a new Si layer. We thus obtain deltas of B in Si. Those allowed us (thanks to their extreme abruptness): (i) to quantify the contribution of the SIMS tool to the enlargement of the peaks [BAB 02, LAU 04] and (ii) to test the capacity of transmission electronic holography to detect the changes of the electrical potential related to these deltas [COO 07]. Thanks to the availability in the Epi Centura of n-type (PH_3) and p-type (B_2H_6) precursors, we could moreover obtain B doping staircases (in the shape of Mayan pyramids), as well as abrupt p-n junctions in Si. These structures have respectively been studied in in cross-sectional scanning capacitance microscopy (SCM) [CIA 03] and in transmission electronic holography [COO 08, COO 09].

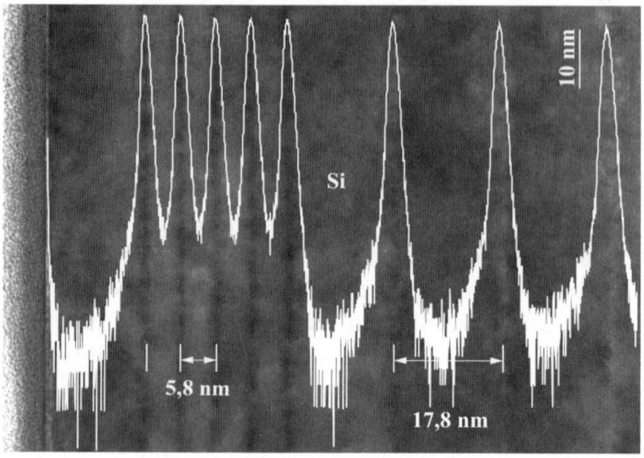

Figure 4.107. *Cross-sectional TEM image and SIMS profile of multiple B deltas in Si*

Moreover, we have "diverted" the RP-CVD reactor from its original use, i.e. epitaxy, in order to deposit Si or Ge nano-crystals (ncs) on an oxidized Si surface functionalized or not [BAR 03, MAZ 03a, MAZ 03b, MAZ 04a]. A method that helps to strictly dissociate the nucleation and the growth of Si ncs was even patented [MAZ 04b]. These nano-crystals have been used later on as storage entities in memory cells, with excellent retention times [MOL 07, SAL 03]. These aspects have not been mentioned in this chapter, because they belong more to the field of poly-Si deposition (in batch-type furnaces) than to epitaxy strictly speaking (i.e. prolongation of the atomic columns of a single-crystal substrate in the deposited layer).

Regarding the various prospects, we could in the long term identify the epitaxy of *in situ* doped Si/SiGe heterostructures for the controlled development of porosities via electrochemical [VYA 02] or photo electrochemical [BAD 03] means. The impact of porous Si substrates on the structural and electronic properties of epitaxial SiGe/Si stacks would also deserve to be studied.

It would also be interesting to explore in more detail the specificities of single-crystal or poly-crystalline SiGeC layers grown at low-temperature thanks to gaseous or liquid precursors such as disilane (Si_2H_6) [DAM 10], trisilane (Si_3H_8) [BAU 07, GOU 09a, TAK 10] or neopentasilane (Si_5H_{12}) [CHU 08]. At a given growth rate, it is indeed possible thanks to these precursors to lower the deposition temperature by several tens to more than a hundred degrees. The thermal budgets used during the fabrication of electronic devices would then be minimized.

It is possible, when adding liquid digermylmethane ($(GeH_3)_2CH_2$) to digermane (Ge_2H_6), to grow at low temperatures thick, smooth Ge layers of excellent crystalline quality (thanks to a surfactant effect of digermylmethane), this without C incorporation [WIS 07]. The possibility of obtaining on a Si (001) substrate either thin GeSi layers (300°C - 470°C), or small Ge rich GeSi islands (T > 470°C) was also shown, thanks to liquid hydrogenated precursors of the $(H_3Ge)_nSiH_{4-n}$ (n = 1 – 4) family [HU 05a, HU 05b].

From the epitaxy and the electronic properties points of view, we can only be impressed by the recent Arizona State University studies on the UHV-CVD of IV-IV heterostructures based on tin. Thanks to the use of SnD_4, of Ge_2H_6 (for GeSnSi), of Si_3H_8 (for SiSn), of $(H_3Ge)_nSiH_{4-n}$ (n = 1 – 3) liquid precursors (for GeSnSi) and of low growth temperatures (typically 250°C-350°C), it is indeed possible to deposit on Si substrates GeSn/SiGeSn heterostructures of excellent structural quality, for Sn concentrations able to go up to 20%! [AEL 04, ROU 05]. Knowing that tin is: (i) a semi-metal (i.e. Eg = -0.4 eV) and (ii) that its lattice parameter is higher than those of Si or of Ge (a_{Sn} = 6.493 Å ⇔ a_{Si} = 5.431 Å and a_{Ge} = 5.658 Å), we clearly glimpse all the band structure engineering opportunities and difficulties resulting from it. We can in particular take advantage of the apparent compliance of $Ge_{1-x}Sn_x$ buffer layers (with x = 0.02 - 0.08), in order to obtain tensily-strained GeSnSi [TOL 06a], Ge [FAN 07] or SiSn [TOL 06b] layers of good structural (and optical) quality, this on a Si wafer. It was notably predicted that highly-tensily strained Ge layers on $Ge_{1-x}Sn_x$ with x> 13% should have a direct band structure (with therefore potentially important photo- and electro-luminescence yields), with an associated energy band gap between 0.35 and 0.6 eV [MEN 04]. These $Ge_{1-x}Sn_x$ buffer layers can (with their lattice parameter between that of germanium and tin) also be used as substrates for the epitaxy of InGaAs or GaAsSb ternary alloys, for telecommunication wavelengths applications [ROU 07]. An excellent summary of the recent results obtained on these IV-IV heterostructures based on tin can be found in [KOU 08]. The epitaxy of such stacks in our industrial RP-CVD reactor remains *a priori* a problem, in particular because of the very low growth temperatures to be used. As an example, the IR pyrometer of the Epi Centura is no longer operational below 330°C.

Properties at room temperature	**Si**	**Ge**	**GaAs**	**InAs**	**InP**	**InSb**
Electrons mobility ($cm^2 V^{-1} s^{-1}$)	1,400	3,900	8,500	40,000	5,400	77,000
Holes mobility ($cm^2 V^{-1} s^{-1}$)	450	1,900	400	500	200	850
Energy band gap (eV)	1.12	0.66	1.42	0.35	1.34	0.17
Lattice parameter (Å)	5.431	5.658	5.653	6.058	5.869	6.749

Table 4.1. *Electrons and hole mobilities, energy band gap and lattice parameters associated with various IV-IV or III-V semiconductors crystallizing in the diamond or sphalerite cubic structures*

We could also think of using III-V heterostructures on IV-IV substrates for the channel engineering of future technology nodes nMOS transistors, in order to profit from the electron mobilities which are much higher than in column IV semiconductors [DEW 08, HUD 07, RAD 09]. As an example, we will find in Table 4.1 the electrons and the holes mobilities, the energy band gap, as well as the lattice parameter characteristic of various sphalerite and diamond semiconductors. The hole mobility is by far the most important in Ge. However, significant electron mobility gains (by a factor of 20 going from Ge to InSb!) could theoretically be obtained with III-V semiconductors as the channel material in nMOS transistors.

These heterostructures could be deposited in metallo-organic CVD [BRA 08, CHI 04], as well as in MBE [TAN 06] on Ge thick layers on Si, on $Si_{1-x}Ge_x$ virtual substrates (typically x = 0.96) or on GeOI substrates. Hetero-junction bipolar transistors (HBT) and high electron mobility transistors based on AlGaAs and InP have recently been fabricated on top of GeSi and GeOI [LEW 07, LIU 09, LUB 08]. They have performances close to those obtained respectively on GaAs and InP substrates. High optical quality III-V heterostructures have otherwise been obtained on these Ge rich substrates, as mentioned in section 4.7. The pseudomorphic growth of heterostructures based on Ga(NAsP) alloys on Si substrates is most interesting for the Si photonics [KUN 10]. We could indeed consider a monolithic integration of

III-V laser sources on IV-IV substrates, instead of using a complex flip-chip method [CAM 07].

As far as prospects are concerned, we should not forget subtle refinements regarding the selective or blanket epitaxy of {n-type doped Si / boron-doped SiGeC / n-type doped Si} heterostructures on patterned substrates, in order to form the core of HBT [CHE 09b, DEC 09, DUT 07, DUV 07]. The scopes of the hetero-epitaxy are indeed tremendous.

4.10. Bibliography

[ABB 06] ABBADIE A., HARTMANN J.M., DI NARDO C., BILLON T., CAMPIDELLI Y, BESSON P., *Microelectronic Engineering,* vol. 83, p. 1986, 2006.

[ABB 07a] ABBADIE A., HARTMANN J.M., BRUNIER F., *ECS Transactions*, vol. 10, no. 1, 2007.

[ABB 07b] ABBADIE A., BEDELL S.W., HARTMANN J.M., SADANA D.K., BRUNIER F., FIGUET C., CAYREFOURCQ I., *Journal of the Electrochemical Society*, vol. 154, H713, 2007.

[ABB 07c] ABBADIE A., HARTMANN J.M.,CHERKASHIN N., DEGUET C., SANCHEZ L., BRUNIER F., LETERTRE F., *ECS Transactions*, vol. 6, no. 4, 263, 2007.

[ABB 08] ABBADIE A., HARTMANN J.M., BESSON P., ROUCHON D., MARTINEZ E., HOLLIGER P., DI NARDO C., LARDIN T., CAMPIDELLI Y., GHYSELEN B., BILLON T., *Applied Surface Science*, vol. 254, p. 6793, 2008.

[AEL 04] AELLA P., COOK C., TOLLE J., ZOLLNER S., CHIZMESHYA A.V.G., KOUVETAKIS J., *Applied Physics Letters*, vol. 84, p. 888, 2004.

[AKA 06a] AKATSU T. *et al.*, *ECS Transactions*, vol. 3, no. 6, 107, 2006.

[AKA 06b] AKATSU T. *et al.*, *Material Science in Semiconductor Processing*, vol. 9, p. 444, 2006.

[AND 03] ANDRIEU F. *et al.*, *Proceedings of the ESSDERC Conference*, p. 267, 2003.

[AND 05a] ANDRE C.L., WILT D.M., PITERA A.J., LEE M.L., FITZGERALD E.A., RINGEL S.A., *Journal of Applied Physics*, vol. 98, 014502, 2005.

[AND 05b] ANDRIEU F. *et al.*, *Proceedings of the SOI Conference*, p. 223, 2005.

[AND 06] ANDRIEU F. *et al.*, *Proceedings of the VLSI Technology Symposium*, p. 168, 2006.

[AND 07] ANDRIEU F. *et al.*, *Proceedings of the VLSI Technology Symposium*, p. 50, 2007.

[ANG 05] ANG K.W., CHUI K.J., BLIZNETSOV V., TANG C.H., DU A., BALASUBRAMIAN N., SAMUDRA G., LI M.F., YEO Y.C., *Applied Physics Letters 86*, 093102, 2005.

[AUG 09] AUGENDRE E. *et al.*, *ECS Transactions*, vol. 25, no. 7, 351, 2009.

[BAB 02] BABOUX N., DUPUY J.C., PRUDON G., HOLLIGER P., LAUGIER F., PAPON A.M., HARTMANN J.M., *Journal of Crystal Growth*, vol. 245, no. 1, 2002.

[BAD 03] BADEL X., LINNROS J., KLEIMANN P., *Electrochemical and Solid-State Letters,* vol. 6, C79, 2003.

[BAR 08] BARNOLA S. *et al.*, *ECS Transactions*, vol. 16, no. 10, 923, 2008.

[BAR 03] BARON T., PELISSIER B., PERNIOLA L., MAZEN F., HARTMANN J.M., ROLLAND G., *Applied Physics Letters*, vol. 83, p. 1444, 2003.

[BAR 07] BARRAL V. *et al.*, *Proceedings of the IEDM Conference*, p. 61, 2007.

[BAS 01] BASCHIR R., CHAO K.-J., KABIR A.E., *Journal of Vacuum Science and Technology B19*, p. 517, 2001.

[BAT 07] BATAIL E. *et al.*, *Proceedings of the ESSDERC Conference*, p. 450, 2007.

[BAT 08] BATAIL E. *et al.*, *Proceedings of the IEDM Conference*, p. 397, 2008.

[BAT 09a] BATUDE P. *et al.*, *Proceedings of the IEDM Conference*, p. 345, 2009.

[BAT 09b] BATUDE P. *et al.*, *Proceedings of the VLSI Technology Symposium,* p. 166, 2009.

[BAU 06] BAUER M., ARENA C., BERTRAM R., TOMASINI P., CODY N., BRABANT P., ITALIANO J., JACOBSON P., WEEKS K.D., US Patent 2006/0234504 A1, 2006.

[BAU 07] BAUER M., MACHKAOUTSAN V., ARENA C., *Semiconductor Sciences and Technology*, vol. 22, S183, 2007.

[BED 04] BEDELL S.W., FOGEL K., SADANA D.K., CHEN H., DOMENICUCCI A., *Applied Physics Letters*, vol. 85, 2493, 2004.

[BER 08] BERNARD E. *et al.*, *Proceedings of the VLSI Technology Symposium*, p. 16, 2008.

[BER 09a] BERNARD E., ERNST T., GUILLAUMOT B., VULLIET N., CORONEL P., SKOTNICKI T., DELEONIBUSS., FAYNOT O., *IEEE Transactions on Electron Devices*, vol. 56, p. 1243 and 1252, 2009.

[BER 09b] BERNARD E., ERNST T., GUILLAUMOT B., VULLIET N., LIM T.Ch., ROZEAU O., DANNEVILLE F., CORONEL P., SKOTNICKI T., *IEEE Electron Device Letters*, vol. 30, p. 148, 2009.

[BID 08] BIDAL G.*et al.*, *Proceedings of the VLSI Technology Symposium*, p. 146, 2008.

[BID 09a] BIDAL G. *et al.*, *Solid-State Electronics*, vol. 53, p. 735, 2009.

[BID 09b] BIDAL G. *et al.*, *Proceedings of the IEDM Conference*, p. 677, 2009.

[BOD 97] BODNAR S., DE BERRANGER E., BOUILLON P., MOUIS M., SKOTNICKI T., REGOLINI J.L., *Journal of Vacuum Science and Technology B15*, p. 712, 1997.

[BOE 08] DE BOER W.B., *ECS Transactions*, vol. 16, no. 10, 13, 2008.

[BOG 04] BOGUMILOWICZ Y., HARTMANN J.M., LAUGIER F., ROLLAND G., BILLON T., RENARD V., OLSHANETSKY E.B., ESTIBALS O., KVON Z.D., PORTAL J.C., *Materials Research Society Symposium Proceedings*, vol. 809, B1.9.1, 2004.

[BOG 05a] BOGUMILOWICZ Y., HARTMANN J.M., ROLLAND G., BILLON T., *Journal of Crystal Growth*, vol. 274, p. 28, 2005.

[BOG 05b] BOGUMILOWICZ Y., HARTMANN J.M., LAUGIER F., ROLLAND G., BILLON T., *Solid State Phenomena*, vol. 108-109, p. 445, 2005.

[BOG 05c] BOGUMILOWICZ Y., HARTMANN J.M., TRUCHE R., CAMPIDELLI Y., ROLLAND G., BILLON T., *Semiconductor Science and Technology*, vol. 20, p. 127, 2005.

[BOG 05d] BOGUMILOWICZ Y., HARTMANN J.M.,CHERKASHIN N., CLAVERIE A., ROLLAND G., BILLON T., *Materials Science and Engineering B 124-125*, p. 113, 2005.

[BOG 05e] BOGUMILOWICZ Y., HARTMANN J.M., TRUCHE R., CAMPIDELLI Y., ROLLAND G., BILLON T., *Semiconductor Science and Technology*, vol. 20, p. 127, 2005.

[BOG 06a] BOGUMILOWICZ Y., HARTMANN J.M., DI NARDO C., HOLLIGER P., PAPON A.M., ROLLAND G., BILLON T., *Journal of Crystal Growth*, vol. 290, p. 523, 2006.

[BOG 06b] BOGUMILOWICZ Y. *et al.*, *ECS Transactions*, vol. 3, no. 7, 1099, 211, 2006.

[BOG 06c] BOGUMILOWICZ Y., HARTMANN J.M., FABBRI J.M., BILLON T., *Semiconductor Science and Technology*, vol. 21, p. 1668, 2006.

[BOR 06] BOREL S., CAUBET V., BILDE J., CHERIF A., ARVET C., VIZIOZ C., HARTMANN J.M., RABILLÉ G., BILLON T., *ECS Transactions*, vol. 3, no. 7, 627, 2006.

[BOU 04] BOUCAUD P., EL KURDI M., HARTMANN J.M., *Applied Physics Letters*, vol. 85, p. 46, 2004.

[BRA 08] BRAMMERTZ G., CAYMAX M., MEURIS M., HEYNS M., MOLS Y., DEGROOTE S., LEYS F., *Thin Solid Films*, vol. 517, p. 148, 2008.

[BRU 00] BRUNHES T., BOUCAUD P., SAUVAGE S., ANIEL F., LOURTIOZ J.M., HERNANDEZ C., CAMPIDELLI Y., KERMARREC O., BENSAHEL D., FAINI G., SAGNES I., *Applied Physics Letters*, vol. 77, p. 1822, 2000.

[CAM 07] VAN CAMPENHOUT J., ROJO-ROMEO P., REGRENY P., SEASSAL C., VAN THOURHOUT D., VERSTUYFT S., DI CIOCCI L., FEDELI J.M., LAGAHE C., BAETS R., *Optics Express*, vol. 15, p. 6744, 2007.

[CAU 06] CAUBET V., BEYLIER C., BOREL S., RENAULT O., *Journal of Vacuum Science and Technology B 24*, p. 2748, 2006.

[CHA 87] CHANG S., UNZICKER D., ANDERSON T.J., *Proceedings of the 10th International Conference on CVD*, vol. 87-8, p. 122, 1987.

[CHE 02] CHEN H., LI Y.K., PENG C.S., LIU H.F., LIU Y.L., HUANG Q., ZHOU J.M., XUE Q.K., *Physics Review B 65*, p. 233303, 2002.

[CHE 09a] CHENG K. *et al.*, *Proceedings of the VLSI Technology Symposium*, p. 212, 2009.

[CHE 06] CHERKASHIN N., HŸTCH M.J., SNOECK E., HÜE F., HARTMANN J.M., BOGUMILOWICZ Y., CLAVERIE A., *Nuclear Instruments and Methods in Physics Research B 253*, p. 145, 2006.

[CHE 09b] CHEVALIER P. *et al.*, *Bipolar/BiCMOS Circuits and Technology Meeting*, p. 1, 2009.

[CHI 04] CHRIQUI Y., LARGEAU L., PATRIARCHE G., SAINT-GIRONS G., BOUCHOULE S., SAGNES I., BENSAHEL D., CAMPIDELLI Y., KERMARREC O., *Journal of Crystal Growth*, vol. 265, p. 53, 2004.

[CHO 06] CHOPRA S., OZTURK M.C., Misra V., MCGUIRE K., MCNEIL L.E., *Applied Physics Letters*, vol. 88, p. 202114, 2006.

[CHU 08] CHUNG K.H., YAO N., BENZIGER J., STURM J.C., SINGH K.K., CARLSON D., KUPPURAO S., *Applied Physics Letters*, vol. 92, no. 113, p. 506, 2008.

[CHU 98] CHURCHILL A.C., ROBBINS D.J., WALLIS D.J., GRIFFIN N., PAUL D.J., PIDDUCK A.J., LEONG W.Y., WILLIAMS G.M., *Journal of Vacuum Science and Technology B 16*, p. 1634, 1998.

[CIA 03] CIAMPOLINI L., BERTIN F., HARTMANN J.M., ROCHAT N., HOLLIGER P., LAUGIER F., CHABLI A., *Materials Science and Engineering B 102*, p. 113, 2003.

[CLA 07] CLAEYS C., SIMOEN E., *Germanium-based Technologies: From Materials to Devices*, Elsevier, Oxford, 2007.

[CLA 06] CLAVELIER L. *et al.*, *ECS Transactions*, vol. 3, no. 7, p. 789, 2006.

[COL 98] COLACE L., MASINI G., GALLUZZI F., ASSANTO G., CAPELLINI G., DI GASPARE L., PELANGE E., EVANGELISTI F., *Applied Physics Letters*, vol. 72, p. 3175, 1998.

[COL 03] COLLAERT N., VERHEYEN P., DE MEYER K., LOO R., CAYMAX M., *Solid-State Electronics*, vol. 47, p. 1173, 2003.

[COL 06] COLLAERT N., ROOYACKERS R., CLEMENTE F., ZIMMERMAN P., CAYREFOURCQ I., GHYSELEN B., SAN K.T., JURCZAK M., BIESEMANS S., *Proceedings of the VLSI Technology Symposium*, p. 64, 2006.

[COO 07] COOPER D., TRUCHE R., RIVALLIN P., HARTMANN J.M., LAUGIER F., BERTIN F., CHABLI A., ROUVIÈRE J.L., *Applied Physics Letters*, vol. 91, p. 143501, 2007.

[COO 08] COOPER D., AILLIOT C., TRUCHE R., BARNES J.P., HARTMANN J.M., BERTIN F., *Journal of Applied Physics,* vol. 104, p. 064513, 2008.

[COO 09] COOPER D., RIVALLIN P., HARTMANN J.M., CHABLI A., DUNIN-BORKOWSKI R.E., *Journal of Applied Physics*, vol. 106, 06 4506, 2009.

[COO 10] COOPER D., BÉCHÉ A., HARTMANN J.M., CARRON V., ROUVIÈRE J.L., *Journal of Applied Physics*, vol. 96, 113508, 2010.

[CRE 06] CRESSLER J.D. (ed.), *Silicon Heterostructure Handbook: Materials, Fabrication, Devices, Circuits and Applications of SiGe and Si Strained-Layer Epitaxy*, CRC Press, Taylor & Francis Group, Boca Raton, 2006.

[CUR 98] CURRIE M.T., SAMAVEDAM S.B., LANGDO T.A., LEITZ C.W., FITZGERALD E.A., *Applied Physics Letters*, vol. 78, p. 1718, 1998.

[CUR 01] CURRIE M.T., LEITZ C.W., LANGDO T.A., TARASCHI G., FITZGERALD E.A., ANTONIADIS D.A., *Journal of Vacuum Science and Technology B 19*, p. 2268, 2001.

[DAM 04a] DAMLENCOURT J.F., WEBER O., RENAULT O., HARTMANN J.M., POGGI C., DUCROQUET F., BILLON T., *Journal of Applied Physics*, vol. 96, p. 5478, 2004.

[DAM 04b] DAMLENCOURT J.F., VANDELLE B., CLUZEL B., CALVO V., HARTMANN J.M., DAVID S., FÉDÉLI J.M., BILLON T., *Electrical Society Proceedings*, vol. 2004-07, p. 1033, 2004.

[DAM 10] DAMLENCOURT J.F., *ECS Transactions*, vol. 28, no. 1, p. 343, 2010.

[DEC 09] DECOUTERE S., VAN HUYLENBROECK S., HEINEMANN B., FOX A., CHEVALIER P., CHANTRE A., MEISTER T.F., AUFINGER K., SCHRÖTER M., *Bipolar / BiCMOS Circuits and Technology Meeting*, p. 16, 2009.

[DEL 04] DELHOUGNE R., ENEMAN G., CAYMAX M., LOO R., MEUNIER-BEILLARD P., VERHEYEN P., VANDERVORST W., DE MEYER K., HEYNS M., *Solid-State Electronics*, vol. 48, p. 1307, 2004.

[DES 07] DESTEFANIS V., MORAND Y., HARTMANN J.M., ROUCHON D., BARBÉ J.C., MERMOUX M., *Applied Surface Science*, vol. 254, p. 1436, 2007.

[DES 08a] DESTEFANIS V., HARTMANN J.M., BOREL S., BENSAHEL D., *Semiconductor Science and Technology*, vol. 23, p. 105019, 2008.

[DES 08b] DESTEFANIS V., HARTMANN J.M., HÜE F., BENSAHEL D., *ECS Transactions*, vol. 16, no. 10, 427, 2008.

[DES 08c] DESTEFANIS V., HARTMANN J.M., HOPSTAKEN M., DELAYE V., BENSAHEL D., *Semiconductor Science and Technology*, vol. 23, 105018, 2008.

[DES 10] DESTEFANIS V. *et al.*, *Semiconductor Science and Technology*, vol. 25, p. 045014, 2010.

[DEW 08] DEWEY G., HUDAIT M.K., LEE K., PILLARISETTY R., RACHMADY W., RADOSAVLJEVIC M., RAKSHIT T., CHAU R., *IEEE Electron Device Letters*, vol. 29, p. 1094, 2008.

[DIM 07] DIMOULAS A., GUSEV E., MCINTYRE P.C., HEYNS M. (eds), *Advanced Gate Stacks for High-Mobility Semiconductors*, Springer, Berlin Heidelberg, 2007.

[DOR 07] DORNEL E., ERNST T., BARBÉ J.C., HARTMANN J.M., DELAYE V., AUSSENAC F., VIZIOZ C., BOREL S., MAFFINI-ALVARO V., ISHEDEN C., FOUCHER J., *Applied Physics Letters*, vol. 91, p. 233502, 2007.

[DUC 04a] DUCROQUET F., ERNST T., WEBER O., HARTMANN J.M., LOUP V., BESSON P., BRÉVARD L., DI MARIA J.L., DELEONIBUS S., *Applied Surface Science*, vol. 224, p. 274, 2004.

[DUC 04b] DUCROQUET F. *et al.*, *Proceedings of the IEDM Conference*, p. 437, 2004.

[DUC 06] DUCROQUET F., HARTMANN J.M., TABONE C., LAFOND D., VIZIOZ C., ERNST T., DELEONIBUS S., *ECS Transactions*, vol. 3, no. 2, 333, 2006.

[DUP 08a] DUPRÉ C., ERNST T., MAFFINI-ALVARO V., DELAYE V., HARTMANN J.M., BOREL S., VIZIOZ C., FAYNOT O., GHIBAUDO G., DELEONIBUS S., *Solid-State Electronics*, vol. 52, p. 519, 2008.

[DUP 08b] DUPRÉ C. *et al.*, *Proceedings of the IEDM Conference*, p. 749, 2008.

[DUT 91] DUTARTRE D., BRÉMOND G., SOUIFI A., BENYATTOU T., *Physics Review B*, vol. 44, p. 11525, 1991.

[DUT 07] DUTARTRE D., LOUBET N., BROSSARD F., VANDELLE B., CHEVALIER P.,CHANTRE A., MONFRAY S., FENOUILLET-BERANGER C., POUYDEBASQUE A., SKOTNICKI T., *Proceedings of the IEDM Conference*, p. 689, 2007.

[DUV 07] DUVERNAY J., BOROT G., CHEVALIER P., DUTARTRE D., PANTEL R.,RUBALDO L.,SCHWARTZMANN T., VANDELLE B., CHANTRE A., *Proceedings of the ESSDERC Conference*, p. 346, 2007.

[EAG 91] EAGLESHAM D.J., CERULLO M., *Applied Physics Letters*, vol. 58, p. 2276, 1991.

[ERD 05] ERDTMANN M., CURRIE M.T., WOICIK J.C., BLACK D., *Materials Research Society Symposium Proceedings*, p. 891, EE12-05, 2005.

[ERN 02] ERNST T. *et al.*, *Proceedings of the VLSI Technology Symposium*, p. 92-93, 2002.

[ERN 03] ERNST T. *et al.*, *Proceedings of the VLSI Technology Symposium*, p. 51-52, 2003.

[ERN 09] ERNST T. *et al.*, *ECS Transactions*, vol. 25, no. 7, 471, 2009.

[FAN 07] FANG Y.Y., TOLLE J., ROUCKA R., CHIZMESHYA A.V.G., KOUVETAKIS J., D'COSTA V., MENÉNDEZ J., *Applied Physics Letters*, vol. 90, p. 061915, 2007.

[FIO 04] FIORENZA J.G. *et al.*, *Semiconductor Science and Technology*, vol. 19, 2004.

[FIT 99] FITZGERALD E.A., CURRIE M.T., SAMAVEDAM S.B., LANGDO T.A., TARASCHI G., YANG V., LEITZ C.W., BULSARA M.T., *Physica Status Solidi (a)*, vol. 171, p. 227, 1999.

[GHY 04] GHYSELEN B. *et al.*, *Solid-State Electronics*, vol. 48, p. 1285, 2004.

[GIA 03] GIANNAKOPOULOS K.P., ROTH S., BURGHAMMER M., FELLOUS C., RICHARD D., DUTARTRE D., *Journal of Applied Physics*, vol. 93, p. 259, 2003.

[GON 08] GONZATTI F., HARTMANN J.M., YCKACHE K., *ECS Transactions*, vol. 16, no. 10, p. 485, 2008.

[GOU 09a] GOUYÉ A., KERMARREC O., HALIMAOUI A., CAMPIDELLI Y., ROUCHON D., BURDIN M., HOLLIGER P., BENSAHEL D., *Journal of Crystal Growth*, vol. 311, p. 3522, 2009.

[GOU 09b] GOUYÉ A., HÜE F., HALIMAOUI A., KERMARREC O., CAMPIDELLI Y., HŸTCH M.J., HOUDELLIER F., CLAVERIE A., BENSAHEL D., *Material Science in Semiconductor Processing*, vol. 12, p. 34, 2009.

[GUP 90] GUPTA P., COON P.A., KOELER B.G., GEORGES S.M., *Journal of Chemical Physics*, vol. 93, p. 2827, 1990.

[HAN 06] HAN J.P. *et al.*, *Proceedings of the IEDM Conference*, p. 59, 2006.

[HAR 02a] HARTMANN J.M., LOUP V., ROLLAND G., HOLLIGER P., LAUGIER F., VANNUFFEL C., SÉMÉRIA M.N., *Journal of Crystal Growth*, vol. 236, p. 10, 2002.

[HAR 02b] HARTMANN J.M., CHAMPAY F., LOUP V., ROLLAND G., SÉMÉRIAM.N., *Journal of Crystal Growth*, vol. 241, p. 93, 2002.

[HAR 02c] HARTMANN J.M., ERNST T., LOUP V., DUCROQUET F., ROLLAND G., LAFOND D., HOLLIGER P., LAUGIER F., SÉMÉRIA M.N., DELEONIBUS S., *Journal of Applied Physics*, vol. 92, p. 2368, 2002.

[HAR 03a] HARTMANN J.M., ABBADIE A., VINET M., CLAVELIER L., HOLLIGER P., LAFOND D., SÉMÉRIA M.N., GENTILE P., *Journal of Crystal Growth*, vol. 257, p. 19, 2003.

[HAR 03b] HARTMANN J.M., LOUP V., ROLLAND G., SÉMÉRIA M.N., *Journal of Vacuum Science and Technology B 21*, p. 2524, 2003.

[HAR 03c] HARTMANN J.M., BERTIN F., ROLLAND G., LAUGIER F., SÉMÉRIA M.N., *Journal of Crystal Growth*, vol. 259, p. 419, 2003.

[HAR 04a] HARTMANN J.M., CLAVELIER L., JAHAN C., HOLLIGER P., ROLLAND G.,BILLON T., DEFRANOUX C., *Journal of Crystal Growth*, vol. 264, p. 36, 2004.

[HAR 04b] HARTMANN J.M., ERNST T., DUCROQUET F., ROLLAND G., LAFOND D., PAPON A.M., TRUCHE R., HOLLIGER P., LAUGIER F., SÉMÉRIA M.N., DELEONIBUS S., *Semiconductor Science and Technology*, vol. 19, p. 593, 2004.

[HAR 04c] HARTMANN J.M. *et al.*, *Semiconductor Science and Technology*, vol. 19, p. 311, 2004.

[HAR 04d] HARTMANN J.M., ABBADIE A., PAPON A.M., HOLLIGER P., ROLLAND G., BILLON T., FÉDÉLI J.M., ROUVIÈRE M., VIVIEN L., LAVAL S., *Journal of Applied Physics*, vol. 95, p. 5905, 2004.

[HAR 04e] HARTMANN J.M., PAPON A.M., HOLLIGER P., ROLLAND G., BILLON T., ROUVIÈRE M., VIVIEN L., LAVAL S., *Material Research Society Symposium Proceedings*, vol. 809, B4.3.1, 2004.

[HAR 05a] HARTMANN J.M., BOGUMILOWICZ Y., ANDRIEU F., HOLLIGER P., ROLLAND G., BILLON T., *Journal of Crystal Growth*, vol. 277, p. 114, 2005.

[HAR 05b] HARTMANN J.M., HOLLIGER P., LAUGIER F., ROLLAND G., SUHM A., ERNST T., BILLON T., VULLIET N., *Journal of Crystal Growth*, vol. 283, p. 57, 2005.

[HAR 05c] HARTMANN J.M., BERTIN F., ROLLAND G., SÉMÉRIA M.N., BRÉMOND G., *Thin Solid Films*, vol. 479, p. 113, 2005.

[HAR 05d] HARTMANN J.M., DAMLENCOURT J.F., BOGUMILOWICZ Y., HOLLIGER P., ROLLAND G., BILLON T., *Journal of Crystal Growth*, vol. 274, p. 90, 2005.

[HAR 06a] HARTMANN J.M., BAUD L., ROLLAND G., FABBRI J.M., BILLON T., *ECS Transactions*, vol. 3, no. 7, 219, 2006.

[HAR 06b] HARTMANN J.M., PAPON A.M., FABBRI J.M., ROLLAND G., BILLON T., JUHEL M., HALIMAOUI A., *ECS Transactions*, vol. 3, no. 7, 489, 2006.

[HAR 06c] HARTMANN J.M., BURDIN M., ROLLAND G., BILLON T., *Journal of Crystal Growth*, vol. 294, p. 288, 2006.

[HAR 07a] HARTMANN J.M., *Journal of Crystal Growth*, vol. 305, p. 113, 2007.

[HAR 07b] HARTMANN J.M., ABBADIE A., GUINCHE Y., HOLLIGER P., ROLLAND G., BUISSON M., DEFRANOUX C., PIERREL F., BILLON T., *Semiconductor Science and Technology*, vol. 22, p. 354, 2007.

[HAR 07c] HARTMANN J.M., ABBADIE A., ROUCHON D., MERMOUX M., BILLON T., *Semiconductor Science and Technology*, vol. 22, p. 362, 2007.

[HAR 08a] HARTMANN J.M., GONZATTI F., FILLOT F., BILLON T., *Journal of Crystal Growth*, vol. 310, p. 62, 2008.

[HAR 08b] HARTMANN J.M., ANDRIEU F., LAFOND D., ERNST T., BOGUMILOWICZ Y., DELAYE V., WEBER O., ROUCHON D., PAPON A.M., *Materials Science and Engineering B 154-155*, p. 76, 2008.

[HAR 08c] HARTMANN J.M., ABBADIE A., ROUCHON D., BARNES J.P., MERMOUX M., BILLON T., *Thin Solid Films 516*, p. 4238, 2008.

[HAR 08d] HARTMANN J.M., BOGUMILOWICZ Y., ABBADIE A., FILLOT F., BILLON T., *Journal of Crystal Growth*, vol. 310, p. 249, 2008).

[HAR 08e] HARTMANN J.M., PAPON A.M., DESTEFANIS V., BILLON T., *Journal of Crystal Growth*, vol. 310, p. 5287, 2008.

[HAR 09a] HARTMANN J.M., PAPON A.M., BARNES J.P., BILLON T., *Journal of Crystal Growth*, vol. 311, p. 3152, 2009.

[HAR 09b] HARTMANN J.M., ABBADIE A., CHERKASHIN N., GRAMPEIX H., CLAVELIER L., *Semiconductor Science and Technology*, vol. 24, p. 055002, 2009.

[HAR 10a] HARTMANN J.M., SANCHEZ L., VAN DEN DAELE W., ABBADIE A., BAUD L., TRUCHE R., AUGENDRE E., CLAVELIER L., CHERKASHIN N., HŸTCH M., CRISTOLOVEANU S., *Semiconductor Science and Technology*, vol. 25, p. 075010, 2010.

[HAR 10b] HARTMANN J.M., ABBADIE A., BARNES J.P., FÉDÉLI J.M., BILLON T., VIVIEN L., *Journal of Crystal Growth*, vol. 312, p. 532, 2010.

[HAR 10c] HARTMANN J.M., PY M., MOREL P.H., ERNST T., PRÉVITALI B., BARNES J.P., VULLIET N., CHERKASHIN N., REBOH S., HŸTCH M., PAILLARD V., *ECS Transactions*, vol. 33, no. 7, 391, 2010.

[HER 99] HERNANDEZ C., CAMPIDELLI Y., SIMON D., BENSAHEL D., SAGNES I., PATRIARCHE G., BOUCAUD P., SAUVAGE S., *Journal of Applied Physics*, vol. 86, p. 1145, 1999.

[HER 03] HERNANDEZ C., CAMPIDELLI Y., BENSAHEL D., United States Patent 6, 537, 370 B1, 2003.

[HIK 09] HIKAVYY A., LOO R., WITTERS L., TAKEOKA S., GEYPEN J., BRIJS B., MERCKLING C., CAYMAX M., DEKOSTER J., *ECS Transactions*, vol. 25, no. 7, 201, 2009.

[HIR 05] HIRASHITA N., SUGIYAMA N., TOYODA E., TAGAKI S., *Applied Physics Letters*, vol. 86, p. 221923, 2005.

[HOL 10] HOLLÄNDER B., BUCA D., MANTL S., HARTMANN J.M., *Journal of the Electrochemical Society*, vol. 157, H643, 2010.

[HU 05a] HU C.W., TSONG I.S.T., D'COSTA V., MENÉNDEZ J., CROZIER P.A., TOLLE J., KOUVETAKIS J., *Applied Physics Letters*, vol. 87, p. 083101, 2005.

[HU 05b] HU C.W., MENÉNDEZ J., TSONG I.S.T., TOLLE J., CHIZMESHYA A.V.G., RITTER R., KOUVETAKIS J., *Applied Physics Letters*, vol. 87, p. 181903, 2005.

[HUA 98] HUANG J., YE Y.Z., LU H., QUE D., *Journal of Applied Physics*, vol. 83, p. 171, 1998.

[HUB 08] HUBERT A. *et al.*, *ECS Transactions*, vol. 13, no. 1, 195, 2008.

[HUD 07] HUDAIT M.K. *et al.*, *Proceedings de la conférence IEDM*, p. 625, 2007.

[HUE 07] HÜE F., HŸTCH M., BENDER H., HARTMANN J.M., CLAVERIE A., *Material Research Symposium Proceedings*, vol. 1026E,1026-C20-04, 2007.

[HUE 08] HÜE F., HŸTCH M., BENDER H., HOUDELLIER F., CLAVERIE A., *Physical Review Letters*, vol. 100, p. 156602, 2008.

[HUE 09] HÜE F., HŸTCH M., HOUDELLIER F., BENDER H., CLAVERIE A., *Applied Physics Letters*, vol. 95, p. 073103, 2009.

[HUT 10a] HUTIN L. *et al.*, *Proceedings of the IEDM Conference*, p. 253, 2010.

[HUT 10b] HUTIN L., LE ROYER C., DAMLENCOURT J.F., HARTMANN J.M., GRAMPEIX H., MAZZOCCHI V., TABONE C., PREVITALI B., POUYDEBASQUE A., VINET M., FAYNOT O., *IEEE Electron Device Letters*, vol. 31, p. 234, 2010.

[ISH 02] ISHIKAGAWA Y., KUMEZAWA M., NURYADI R., TABE M., *Applied Surface Science*, vol. 190, p. 11, 2002.

[ITO 95] ITOH S., NAKAMURA T., NISCHIKAWA S., *Journal of Applied Physics*, vol. 78, p. 2716, 1995.

[JAH 05] JAHAN C., FAYNOT O., TOSTI L., HARTMANN J.M., *Journal of Crystal Growth*, vol. 280, p. 530, 2005.

[JOS 07] JOSHIL S. *et al.*, *Proceedings of the DRC Conference*, p. 53, 2007.

[KAM 97] KAMINS T.I., CARR E.C., WILLIAM R.S., ROSNER S.J., *Journal of Applied Physics*, vol. 81, p. 211, 1997.

[KAM 98] KAMINS T.I., BRIGGS G.A.D., STANLEY WILLIAMS R., *Applied Physics Letters*, vol. 73, p. 1862, 1998.

[KAM 03] KAMINS T.I., MEIDEIROS-RIBEIRO G., OHLBERG D.A.A., STANLEY WILLIAMS R., *Journal of Applied Physics*, vol. 94, p. 4215, 2003.

[KAS 00] KASPAR E., LYUTOUICH K., *Properties of Silicon Germanium and SiGe:Carbon*, INSPEC, The Institution of Electrical Engineers, London, 2000.

[KER 04] KERMARREC O., CAMPIDELLI Y., BENSAHEL D., *Journal of Applied Physics*, vol. 96, p. 6175, 2004.

[KIM 06] KIMURA Y., SUGII N., KIMURA S., INUI K., HIRASAWA W., *Applied Physics Letters*, vol. 88, p. 031912, 2006.

[KOL 09] KOLAHDOUZ M., MARESCA L., OSTLING M., RILEY D., WISE R., RADAMSON H.H., *Solid-State Electronics 53*, 858, 2009.

[KON 97] KONGETIRA P., NEUDECK G., TAKOUDIS C.G., *Journal of Vacuum Science and Technology B 15*, 1902, 1997.

[KOU 08] KOUVETAKIS J., TOLLE J., ROUCKA R., D'COSTA V., FANG Y.Y., CHIZMESHYA A.V.G., MENÉNDEZ J., *ECS Transactions*, vol. 16, no. 10, 807, 2008.

[KNU 94] KNUTSON K.L., CARR R.W., LIU W.H., CAMPBELL S.A., *Journal of Crystal Growth*, vol. 140, p. 191, 1994.

[KUN 10] KUNERT B., VOLZ K., STOLZ W., *ECS Transactions*, vol. 28, no. 5, 219, 2010.

[KUR 02a] EL KURDI M., BOUCAUD P., SAUVAGE S., KERMARREC O., CAMPIDELLI Y., BENSAHEL D., SAINT-GIRONS G., SAGNES I., *Applied Physics Letters*, vol. 80, p. 509, 2002.

[KUR 02b] EL KURDI M., BOUCAUD P., SAUVAGE S., FISHMAN G., KERMARREC O., CAMPIDELLI Y., BENSAHEL D., SAINT-GIRONS G., SAGNES I., PATRIARCHE G., *Journal of Applied Physics*, vol. 92, p. 1858, 2002.

[KWO 06] KWON O., BOECKL J.J., LEE M.L., PITERA A.J., FITZGERALD E.A., RINGEL S.A., *Journal of Applied Physics*, vol. 100, p. 013103, 2006.

[LAU 04] LAUGIER F., HARTMANN J.M., MORICEAU H., HOLLIGER P., TRUCHE R., DUPUY J.C., *Applied Surface Science*, vol. 231-232, p. 668, 2004.

[LEE 04] LEE M.L., PITERA A.J., FITZGERALD E.A., *Journal of Vacuum Science and Technology B 22*, p. 158, 2004.

[LEI 01] LEITZ C.W., CURRIE M.T., KIM A.Y., LAI J., ROBBINS E., FITZGERALD E.A., *Journal of Applied Physics*, vol. 90, p. 2730, 2001.

[LEI 02] LEITZ C.W., CURRIE M.T., LEE M.L., CHENG Z.Y., ANTONIADIS D.A., FITZGERALD E.A., *Journal of Applied Physics*, vol. 92, p. 3745, 2002.

[LEW 07] LEW K.L., YOON S.F., LOKE W.K., TANATO H., DOHRMAN C.L., ISAACSON M., FITZGERALD E.A., *Journal of Vacuum Science and Technology B 25*, p. 902, 2007.

[LIU 09] LIU W.K. *et al.*, *Journal of Crystal Growth*, vol. 311, p. 1979, 2009.

[LIU 10] LIU J., SUN X., CAMACHO-AGUILERA R., KIMERLING L.C., MICHEL J., *Optics Express*, vol. 35, p. 679, 2010.

[LOO 01] LOO R., MEUNIER-BEILLARD P., VANHAEREN D., BENDER H., CAYMAX M., VANDERVORST W., DENTEL D., GORYLL M., VESCAN L., *Journal of Applied Physics*, vol. 90, p. 2565, 2001.

[LOU 02] LOUP V., HARTMANN H.M., ROLLAND G., HOLLIGER P., LAUGIER F., VANNUFFEL C., SÉMÉRIA M.N., *Journal of Vacuum Science and Technology B 20*, p. 1048, 2002.

[LOU 03a] LOUP V., HARTMANN J.M., ROLLAND G., HOLLIGER P., LAUGIER F., SÉMÉRIA M.N., *Journal of Vacuum Science and Technology B 21*, p. 246, 2003.

[LOU 03b] LOUP V., HARTMANN J.M., ROLLAND G., HOLLIGER P., LAUGIER F., LAFOND D., SÉMÉRIA M.N., BESSON P., GENTILE P., *Semiconductor Science and Technology*, vol. 18, p. 352, 2003.

[LUA 99] LUAN H.C., LIM D.R., LEE K.K., CHEN K.M., SANDLAND J., WADA K., KIMERLING L.C., *Applied Physics Letters*, vol. 75, p. 2909, 1999.

[LUB 08] LUBYSHEV D., FASTENAU J.M., WU Y., LIU W.K., BULSARA M.T., FITZGERALD E.A., HOKE W.E., *Journal of Vacuum Science and Technology B 26*, p. 1115, 2008.

[MAR 09] MARRIS-MORINI D. *et al.*, *Proceedings of the IEEE 97*, p. 1199, 2009.

[MAS 03] MASAROTTO L., HARTMANN J.M., BRÉMOND G., ROLLAND G., PAPON A.M., SÉMÉRIA M.N., *Journal of Crystal Growth*, vol. 255, p. 8, 2003.

[MAZ 03a] MAZEN F., BARON T., HARTMANN J.M., BRÉMOND G., SÉMÉRIA M.N., *Journal of Crystal Growth*, vol. 255, p. 250, 2003.

[MAZ 03b] MAZEN F., BARON T., PAPON A.M., TRUCHE R., HARTMANN J.M., *Applied Surface Science*, vol. 214, p. 359, 2003.

[MAZ 04a] MAZEN F., MOLLARD L., BARON T., DECOSSAS S., HARTMANN J.M., *Microelectronic Engineering*, vol. 73-74, p. 632, 2004.

[MAZ 04b] MAZEN F., BARON T., HARTMANN J.M., SÉMÉRIAM.N., US Patent 6, 946, 369 B2 (20 September 2005) and European Patent 1 426 328 A2, 2004.

[MEN 04] MENÉNDEZ J., KOUVETAKIS J., *Applied Physics Letters*, vol. 85, p. 1175, 2004.

[MEN 01] MENON C., BENTZEN A., RADAMSON H.H., *Journal of Applied Physics*, vol. 90, p. 4805, 2001.

[MEN 03] MENON C., LINDGREN A.-C., PERSSON P.O.A., HULTMAN L., RADAMSON H.H., *Journal of the Electrochemical Society*, vol. 150, G253, 2003.

[MER 10] MERMOUX M., CRISCI A., BAILLET F., DESTEFANIS V., ROUCHON D., PAPON A.M., HARTMANN J.M., *Journal of Applied Physics*, vol. 107, p. 013512, 2010.

[MI 95] MI J., WARREN P., LETOURNEAU P., JUDELEWICZ M., GAILHANOU M., DUTOIT M., DUBOIS C., DUPUY J.C., *Applied Physics Letters*, vol. 67, p. 259, 1995.

[MIS 07] MISTRY K. *et al.*, *Proceedings of the IEDM Conference*, p. 247, 2007.

[MIT 09] MITARD J. *et al.*, *Proceedings of the VLSI Technology Symposium*, p. 82, 2009.

[MIT 97] MITCHELL T.O., HOYT J.L., GIBBONS J.F., *Applied Physics Letters*, vol. 71, p. 1688, 1997.

[MIY 99] MIYANO K., MIZUSHIMA I., OHUCHI K., HOKAZANO A., TSUNASHIMA Y., *Japanese Journal of Applied Physics*, vol. 38, p. 2419, 1999.

[MO 90] MO Y.W., SAVAGE D.E., SCHWARTSENTRUBER B.S., LAGALLY M., *Physical Review Letters*, vol. 65, p. 1020, 1990.

[MOL 07] MOLAS G. *et al.*, *Proceedings of the IEDM Conference*, p. 453, 2007.

[MON 07] MONFRAY S.*et al.*, *Proceedings of the IEDM Conference*, p. 693, 2007.

[MON 10] MONFRAY S. *et al.*, *Solid-State Electronics*, vol. 54, p. 90, 2010.

[MUN 08a] MUNGUÍA J., BREMOND G., BLUET J.M., HARTMANN J.M., MERMOUX M., *Applied Physics Letter*, vol. 93, p. 102101, 2008.

[MUN 08b] MUNGUÍA J., BLUET J.M., BAIRA M., MARTY O., BREMOND G., HARTMANN J.M., MERMOUX M., *Applied Physics Letter*, vol. 93, p. 191913, 2008.

[NGU 07] NGUYEN Q.T., DAMLENCOURT J.F., VINCENT B., CLAVELIER L., MORAND Y., GENTIL P., CRISTOLOVEANU S., *Solid-State Electronics*, vol. 51, p. 1172, 2007.

[OKA 07] OKAMOTO S. *et al.*, *Proceedings of the IEDM Conference*, p. 277, 2007.

[ONO 95] ONO Y., NAGASE M., TABE M., TAKAHASHI Y., *Japanese Journal of Applied Physics*, vol. 34, p. 1728, 1995.

[OSM 09] OSMOND J., VIVIEN L., FÉDÉLI J.M., MARRIS-MORINI D., CROZAT P., DAMLENCOURT J.F., CASSAN E., LECUNFF Y., *Applied Physics Letter*, vol. 95, p. 151116, 2009.

[PAN 06] PAN J.W., LIU P.W., CHANG T.Y., CHIANG W.T., TSAI C.H., LIN Y.H., TSAI.C.T., MA G.H., CHIEN S.C., SUN S.W., *Proceedings of the IEDM Conference*, p. 1, 2006.

[POU 08] POUYDEBASQUE A. *et al.*, *Proceedings of the VLSI-TSA Symposium*, p. 16, 2008.

[POU 09] POUYDEBASQUE A., ROMANJEK K., LE ROYER C., TABONE C., PREVITALI B., ALLAIN F.,AUGENDRE E.,HARTMANN J.M., GRAMPEIX H., VINET M., *IEEE Transactions on Electron Devices*, vol. 56, p. 3240, 2009.

[RAD 09] RADOSAVLJEVIC M. *et al.*, *Proceedings of the IEDM Conference*, p. 319, 2009.

[REE 96] REEBER R.R., WANG K., *Materials Chemistry and Physics*, vol. 46, p. 259, 1996.

[RIM 03] RIM K.*et al.*, *Proceedings of the IEDM Conference*, p. 49, 2003.

[ROB 91] ROBBINS D.J., GLASPER J.L., GULLIS A.G., LEONG W.Y., *Journal of Applied Physics*, vol. 69, p. 3729, 1991.

[ROB 92] ROBBINS D.J., GASPER J.L., CULLIS A.G., LEONG W.Y., *Journal of Applied Physics*, vol. 69, p. 3729, 1992.

[ROM 09] ROMANJEK K. *et al.*, *Microelectronics Engineering*, vol. 86, p. 1585, 2009.

[ROU 08] ROUCHON D., HARTMANN J.M., CRISCI A., MERMOUX M., *ECS Transactions*, vol. 16, no. 10, 181, 2008.

[ROU 05] ROUCKA R., TOLLE J., COOK C., CHIZMESHYA A.V.G., KOUVETAKIS J., D'COSTA V., MENÉNDEZ J., CHEN Z.D., ZOLLNER S., *Applied Physics Letters*, vol. 86, p. 191192, 2005.

[ROU 07] ROUCKA R., TOLLE J., FORREST B., KOUVETAKIS J., D'COSTA V., MENÉNDEZ J., *Journal of Applied Physics*, vol. 101, p. 013518, 2007.

[ROU 06] ROUVIÈRE M., VIVIEN L., LE ROUX X., MANGENEY J., CROZAT P., HOARAU C., CASSAN E., PASCAL S., LAVAL S., FÉDÉLI J.M., DAMLENCOURT J.F., HARTMANN J.M., KOLEV S., *Applied Physics Letters*, vol. 87, p. 231109, 2005.

[ROY 08] LE ROYER C.*et al.*, *Solid-State Electronics*, vol. 52, 1285, 2008.

[RUC 98] RÜCKER H., HEINEMANN B., RÖPKE W., KURPS R., KRÜGER D., LIPPERT G., OSTEN H.J., *Applied Physics Letters*, vol. 73, p. 1682, 1998.

[SAK 97] SAKAI A., TATSUMI T., AOYAMA K *Applied Physics Letters*, vol. 71, p. 3510, 1997.

[SAL 03] DE SALVO B. *et al.*, *Proceedings of the IEDM Conference*, p. 26.1.1, 2003.

[SAL 08] SALVETAT T., DESTEFANIS V., BOREL S., HARTMANN J.M., KERMARREC O., CAMPIDELLI Y., *ECS Transactions*, vol. 16, no. 10, 439, 2008.

[SAR 09] SARACCO E. *et al.*, *ECS Transactions*, vol. 19, no. 1, 207, 2009.

[SCH 97] SCHÄFFLER F., *Semiconductor Science and Technology*, vol. 12, p. 1515, 1997.

[SHA 05] SHANG H. e*t al.*, *Proceedings of the VLSI Technology Symposium*, p. 78, 2005.

[SHE 05] SHERAW C.D. *et al.*, *Proceedings of the VLSI Technology Symposium*, p. 12, 2005.

[SIN 99] SINGH D.V., RIM K., MITCHELL T.O., HOYT J., GIBBONS J.F., *Journal of Applied Physics*, vol. 85, p. 978 and 985, 1999.

[SIN 89] SINNIAH K., SHERMAN M.G., LEWIS L.B., WEINBERG W.H., YATES Jr. J.T., JANDA K.C., *Physical Review Letters*, vol. 62, p. 567, 1989.

[SUE 00] SUE K.Y., LEE H.H., *Journal of Applied Physics*, vol. 88, p. 404, 2000.

[SUG 05] SUGII N. *et al.*, *Proceedings of the VLSI Technology Symposium*, p. 54, Kyoto, Japan, 2005.

[SUN 09] SUN X., LIU J., KIMERLING L.C., MICHEL J., *Applied Physics Letters*, vol. 95, p. 011911, 2009.

[SUR 84] SURNEV L., TIKHOV M., *Surface Science,* vol. 138, p. 40, 1984.

[TAC 09] TACHI K. *et al.*, *Proceedings of the IEDM Conference*, p. 313, 2009.

[TAC 10] TACHI K. *et al.*, *Proceedings of the ESSDERC Conference*, p. 368, 2010.

[TAK 10] TAKEUCHI S., NGUYEN N.D., GOOSENS J., CAYMAX M., LOO R., *Thin Solid Films*, vol. 518, S18, 2010.

[TAN 06] TANOTO H., YOON S.F., LOKE W.K., FITZGERALD E.A., DOHRMAN C., NARAYANAN B., DOAN M.T., TUNG C.H., *Journal of Vacuum Science and Technology B 24*, p. 152, 2006.

[THE 05a] THEAN A.V.Y. *et al.*, *Proceedings of the VLSI Technology Symposium*, p. 134n Kyoton Japan, 2005.

[THE 05b] THEAN A.V.Y. *et al.*, *Proceedings of the IEDM Conference*, p. 515, 2005.

[THE 06] THEAN A.V.Y. *et al.*, *Proceedings of the VLSI Technology Symposium*, Honolulu, p. 164, 2006.

[THO 03] THOMAS S.G. *et al.*, *Journal of Electronic Materials*, vol. 32, p. 976, 2003).

[THO 06] THOMPSON S.E., GUANGYU S., YOUN SUNG C., NISHIDA T., *IEEE Transactions on Electron Devices*, vol. 53, p. 1010, 2006.

[TOL 06a] TOLLE J., ROUCKA R., CHIZMESHYA A.V.G., KOUVETAKIS J., D'COSTA V., MENÉNDEZ J., *Applied Physics Letters*, vol. 88, p. 252112, 2006.

[TOL 06b] TOLLE J., CHIZMESHYA V.G., FANG Y.Y., KOUVETAKIS J., D'COSTA V., HU C.W., MENÉNDEZ J., TSONG I.S.T., *Applied Physics Letters*, vol. 89, p. 231924, 2006.

[VIN 05] VINET M. *et al.*, *IEEE Electron Device Letters*, vol. 26, p. 317, 2005.

[VIN 10] VINET M., LE ROYER C., BATUDE P., DAMLENCOURT J.F., HARTMANN J.M., HUTIN L., ROMANJEK K., POUYDEBASQUE A., THOMAS O., *International Journal of Nanotechnology*, vol. 7, p. 304, 2010.

[VIV 07] VIVIEN L. *et al.*, *Optics Express*, vol. 15, p. 9843, 2007.

[VIV 09] VIVIEN L., OSMOND J., FÉDÉLI J.M., MARRIS-MORINI D., CROZAT P., DAMLENCOURT J.F., CASSAN E., LECUNFF Y., LAVAL S., *Optics Express*, vol. 17, p. 6252, 2009.

[VYA 02] VYATKIN A., STARKOV V., TZEITLIN V., PRESTING H., KONLE J., KÖNIG U., *Journal of the Electrochemical Society*, vol. 149, G70, 2002.

[WAL 86] WALLE C.G., MARTIN R.M., *Physical Review B 34*, p. 5621, 1986.

[WEB 03] WEBER O. *et al.*, *Proceedings of the ESSDERC Conference*, p. 271, 2003.

[WEB 04a] WEBER O. *et al.*, *Proceedings of the VLSI Technology Symposium*, p. 42-43, 2004.

[WEB 04b] WEBER O. *et al.*, *Proceedings of the IEDM Conference*, p. 867, 2004.

[WEB 05] WEBER O. *et al.*, *Proceedings of the IEDM Conference*, p. 143, 2005.

[WES 04] WESTHOFF R., CARLIN J., ERDTMANN M., LANGDO T.A., LEITZ C.W., YANG V., PETROCELLI K., BULSARA M.T., FITZGERALD E.A., VINEIS C.J., *Electrochemical Society Proceedings*, vol. 2004-07, p. 589, 2004.

[WHA 98] WHALL T.E., PARKER C.H.E., *Journal of Physics D: Applied Physics*, vol. 31, p. 1397, 1998.

[WIS 07] WISTLEY M.A., FANG Y.Y., TOLLE J., CIZMESHYA A.V.G., KOUVETAKIS J., *Applied Physic Letters*, vol. 90, 082, 108, 2007.

[YAN 00] YANG J., NEUDECK G.W., DENTON J.P., *Applied Physic Letters*, vol. 77, p. 4034, 2000.

[YAN 08] YANG B. *et al.*, *Proceedings of the IEDM Conference*, 2008.

[YIN 06] YIN H. *et al.*, *Proceedings of the VLSI Technology Symposium*, Honolulu, USA, p. 94, 2006.

[ZHA 06] ZHANG P., ISTRATOV A.A., WEBER E.R., KISIELOWSKI C., HE H., NELSON C., SPENCE J.C.H., *Applied Physics Letters*, vol. 89, p. 161907, 2006.

List of Authors

Annie BAUDRANT
CEA LETI
Grenoble
France

Jean-Jacques GANEM
Institut des nanosciences
CEMIP
Paris
France

Jean-Jacques GROB
InESS
Strasbourg
France

Jean-Michel HARTMANN
CEA LETI
Grenoble
France

Daniel MATHIOT
InESS
Strasbourg
France

Isabelle TRIMAILLE
CEMIP
Paris
France

Index

V, W